W9-CPD-462

Airway Smooth Muscle in Health and Disease

Airway Smooth Muscle in Health and Disease

Edited by

Ronald F. Coburn

University of Pennsylvania School of Medicine
Philadelphia, Pennsylvania

Plenum Press • New York and London

Library of Congress Cataloging in Publication Data

Airway smooth muscle in health and disease / edited by Ronald F. Coburn.
   p.   cm.
   Includes bibliographies and index.
   ISBN 0-306-43120-3
   1. Respiratory organs — Muscles — Physiology. 2. Smooth muscle — Physiology. 3. Airway (Medicine) — Physiology. I. Coburn, Ronald F., 1931-   .
   [DNLM: 1. Airway Resistance. 2. Muscle, Smooth — physiology. 3. Muscle, Smooth — physiopathology. 4. Respiratory System — physiology. 5. Respiratory System — physiopathology. WF 102 A2983]
QP121.A57   1989
612'.21 — dc19
DNLM/DLC
for Library of Congress
89-3462
CIP

Contributors

Kenji Baba Department of Physiology, University of Pennsylvania School of Medicine, Philadelphia, Pennsylvania 19104-6085

Peter J. Barnes Department of Clinical Pharmacology, Cardiothoracic Institute, Brompton Hospital, London SW3 6HP, England

Carl B. Baron Department of Physiology, University of Pennsylvania School of Medicine, Philadelphia, Pennsylvania 19104-6085

Ronald F. Coburn Department of Physiology, University of Pennsylvania School of Medicine, Philadelphia, Pennsylvania 19104-6085

Primal de Lanerolle Department of Physiology and Biophysics, College of Medicine, University of Illinois at Chicago, Chicago, Illinois 60680

Giorgio Gabella Department of Anatomy, University College, London, London WC1E 6BT, England

Douglas W. P. Hay Department of Pharmacology, Research and Development, Smith, Kline & French Laboratories, King of Prussia, Pennsylvania 19406-0939

James C. Hogg Pulmonary Research Laboratory, St. Paul's Hospital, University of British Columbia, Vancouver, British Columbia, Canada V6Z 1Y6

Ruth Jacobs Allergic Diseases Section, National Institute of Allergy and Infectious Diseases, National Institutes of Health, Bethesda, Maryland 20205

David B. Jacoby Pulmonary Division, Department of Medicine, University of Maryland, Baltimore, Maryland 21201

Michael Kaliner Allergic Diseases Section, National Institute of Allergy and Infectious Diseases, National Institutes of Health, Bethesda, Maryland 20205

Michael I. Kotlikoff Department of Animal Biology, School of Veterinary Medicine and Cardiovascular–Pulmonary Division, Department of Medicine, University of Pennsylvania School of Medicine, Philadelphia, Pennsylvania 19104-6046

Jay A. Nadel Cardiovascular Research Institute, Moffitt Hospital, University of California, San Francisco, San Francisco, California 94143-0130

Sami I. Said Department of Medicine, University of Illinois College of Medicine, Chicago, Illinois 60612

Joseph F. Souhrada John B. Pierce Foundation Laboratory and School of Medicine, Yale University, New Haven, Connecticut 06519

Tadao Tomita Department of Physiology, School of Medicine, Nagoya University, Nagoya 466, Japan

John G. Widdicombe Department of Physiology, St. George's Hospital Medical School, London SW17 0RE, England

Preface

I organized this book because there is a need to put together in book form recent advances in our knowledge of how airway smooth muscle works in health and in disease. After a period when it seemed that progress was very slow, there has been in the past few years an incredibly rapid gathering of knowledge in this area. In particular, our understanding has improved regarding the cascades of events that follow the initial binding of agonist to plasma membrane receptors and that lead to the cross-bridge movements that determine contraction. This advance in our knowledge was stimulated by use of single- and whole-cell channel recordings of plasma membrane currents and by description of the β-receptor–GTP-binding protein–adenylate cyclase–cAMP coupling system, which serves as a model for other coupling mechanisms. The discovery of the receptor-activated inositol phospholipid transduction system has greatly stimulated research and led to advances in our understanding of mechanisms involved in smooth muscle contraction. Major advances were also triggered by the development of indicators for measuring free cytosolic calcium concentration and starting the unraveling of the events involved in Ca^{2+}-dependent activation of contractile proteins. Although most of the studies that led to our current understanding of these areas were performed on nonairway smooth muscle, these studies usually add to our understanding of airway smooth muscle, and there is an enlarging body of data that have been obtained on airway smooth muscle. In addition, recent years have seen advances in our knowledge of the neural signals that control activation of airway smooth muscle, stimulated by the development of techniques for measurement of neuropeptides and other transmitters, as well as techniques for investigation of the neurons that innervate airway smooth muscle.

This book is organized to emphasize this new information. It also emphasizes our current concepts of how smooth muscle function is altered during disease. Major advances have been made in the understanding of asthmatic airway smooth muscle hyperreactivity and hyperreactivity induced by non-allergic mechanisms. The book discusses the multiple approaches that have been used in this area.

All the contributors to this volume have worked with airway smooth muscle, either normal or diseased, or have studied airway function. The goal of this book is to document our state of knowledge at this time about normal and abnormal airway smooth muscle function.

Ronald F. Coburn, M.D.

Philadelphia

Contents

Chapter 5
Cell-Surface Receptors in Airway Smooth Muscle   77

Peter J. Barnes

Chapter 6
Cellular Control Mechanisms in Airway Smooth Muscle 99

Primal de Lanerolle

Chapter 7
Transduction and Signaling in Airway Smooth Muscle 127

Carl B. Baron

Chapter 8
Electrical Properties of Airway Smooth Muscle **151**

Tadao Tomita

Chapter 9
Ion Channels in Airway Smooth Muscle **169**

Michael I. Kotlikoff

Chapter 10
Coupling Mechanisms in Airway Smooth Muscle **183**

Ronald F. Coburn and Kenji Baba

Chapter 11
Postulated Mechanisms Underlying Airway Hyperreactivity: Relationship to Supersensitivity 199

Douglas W. P. Hay

Chapter 12
Airway Epithelial Metabolism and Airway Smooth Muscle Hyperresponsiveness 237

David B. Jacoby and Jay A. Nadel

Chapter 13
Airway Hyperreactivity: Relationship to Disease States 267

James C. Hogg

Chapter 14
Current Concepts of the Pathophysiology of Allergic Asthma 277

Ruth Jacobs and Michael Kaliner

Chapter 15

Exercise-Induced Bronchoconstriction **301**

Joseph F. Souhrada

Chapter 1

Structure of Airway Smooth Muscle and Its Innervation

Giorgio Gabella
Department of Anatomy
University College, London
London WC1E 6BT, England

I. NERVE SUPPLY TO TRACHEA AND BRONCHI

The musculature of the trachea and bronchi is well supplied with nerves; its activity is under myogenic control as well as under nervous and hormonal control. The main source of afferent fibers to trachea and bronchi are nerve cells in the nodose ganglion whose axons reach the trachea through the recurrent laryngeal nerve; whether there are also some afferent fibers from dorsal root ganglia remains to be proved—a possibility suggested by Dalsgaard and Lundberg (1984). The vagal (afferent) fibers represent the largest nerve supply to the bronchial and tracheal muscles of the mouse, as judged by the extent of the loss of intramuscular nerve endings after vagotomy (Pack *et al.*, 1984). The main source of efferent fibers are ganglion cells in the tracheal and bronchial ganglia (see Chapter this volume) and possibly in some other of the many ganglia associated with mediastinal organs. These neurons are driven (exclusively?) by vagal preganglionic nerve fibers and are part of the parasympathetic outflow; they include excitatory and inhibitory neurons. In the ferret, axons projecting from these neurons to the tracheal muscle have been traced after intracellular injection of horseradish peroxidase (HRP) (Kalia *et al.*, 1983). Efferent vagal fibers are preganglionic and do not reach the musculature, but they all end within the ganglia. There are on average 235 neurons in the ganglionated plexus of the trachea of the mouse (Chiang and Gabella, 1986), but there can be as many as 4000 neurons in the ferret (Baker *et al.*, 1986). Other efferent fibers (postganglionic) are issued by ganglion neurons of the sympathetic chain, mainly in the superior cervical ganglion in the case of the cervical portion of the trachea of the guinea pig (Smith and Satchell, 1985) but probably in the stellate ganglion

1

as well. All the nerves leading into the trachea and bronchi are mixed; i.e., they contain a mixture of afferent and efferent fibers and a mixture of parasympathetic preganglionic and sympathetic postganglionic fibers. So, for example, adrenergic fibers are present in the recurrent laryngeal nerve of the guinea pig, and in some preparations the sympathetic trunk contains cholinergic parasympathetic fibers (Blackman and McCraig, 1982).

II. STRUCTURE OF THE MUSCULATURE

As a whole, the tracheal muscle is a narrow thin band of smooth muscle running transversely within the dorsal (membranous) wall of the trachea. The muscle is organized into bundles by thin septa of connective tissue carrying blood vessels. The bundles split and merge repeatedly; thus, the muscle maintains its continuity along the length of the trachea. The bundles are compact (Fig. 1) and run approximately parallel to each other; however, at the ends of the muscle, they converge and insert to the cartilage arches. Details of this insertion vary between species: in the rat the muscle is attached to the tips of the arches, whereas in the guinea pig the insertion occurs on the concave surface of the cartilages at some distance from the tips. Either a full layer or bundles of longitudinal musculature are found in many preparations, but not consistently. In the human the longitudinal musculature is more developed in the thoracic than in the cervical portion of the trachea (Wailoo and Emery, 1980).

In bronchi and bronchioli, the musculature is a system of bundles arranged as geodesic lines within the wall; their precise relationship with the cartilage is not well understood. There are many references in the old literature as to the amounts of muscle tissue present in the intrapulmonary airways (see Daly and Hebb, 1966), including the observation that the musculature in the guinea pig is more prominent than in other species (the guinea pig also has a distinct layer of smooth muscle in the pleura). However, there are no proper comparative studies of this question.

III. NERVE-MEDIATED RESPONSES

Cholinergic fibers, issued by tracheal and bronchial ganglion neurons (parasympathetic postganglionic fibers), and operating via muscarinic receptors, are excitatory and are the principal nerve input to the tracheobronchial muscles in all species investigated (Widdicombe, 1963; Foster, 1964).

A small but distinct α-adrenergic excitatory response is present *in vitro* in the trachea of the dog (Suzuki *et al.*, 1976), human (Kneussl and Richardson, 1978), and rat (Vornanen, 1982) and *in vivo* in the rabbit (Mustafa *et al.*, 1982) but is absent in the guinea pig trachea (Yip *et al.*, 1981). In the latter species, a

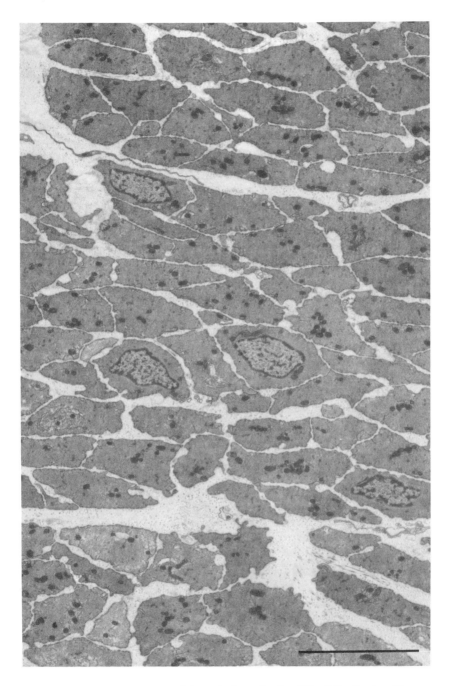

FIGURE 1. Transverse section of the tracheal muscle of a rabbit. Calibration bar: 10 μm.

noncholinergic nonadrenergic excitatory nerve response is recorded (Szolcsanyi and Bartho, 1982).

Two nerve-mediated inhibitory responses have been uncovered. The more conspicuous one is mediated by adrenergic nerve endings acting on β-adrenergic receptors (Foster, 1964). This response is well documented *in vitro* (Fleisch *et al.*, 1970; Suzuki *et al.*, 1976) and *in vivo* [e.g., in the cervical trachea of the guinea pig (Yip *et al.*, 1981)]. A relaxation of lesser extent but of similar time course is obtained by stimulation of the sympathetic trunks and is abolished by hexamethonium. This result indicates that the trunk contains preganglionic fibers that synapse within ganglia (mainly the stellate ganglion) (Yip *et al.*, 1981).

The second inhibitory response has been demonstrated *in vitro* in the trachea of several species and is insensitive to adrenergic and cholinergic blockers (Coburn and Tomita, 1973; Coleman and Levy, 1974; Richardson and Beland, 1976). It is absent in the trachea of the dog (Suzuki *et al.*, 1976). *In vivo* this response has been observed with transmural stimulation or with stimulation of the vagus nerve in cat bronchi (Irvin *et al.*, 1980; Diamond and O'Donnell, 1980) and in guinea pig trachea (Chesrown *et al.*, 1980; Yip *et al.*, 1981)—more readily in the thoracic than in the cervical portion of the latter. Since the response to vagal nerve stimulation is inhibited by hexamethonium (Yip *et al.*, 1981), the inhibitory fibers are not the vagal fibers themselves but fibers that originate from local (tracheal) ganglion cells that are synaptically driven by vagal fibers. The presence of bronchodilator fibers in the vagus nerve, in addition to the principal, bronchoconstrictor, fibers, has often been noted (Widdicombe, 1963). Nerve fibers containing neuropeptides such as vasoactive intestinal polypeptide (VIP) and peptide histidine isoleucine are found in the musculature of the trachea and to a lesser extent in that of bronchi (Uddman *et al.*, 1978; Dey *et al.*, 1981; Lundberg and Saria, 1987).

Afferent mechanisms involving mainly the vagus nerve are well established (Sant Ambrogio, 1982). The sensory receptors include chemoreceptors, mechanoreceptors, and stretch receptors (the so-called rapidly adapting receptors and slowly adapting receptors; the latter outnumber the former by a factor of 10 (Widdicombe, 1954)). In the trachea of the dog, cat, and rabbit, the slowly adapting stretch receptors are probably entirely located within the trachealis muscle itself (Bartlett *et al.*, 1976; Mortola and Sant Ambrogio, 1979). Other afferent mechanisms are mediated by nerve fibers that travel in the sympathetic nerves and through the stellate ganglion (Holmes and Torrance, 1959; Knight *et al.*, 1981).

Finally, there are motor responses that are mediated not by efferent nerves but by afferent fibers, through a mechanism known as the axon reflex. In the trachea of the guinea pig and rat, axon reflexes involving vagal afferent axons enhance vascular permeability and produce contraction (Lundberg and Saria, 1982; Lundberg *et al.*, 1984a); the latter response can also be induced in human bronchi *in vitro* (Lundberg *et al.*, 1983).

IV. DISTRIBUTION OF INTRAMUSCULAR NERVES

Nerve fibers penetrate the muscle mainly from its adventitial side and are gathered into small bundles linked to the meshes of the tracheal ganglionated plexus and to perivascular nerves. Fibers of different types are grouped in the same bundle. The nerve bundles are found mostly within the intramuscular septa, rather than inside muscle bundles. The nerve bundles lie approximately parallel to the muscle, and this is also true of the bronchial musculature, where the orientation of the muscle is more variable; the nerve bundles branch and merge, forming a fine-meshed net, the pattern of which is still unknown. It is unlikely that this resembles the pattern of a tree, because the bundles appear to form a closed mesh, and the true anatomical endings are only short projections emerging from the meshwork. Many of the endings, in a functional sense, i.e., the sites of transmitter release, are varicosities of axons that do not leave the nerve bundle they inhabit. There are no data on the innervation of the longitudinal musculature of the trachea, when present.

Nerve bundles do not show a preferential distribution across the thickness of the tracheal muscle. Only in the ferret has it been found that the great majority of axons lie at the adventitial surface of the muscle or very close to it (Basbaum *et al.*, 1984), an arrangement of the innervation that more closely resembles that of blood vessels than that of other viscera.

V. HISTOCHEMICAL TYPES OF NERVE FIBERS

Cholinergic fibers, tentatively identified by acetylcholinesterase (AChE) histochemistry, have been localized in the bronchial muscle of several species; the AChE reaction is also positive in the tracheal and bronchial ganglion neurons (Fillenz, 1970; Smith and Taylor, 1971; El-Bermani and Grant, 1975; Baker *et al.*, 1986).

There are abundant adrenergic fibers in the bronchial muscles of the cat (Dahlstrom *et al.*, 1966; Silva and Ross, 1974), of the goat, sheep, pig, and calf (Mann, 1971), and of the dog (Knight *et al.*, 1981), but few in the bronchial muscles of rabbit (Mann, 1971) and humans (Pack and Richardson, 1984; Laitinen, 1985). The trachea of the rabbit (Mann, 1971) and of the dog (Suzuki *et al.*, 1976) has few fluorescent fibers, and most of these are perivascular. No fluorescent fibers are found in the bovine tracheal muscle (Cameron *et al.*, 1983), whereas in the cat trachea the adrenergic innervation is well developed (Silva and Ross, 1974). In the guinea pig tracheal muscle, adrenergic fibers are readily found (by fluorescence microscopy) in the cervical portion, whereas they are scanty in the thoracic portion (Coburn and Tomita, 1973; O'Donnell and Saar, 1973; O'Donnell *et al.*, 1978) and are rare in the bronchiole muscles (O'Donnell *et al.*, 1978). The uptake of tritiated norepinephrine by the guinea pig trachea has a similar gradient (Foster and O'Donnell, 1972), and the pharmacological effect of β-blockers on the relaxation elicited by transmural stimulation is markedly greater in the cervical than in the thoracic trachea of this species (Coburn and Tomita, 1973). However, no difference was noted electron microscopically in the frequency of adrenergic endings in the two parts of the trachea (Hoyes and Barber, 1980; Jones et al., 1980; see also Section VII). Histochemically, adrenergic fibers are always less numerous than the fibers interpreted as cholinergic, although their distribution is the same (Silva and Ross, 1974).

Adrenergic fibers originate from neurons in the sympathetic chain, whereas adrenergic neurons are not found in the tracheal or bronchial ganglia (Fillenz, 1970; Jacobowitz et al., 1973). Interestingly, however, in some animals, such as the calf, adrenergic fibers are seen by fluorescence microscopy to be closely associated with some bronchial ganglion neurons (Jacobowitz et al., 1973). Knight (1980) made a similar observation in the cat and extended it to the electron microscopic level. The adrenergic endings were described as forming numerous synaptic complexes with dendrites. The possibility of an adrenergic modulation of the parasympathetic transmission to the airways has been discussed by many investigators and it has recently been shown that norepinephrine inhibits cholinergic (nicotinic) transmission in tracheal ganglia of the ferret by acting on α-adrenoceptors (Baker et al., 1983).

Other chemical classes of nerve fibers are recognized by immunohistochemistry for peptides, which include substance P (Cadieux et al., 1986) and calcitonin gene-related peptide (Lundberg et al., 1984b), both substances probably associated with afferent vagal fibers and VIP (Matsuzaki et al., 1980; Dey et al., 1981; Laitinen, 1985) (see Chapter 4, this volume). No substance P immunofluorescence is detectable in the human trachea and bronchi (Laitinen, 1985). Other chemicals localized in airway ganglia and intramuscular nerves include the enzyme neuron-specific enolase and the calcium-binding protein S-100 (Sheppard et al., 1983). The latter substance occurs in satellite cells of the ganglia and in the Schwann cells of intramuscular nerve bundles.

VI. DENSITY OF INNERVATION

The precise functional significance of the various nerve fibers within a smooth muscle is difficult to assess. Specialized junctions between nerve endings and muscle cells are not common, and transmission can probably also occur where minimal anatomical specialization is apparent (Fig. 2). The distances between nerve endings and muscle cells vary enormously, but it is not established how the width of the gap relates to transmission and its efficiency. Some varicosities lie so far away from the nearest muscle cell that they are unlikely to influence its contraction. Some intramuscular axons may simply be in transit to other parts of the muscle. It seems useful to distinguish intrafascicular and

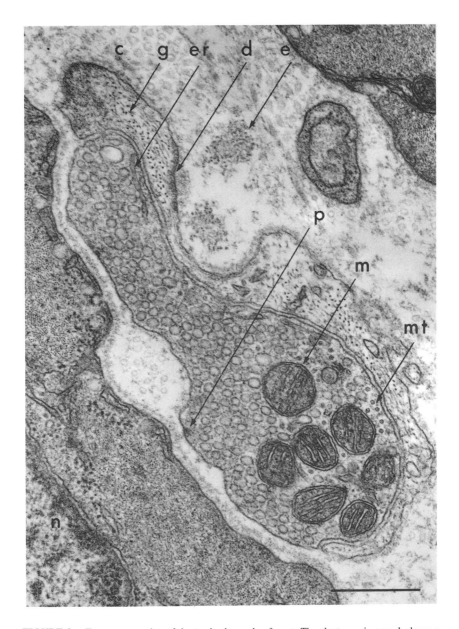

FIGURE 2. Transverse section of the tracheal muscle of a rat. The electron micrograph shows a nerve ending packed with agranular vesicles, mitochondria (m), endoplasmic reticulum (er), and microtubules (mt). The nerve is partly covered by the slender process of a Schwann cell, which contains gliofilaments (g) and membrane-bound densities (d) forming junctions with the stroma. The narrow and regular nerve–muscle cleft is occupied by a basal lamina. A few dense projections (p) are attached to the prejunctional membrane. The muscle cell profile at the bottom left corner shows part of its nucleus (n), caveolae, smooth sarcoplasmic reticulum, and actin and myosin filaments. In the intercellular space, there are collagen fibrils (c) and bundles of microfibrils, probably of elastic nature (e). Calibration bar: 0.5 μm.

interfascicular nerve bundles, as Hoyes and Barber (1980) have done; however, a precise distinction is generally difficult to make. A rough estimate of the density of innervation is usually expressed as number of nerve bundles, axons, varicose axons, and muscle cells on transverse sections of a muscle, i.e., the ratio of axon profiles to muscle cell profiles.

The density of innervation of the tracheal muscle varies greatly between animal species. In humans there is less than one axon per 100 muscle cell profiles (Daniel *et al.*, 1986), whereas in the dog, the ratio is 3.5 : 100 (and the ratio for varicose axons is about 2 : 100) (Kannan and Daniel, 1980). In the guinea pig there are about 19 varicosities per 100 muscle cell profiles in the cervical trachea and 13 per 100 in the thoracic trachea (Jones *et al.*, 1980). Lower values (6.7 and 7.0, respectively) were obtained by Hoyes and Barber (1980), who included in the counts only intrafascicular axons. Further differences between species are shown in Table I (G. Gabella and C. H. Chiang, unpublished results). The range of the axonal densities in the tracheal muscles of mouse, rat, rabbit, and sheep spans more than one order of magnitude. As the volume of the muscle increases, its innervation tends to become sparser. Interestingly, nerve density may also vary along the length of the airways. Whereas there is little or no variability along the trachea of the few species studied, in humans the density of innervation of the muscles of the bronchi of the 4th to 7th order is 18 axons per 100 muscle cells, i.e., about 20 times greater than in the trachea (Daniel *et al.*, 1986).

VII. TYPES OF NERVE ENDINGS

Intramuscular nerve fibers are varicose; they are mostly found grouped into small bundles. Although occasional nerve endings occur singly and almost com-

TABLE 1
Density of Innervation and Frequency of Gap Junctions in the Trachealis Muscle of Mouse, Rat, Rabbit, and Sheep

Animal	Number of muscle cell profiles examined	Number of axons[a]	Number of varicose axons[a]	Number of gap junctions[a]
Mouse 1	280	35	15	10.7
Mouse 2	332	49	21	10.3
Rat 1	318	74	27	6.3
Rat 2	225	43	21	9.8
Rabbit 1	621	17	5	11.9
Rabbit 2	636	15	8	8.3
Sheep 1	1085	4	1	1.8

[a]Per 100 muscle cell profiles.

pletely devoid of a Schwann cell wrapping, most varicosities are found within small nerve bundles composed of two or more varicose axons and a Schwann cell. Many investigators have confidently identified cholinergic and adrenergic nerve endings of airway musculature on the basis of vesicle content. Cholinergic endings mainly contain agranular vesicles, ranging in size from 40 to 70 nm, accompanied by a few large granular vesicles, mitochondria, endoplasmic reticulum, and microtubules (see Fig. 2). In all respects, these endings do not differ from endings interpreted as cholinergic in other visceral muscle. However, both their ultrastructure and their distribution are poorly characterized. It is doubtful that all the endings containing a high proportion of small agranular vesicles are actually cholinergic.

Adrenergic endings mainly contain small granular vesicles (the granularity is often difficult to preserve with standard fixatives and is always enhanced after injection of the adrenergic false transmitter 5-hydroxydopamine) plus a small number of large granular vesicles, mitochondria, and microtubules (Silva and Ross, 1974; Jones et al., 1980; Knight et al., 1981; Daniel et al., 1986). Endings of this type are absent in the bovine tracheal muscle (Cameron et al., 1983).

Other types of nerve ending have been observed; for example, Knight and co-workers (1980, 1981) reported type II axons containing small, round, or irregular vesicles of 20 to 60-nm diameter, small flat agranular vesicles, and a few large dense-core vesicles of 70 to 100-nm diameter. Whether these structures represent a separate type of ending awaits confirmation. A type 2 nerve ending described by Hoyes and Barber (1980) in the guinea pig tracheal muscle contains mainly large dense-core vesicles. These endings are much fewer in number than the type 1 (cholinergic and adrenergic) endings described by the same workers and are interpreted as releasing a nonadrenergic noncholinergic inhibitory transmitter.

The ultrastructural identification of afferent nerve endings is extremely difficult in all viscera, and the airways are no exception. However, the endings of the vagal fibers in the airways contain the peptide substance P (Lundberg and Saria, 1982); it should therefore be possible to identify these endings by immunocytochemistry. The intramuscular nerve endings that contain numerous mitochondria and small numbers of vesicles observed in the guinea pig and mouse trachea are regarded as afferent endings (Hoyes and Barber, 1980; Pack et al., 1984). These endings are relatively few in number compared with the other types, and they are more common within muscle bundles than between muscle bundles. In the bronchial walls of the rat, the nerve endings (some of which are intramuscular) are very elongated, devoid of Schwann cell covering, and rich in mitochondria (Düring et al., 1974). Endings of this type have been labeled free lanceolate terminals and are thought to be specialized sensory endings (mechanoreceptors) in contact with the basal lamina of muscle cells or with elastic and collagen fibers (Düring et al., 1974).

VIII. NEUROMUSCULAR JUNCTIONS

Close contacts between nerve endings (varicosities) and muscle cells are rare in the airway muscles of all species studied. Because of the variability of the width of the gap and the lack of prominent structural specializations, it is difficult to define what constitutes a neuromuscular junction. In the dog trachea, the closest that a nerve ending approaches a muscle cell is 140 nm (Kannan and Daniel, 1980); in humans it is 100 nm, but most of the endings lie >1 μm away from the nearest muscle cell (Daniel et al., 1986). In animal species such as the mouse and the rat, the neuromuscular cleft is <50 nm in ~ 5% of all nerve endings (G. Gabella and C. H. Chiang, unpublished results). A few close neuromuscular contacts (with a cleft of <20 nm) were seen in the tracheal muscle of the guinea pig (Jones et al., 1980), monkey (Knight et al., 1981), and rat (see Fig. 2). Postjunctional structural specialization are usually not apparent.

IX. SMOOTH MUSCLE CELLS

Smooth muscle cells of the airways resemble other visceral muscle cells, such as those of the gut, ultrastructurally. In a detailed morphometric study of bovine trachea, Cameron et al. (1982) calculated a muscle cell length of 0.8 mm and a cell volume of ~3000 μm^3. The surface of the cells is considerable (~ 7400 μm^2); thus, the surface to volume ratio (square microns of cell surface per cubic micron of cell volume) is relatively high, i.e., 2.50 : 1 (Cameron et al., 1982). The same situation is found in the tracheal muscle cells of other species because of the irregular contour of the cells and the abundant cell processes and invaginations (see Fig. 1). The cell membrane is studded with caveolae, arranged in rows, as in the case in other visceral muscle cells.

The muscle cells are fully coated by a basal lamina, except in the areas forming gap junctions. Its appearance is unremarkable in all species examined thus far. In the bronchial musculature of humans, however, the basal lamina is usually markedly thickened (Fig. 3A); the muscle cells are coated by a thick halo of amorphous material in which collagen fibrils and microfibrils are sometimes embedded (Fig. 3B). How this conspicuous structure affects transport to and from the cell is unknown. Between the muscle cells, as elsewhere in the wall of trachea and bronchi, there are abundant collagen and elastic fibers as well as eleunin–oxytalan fibers (Böck and Stockinger, 1984).

X. GAP JUNCTIONS

Gap junctions permit free and direct flow of ions between cells to occur and thus electrically couple the cells. Electrical coupling is a crucial elec-

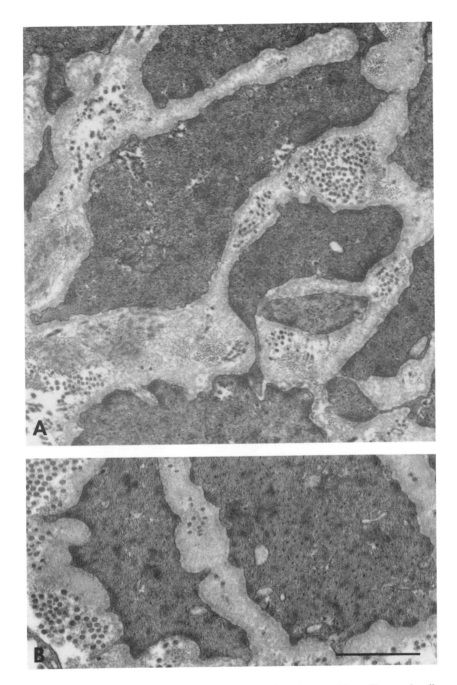

FIGURE 3. Transverse section of bronchial musculature of two human subjects. The muscle cells are surrounded by a thick layer of amorphous material of medium electron density (A). Within this material and in the remainder of the intercellular space, collagen fibrils and bundles of microfibrils can be seen (B). Calibration bar: 1 µm.

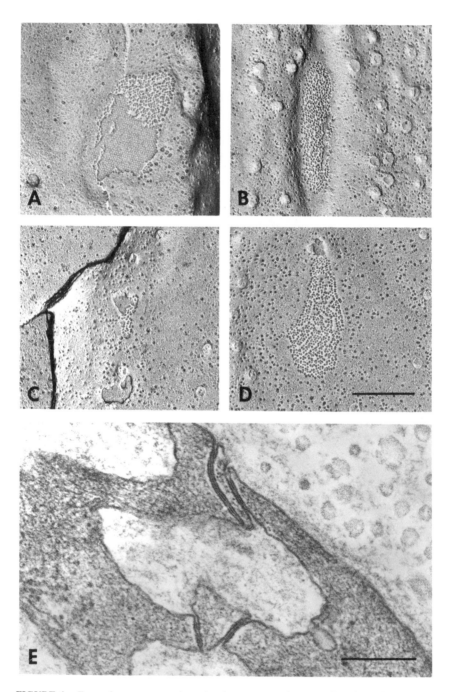

FIGURE 4. Freeze-fracture preparations showing representative examples of gap junctions in muscle cells of the tracheal muscle in four animal species. (A) Mouse. (B) Rat. (C) guinea pig. (D) Rabbit. (E) Section through gap junctions formed by processes of (probably) three muscle cells in the tracheal muscle of a rabbit. Calibration bars: 0.25 μm.

trophysiological characteristic of smooth muscles, but a precise correlation of the occurrence and extent of electrical coupling with the occurrence and density of gap junctions has not been demonstrated. Since some smooth muscles are devoid of gap junctions (Daniel *et al.*, 1976; Gabella, 1981), it is possible that other structures, in addition to gap junctions, support electrical coupling. Moreover, gap junctions may serve other roles by providing other chemical couplings in addition to ionic coupling.

Gap junctions are observed between muscle cells in the tracheal muscle. In humans, 2.7 gap junctions per 100 muscle cell profiles were counted (Daniel *et al.*, 1986) and in the cow about 8 (Cameron *et al.*, 1982); in the latter species, there are about 145 gap junctions per muscle cell, occupying about 0.29% of the cell surface. Other quantitative values on the frequency of gap junctions in mouse, rat, rabbit, and sheep are shown in Table I. Gap junctions are described as rare in the trachea of the dog (Suzuki *et al.*, 1976) and the guinea pig (Richardson and Ferguson, 1979; Jones *et al.*, 1980). In freeze-fracture preparations of tracheal muscle, gap junctions are rare and of minute size in the guinea pig, whereas they are large and common in the mouse, rat, and rabbit (Fig. 4). There is clearly a great variation in this parameter among species, and from the little evidence at hand it seems that gap junctions are not more abundant when innervation is less dense. The muscle cells themselves do not differ in any marked way from other visceral muscle cells.

ACKNOWLEDGMENTS. An earlier version of this chapter appeared in *Annu. Rev. Physiol.* **49**:583–594 (1987). The work discussed in this paper was supported by grants from the Wellcome Trust and the National Institute of Health.

XI. REFERENCES

Baker, D. G., Basbaum, C. B., Herbert, D. A., and Mitchell, R. A., 1983, Transmission in airway ganglia of ferrets: Inhibition by norepinephrine, *Neurosci. Lett.* **41**:139–143.

Baker, D. G., McDonald, D. M., Basbaum, C. B., and Mitchell, R. A., 1986, The architecture of nerves and ganglia of the ferret trachea as revealed by acetylcholinesterase histochemistry, *J. Comp. Neurol.* **246**:513–526.

Bartlett, D., Jr., Jefferey, P., Sant 'Ambrogio, G., and Wise, J. C. M., 1976, Location of stretch receptors in the trachea and bronchi of the dog, *J. Physiol. (Lond.)* **258**:409–420.

Basbaum, C. B., Grillo, M. A., and Widdicombe, J. H., 1984, Muscarinic receptors: Evidence for a nonuniform distribution in tracheal smooth muscle and exocrine glands, *J. Neurosci.* **4**:508–520.

Blackman, J. G., and McCraig, D. J., 1982, Studies on an isolated innervated preparation of guinea-pig trachea, *Br. J. Pharmacol.* **80**:703–710.

Böck, P., and Stockinger, L., 1984, Light and electron microscopic identification of elastic, elaunin and oxytalan fibers in human tracheal and bronchial mucosa, *Anat. Embryol.* **170**:145–153.

Cadieux, A., Springall, D. R., Mulderry, P. K., Rodrigo, J., Ghatei, M. A., *et al.*, 1986, Occurrence, distribution and ontogeny of CGRP immunoreactivity in the rat lower respiratory tract: Effect of capsaicin treatment and surgical denervations, *Neuroscience* **19**:605–627.

Cameron, A. R., Bullock, C. G., and Kirkpatrick, C. T., 1982, The ultrastructure of bovine tracheal smooth muscle, *J. Ultrastruct. Res.* **81:**290–305.

Cameron, A. R., Johnston, C. F., Kirkpatrick, C. T., and Kirkpatrick, M. C. A., 1983, The quest for the inhibitory neurotransmitter in bovine tracheal smooth muscle, *Q. J. Exp. Physiol.* **68:** 413–426.

Chesrown, S. E., Venugopalan, C. S., Gold, W. M., and Drazen, J. M., 1980, In vivo demonstration of nonadrenergic inhibitory innervation of the guinea pig trachea, *J. Clin. Invest.* **65:**314–320.

Chian, C.-H., and Gabella, G., 1986, Quantitative study of the ganglion neurons of the mouse trachea, *Cell Tissue Res.* **100:**243–252.

Coburn, R. F., and Tomita, T., 1973, Evidence for noradrenergic inhibitory nerves in the guinea pig trachealis muscle, *Am. J. Physiol.* **224:**1072–1080.

Coleman, R. A., and Levy, G. P., 1974, A non-adrenergic inhibitory nervous pathway in guinea-pig trachea, *Br. J. Pharmacol.* **52:**167–174.

Dahlstrom, A., Fuxe, K., Hokfelt, T., and Norberg, K.-A., 1966, Adrenergic innervation of the bronchial muscle of the cat, *Acta Physiol. Scand.* **66:**507–508.

Dalsgaard, C.-J., and Lundberg, J. M., 1984, Evidence for a spinal afferent innervation of the guinea pig lower respiratory tract as studied by the horseradish peroxidase technique, *Neurosci. Lett.* **45:**117–122.

Daly, D. I., and Hebb, C., 1966, *Pulmonary and Bronchial Vascular Systems,* Edward Arnold, London.

Daniel, E. E., Daniel, V. P., Duchon, G., Garfield, R. E., Nichols, M., Malhotra, S. K., and Oki, M., 1976, Is the nexus a necessary for cell-to-cell coupling in smooth muscle?, *J. Membr. Biol.* **208:**207–239.

Daniel, E. E., Kannan, M., Davis, C., and Posey-Daniel, V., 1986, Ultrastructural studies on the neuromuscular control of human tracheal and bronchial muscle, *Respir. Physiol.* **63:**109–128.

Dey, D. D., Shannon, W. A., and Said, S. I., 1981, Localization of VIP-immunoreactive nerves in airways and pulmonary vessels of dogs, cats, and human subjects, *Cell Tissue Res.* **220:**231–238.

Diamond, L., and O'Donnell, M., 1980, A nonadrenergic vagal inhibitory pathway to feline airways, *Science* **208:**185–188.

Düring, v. M., Andres, K. H. and Iravani, J., 1974, The fine structure of the pulmonary stretch receptor in the rat, *Z. Anat. Entwicklungsgesch.* **143:**215–222.

El-Bermani, A.-W., and Grant, M., 1975, Acetylcholinesterase-positive nerves of the rhesus monkey bronchial tree, *Thorax* **30:**162–170.

Fillenz, M., 1970, Innervation of pulmonary and bronchial blood vessels of the dog, *J. Anat. (Lond.)* **106:**449–461.

Fleisch, J. H., Maling, H. M., and Brodie, B. B., 1970, Evidence for existence of alpha-adrenergic receptors in the mammalian trachea, *Am. J. Physiol.* **218:**596–599.

Foster, R. W., 1964, A note on the electrically transmurally stimulated isolated trachea of the guinea-pig, *J. Pharm. Pharmacol.* **16:**125–128.

Foster, R. W., and O'Donnell, S. R., 1972, Some evidence of the active uptake of noradrenaline in the guinea-pig isolated trachea, *Br. J. Pharmacol.* **45:**71–82.

Gabella, G., 1981, Structure of smooth muscles, in: *Smooth Muscle: An Assessment of Current Knowledge* (E. Bülbring, A. F. Brading, A. W. Jones, and T. Tomita, eds.), pp. 1–46, Edward Arnold, London.

Holmes, R., and Torrance, R. W., 1959, Afferent fibres of the stellate ganglion, *Q. J. Exp. Physiol.* **44:**271–281.

Hoyes, A. D., and Barber, P., 1980, Innervation of the trachealis muscle in the guinea-pig: A quantitative ultrastructural study, *J. Anat. (Lond.)* **130:**789–800.

Irvin, C. G., Boileau, R., Tremblay, J., Martin, R. R., and Macklem, P. T., 1980, Bronchodilatation: Noncholinergic, nonadrenergic mediation demonstrated in vivo in the cat, *Science* **207**: 791–792.

Jacobowitz, D., Kent, K. M., Fleisch, J. H., and Cooper, T., 1973, Histofluorescence study of catecholamine-containing elements in cholinergic ganglia from the calf and dog lung, *Proc. Soc. Exp. Biol. Med.* **144**:464–466.

Jones, T. R., Kannan, M. S., and Daniel, E. E., 1980, Ultrastructural study of guinea pig tracheal smooth muscle and its innervation, *Can. J. Physiol. Pharmacol.* **58**:974–983.

Kalia, M., Cameron, A. R., and Coburn, R. F., 1983, Morphological characteristics of physiologically identified parasympathetic nerve fibers in ferret trachealis muscle, *Fed. Proc.* **42**:999.

Kannan, M. S., and Daniel, E. E., 1980, Structural and functional study of canine tracheal smooth muscle, *Am. J. Physiol.* **238**:C27–C33.

Kneussl, M. P., and Richardson, J. P., 1978, Alpha-adrenergic receptors in human and canine tracheal and bronchial smooth muscle, *J. Appl. Physiol.* **45**:307–311.

Knight, D. S., 1980, A light and electron microscopic study of feline intrapulmonary ganglia, *J. Anat. (Lond.)* **131**:413–428.

Knight, D. S., Hyman, A. L., and Kadowitz, P. J., 1981, Innervation of intrapulmonary airway smooth muscle of the dog, monkey and baboon, *J. Auton. Nerv. Syst.* **3**:31–43.

Laitinen, A., 1985, Autonomic innervation of the human respiratory tract as revealed by histochemical and ultrastructural methods, *Eur. J. Respir. Dis.* (suppl. 140) **66**:1–42.

Lundberg, J. M., and Saria, A., 1982, Bronchial smooth muscle contraction induced by stimulation of capsaicin-sensitive sensory neurons, *Acta Physiol. Scand.* **116**:473–476.

Lundberg, J. M., and Saria, A., 1987, Polypeptide-containing neurons in airway smooth muscle, *Annu. Rev. Physiol.* **49**:557–572.

Lundberg, J. M., Martling, C.-R., and Saria, A. 1983, Substance P and capsaicin-induced contraction of human bronchi, *Acta Physiol. Scand.* **119**:49–53.

Lundberg, J. M., Brodin, E., Hua, X., and Saria, A., 1984a, Vascular permeability changes and smooth muscle contraction in relation to capsaicin-sensitive substance P afferents in the guinea-pig, *Acta Physiol. Scand.* **120**:217–227.

Lundberg, J. M., Hokfelt, T., Martling, C. R., Saria, A., and Cuello, C., 1984b, Substance P-immunoreactive sensory nerves in the lower respiratory tract of various mammals including man, *Cell Tissue Res.* **235**:251–261.

Mann, S. P., 1971, The innervation of mammalian bronchial smooth muscle: The localization of catecholamines and cholinesterases, *Histochem. J.* **3**:319–331.

Matsuzaki, Y., Hamasaki, Y., and Said, S. I., 1980, Vasoactive intestinal peptide: A possible transmitter of adrenergic relaxation of guinea pig airways, *Science* **210**:1252-1253.

Mortola, J. P., and Sant 'Ambrogio, G., 1979, Mechanics of the trachea and behaviour of its slowly adapting stretch receptors, *J. Physiol. (Lond.)* **286**:577–590.

Mustafa, K. Y., Elkhawad, A. O., Bicik, V., Mardini, I. A., and Thulesius, O., 1982, Adrenergic and cholinergic induced contractions of tracheal smooth muscle in the rabbit as demonstrated by a new in vivo methods, *Acta Physiol. Scand.* **114**:129–134.

O'Donnell, S. R., and Saar, N., 1973, Histochemical localization of adrenergic nerves in the guinea-pig trachea, *Br. J. Pharmacol.* **47**:707–710.

O'Donnell, S. R., Saar, N., and Wood, L. J., 1978, The density of adrenergic nerves at various levels in the guinea-pig lung, *Clin. Exp. Pharmacol. Physiol.* **5**:325–332.

Pack, R. J., Al-Ugaily, L. H., and Widdicombe, J. G., 1984, The innervation of the trachea and extrapulmonary bronchi of the mouse, *Cell Tissue Res.* **338**:61–68.

Pack, R. J., and Richardson, P. S., 1984, The aminergic innervation of the human bronchus: A light and electron microscopic study, *J. Anat. (Lond.)* **138**:493–502.

Richardson, J., and Beland, J., 1976, Nonadrenergic inhibitory nervous system in human airways, *J. Appl. Physiol.* **41**:764–771.

Richardson, J. B., and Ferguson, C. C., 1979, Neuromuscular structure and function in the airways, *Fed. Proc.* **38:** 202–208.

Sant 'Ambrogio, G., 1982, Information arising from the tracheobronchial tree of mammals, *Physiol. Rev.* **62:**531–569.

Sheppard, M. N., Kurian, S. S., Henzen-Longmans, S. C., Michetti, F., Cocchia, D., Cole, P., Rush, R. A., Marangos, P. J., Bloom, S. R., and Polak, J. M., 1983, Neurone-specific enolase and S-100: New markers for delineating the innervation of the respiratory tract in man and other mammals, *Thorax* **38:**333–340.

Silva, D. G., and Ross, G., 1974, Ultrastructural and fluorescence histochemical studies on the innervation of the tracheo bronchial muscle of normal cats and cats treated with 6-hydroxydopamine, *J. Ultrastruc. Res.* **47:**310–328.

Smith, R. V., and Satchell, D. G., 1985, Extrinsic pathways of the adrenergic innervation of the guinea-pig trachealis muscle, *J. Auton. Nerv. Syst.* **14:**61–73.

Smith, R. V., and Taylor, I. M., 1971, Observations on the intrinsic innervation of trachea, bronchi and pulmonary vessels in the sheep, *Acta Anat.* **80:**1–13.

Suzuki, H., Morita, K., and Kuriyama, H., 1976, Innervation and properties of the smooth muscle of the dog trachea, *Jpn. J. Physiol.* **26:**303–320.

Szolcsanyi, J., and Bartho, L., 1982, Capsaicin-sensitive non-cholinergic excitatory innervation of the guinea-pig tracheobronchial smooth muscle, *Neurosci. Lett.* **34:**247–251.

Uddman, R., Alumets, J., Densert, O., Hakanson, R., and Sundler, F., 1978, Occurrence and distribution of VIP nerves in the nasal mucosa and tracheobronchial wall, *Acta. Otolaryngol. (Stockh.)* **86:**443–448.

Vornanen, M., 1982, Adrenergic responses in different sections of rat airways, *Acta Physiol. Scand.* **114:**587–591.

Wailoo, M., and Emery, J. L., 1980, Structure of the membranous trachea in children, *Acta Anat. (Basel)* **106:**254–261.

Widdicombe, J. G., 1954, Receptors in the trachea and bronchi of the cat, *J. Physiol. (Lond.)* **123:** 71–104.

Widdicombe, J. G., 1963, Regulation of tracheobronchial smooth muscle, *Physiol. Rev.* **43:**1–37.

Yip, Palomini, B., and Coburn, R. F., 1981, Inhibitory innervation to the guinea pig trachealis muscle, *J. Appl. Physiol.* **50:**374–382.

Chapter 2

Integration of Neural Inputs in Peripheral Airway Ganglia

Ronald F. Coburn

Department of Physiology
University of Pennsylvania School of Medicine
Philadelphia, Pennsylvania 19104-6085

I. INTRODUCTION

The goal of this chapter is to review information related to neural control of airway smooth muscle, and integration of neural inputs to airway smooth muscle. Neural inputs to airway smooth muscle can be controlled and modulated at the level of the central nervous system (CNS), the peripheral airway ganglia, and at the neuromuscular junction. Mechanisms of integration of neural inputs at the level of the peripheral airway parasympathetic ganglia and at the neuromuscular junction are discussed. Control and modulation at the level of the CNS are not discussed.

II. NEURAL INPUT TO AIRWAY SMOOTH MUSCLE: HOW THE SYSTEM IS WIRED

Control of airway smooth muscle tension in both humans and most animal species is dominated by excitatory neural inputs transmitted by cholinergic motor nerves (Coburn, 1987). Preganglionic cholinergic motor neurons synapse on neurons in peripheral airway ganglia that provide axons innervating smooth muscle cells. The other known neural inputs controlling airway smooth muscle are both inhibitory and occur by means of adrenergic and nonadrenergic noncholinergic neurons. Chapter 1 discusses, in detail, what we know about these neural inputs to airway smooth muscle. Sympathetic nerve terminals directly innervate airway smooth muscle in some species (Cabezas *et al.*, 1971; Coburn and Tomita, 1973; Knight, 1980; Yip *et al.*, 1981; Leff *et al.*, 1983) but have no direct innervation in other species, including humans (Mann, 1971; Cameron *et al.*, 1982; Laitinen, 1985). In the same species, one airway segment can be richly endowed with catecholamine fluorescence neurons, and another segment can be completely devoid of these neurons (O'Donnell and Saar, 1973; Coburn and Tomita, 1973). In some species, catecholamine fluorescence is clearly seen within neurons synapsing on airway ganglia (Dahlstrom *et al.*, 1966; Jacobowitz *et al.*, 1973; Knight, 1980); however, in other species catecholamine fluorescence is not observed within peripheral airway ganglia (Jacobowitz *et al.*, 1973).

There is evidence of a strong inhibitory innervation to airway smooth muscle mediated by nonadrenergic noncholinergic neurons. Field stimulation of isolated airway muscle strips from guinea pig, cat, baboon, rabbit, and human airways evokes nonadrenergic noncholinergic relaxations (Coburn and Tomita, 1973; Coleman and Levy, 1974; Richardson and Beland, 1976; Russell, 1978; Middendorf and Russell, 1980; Altier and Diamond, 1985, 1986). Presynaptic nonadrenergic noncholinergic inhibitory neurons run in the vagus nerve (Chesrown *et al.*, 1980; Yip *et al.*, 1981). Hexamethonium inhibits vagus stimulation-evoked relaxation (Yip *et al.*, 1981), indicating that vagal presynaptic inhibitory fibers synapse in the peripheral airway ganglia rather than directly innervating smooth muscle cells. These vagal presynaptic inhibitory neurons, like cholinergic presynaptic excitatory neurons, release acetylcholine (ACh), which activates postsynaptic neurons by a cholinergic nicotinic mechanism. It is not known whether postsynaptic neurons containing inhibitory neurotransmitters also contain the excitatory neurotransmitter, ACh. Vasoactive intestinal peptide (VIP) and ACh have been identified in the same neurons (i.e., are co-transmitters) (Lundberg and Hokfelt, 1983) in neurons innervating blood vessels and secretory glands. Whether VIP is the nonadrenergic neurotransmitter present in postganglionic inhibitory motor nerves is controversial. The guinea pig trachea is a classic preparation for study of inhibitory innervation to airway smooth muscle, and more is known about inhibitory innervation in this species than in other species. In this species, field stimulation-evoked relaxations are associated with the release of VIP into the media, and relaxations are inhibited by prolonged incubation of the preparation with VIP antibodies (Matsuzaki *et al.*, 1980; Goyal *et al.*, 1980). Field stimulation-evoked nonadrenergic guinea pig trachealis muscle relaxations have been shown to occur without a membrane potential change (Boyle *et al.*, 1987). Exogenous VIP causes an electroneutral relaxation of this muscle, over the range of relaxations seen with field stimulation (Greene and Coburn, 1988). VIP has been shown to be present in smooth muscle neurons (Dey *et al.*, 1981; Uddman *et al.*, 1978). Thus, evidence supporting VIP as an inhibitory neurotransmitter in this species seems to be gaining strength. Nonadrenergic relaxations are not dependent on the presence of the epithelium, not caused by stimulation of sensory neurons, and occur independently of cholinergic mechanisms (Altier and Diamond, 1985, 1986).

From the above considerations, we can make tentative conclusions about the basic wiring of neurons that provide neural input to airway smooth muscle cells (Fig. 1): (1) excitatory input is directed to smooth muscle cells, via synapses on peripheral airway parasympathetic ganglion neurons; (2) nonadrenergic inhibitory neural input also operates via a synapse at peripheral ganglia; and (3) sympathetic nerves can directly innervate smooth muscle cells or synapse on neurons in the peripheral ganglia, depending on the species. There is now increasing evidence that other transmitters or modulators may be involved in the basic "wiring" described above. Substance P is a most likely candidate. Field stimulation-evoked noncholinergic contractions of the guinea pig bronchial smooth muscle are a result of release of substance P from SP neurons (Lundberg and Saria, 1982; Andersson and Grundstrom, 1983). In addition, substance P is released from smooth muscle neurons as part of the action of contracting drugs (Bloomquist and Kream, 1988). It is not known whether substance P is a physiological excitatory neurotransmitter in airway smooth muscle of any species. The basic wiring within individual peripheral airway ganglia or an individual ganglion is not known (for further discussion, see Section IV).

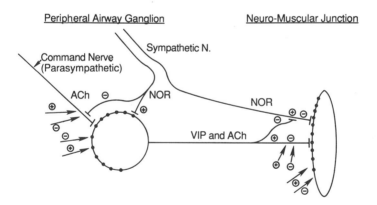

FIGURE 1. Neurotransmission schema for peripheral airway ganglia and the neuromuscular junction. Regarding the peripheral airway ganglion, only one type of ganglion cell, containing both ACh and VIP as excitatory and inhibitory neurotransmitter, is drawn. Although there are at least two different types of neurons in peripheral airway ganglia, classified on the basis of electrophysiological properties, the roles of the different neurons are obscure. Although it is possible that vasoactive intestinal peptide (VIP) and acetylcholine (ACh) are present in different neurons, the evidence seems to suggest co-transmitters in single neurons. Sympathetic neurons are drawn to synapse at both the ganglion and the neuromuscular junction. Plus (+) and minus (−) arrows indicate inputs that modulate effects of ACh at the nicotinic synapse, as discussed in the text. There is little evidence for these inputs (except for norepinephrine) in airway ganglia, and we speculate this possibility on the basis of studies on other autonomic ganglia. Regarding the neuromuscular junction, arrows indicate similar modulatory mechanisms as described for the ganglion, exerted both prejunctionally and postjunctionally. Plus (+) and minus (−) for the NOR effect reflect binding to α- and β-receptors.

III. BASIC PROPERTIES OF PERIPHERAL AIRWAY GANGLIA

Research on the function of peripheral airway ganglia has been stimulated by the development of isolated preparations that permit study using electrophysiological techniques (Cameron and Coburn, 1981; Baker *et al.*, 1983; Fowler and Weinreich, 1986). These preparations also permit intracellular microinjection of fluorescent markers or horseradish peroxidase (HRP), for the study of neuron morphology and circuitry. There have been advances in our knowledge about the anatomy and histology of airway ganglia from different species (Coburn, 1984; Baluk *et al.*, 1985; Baker *et al.*, 1986; Coburn and Kalia, 1986; see also Chapter 1, this volume). Although we have a better understanding of how peripheral airway ganglia operate, the field is still in its infancy, compared with the study of other autonomic ganglia. So far, scant information is available about the basic biophysical properties of neurons present in peripheral airway ganglia, mechanisms of modulation of neurotransmission, intracellular control mechanisms, and so forth. Airway ganglion cells have not yet been put in culture for the study of ion channels or other areas.

Electrophysiological peripheral airway ganglia experiments have been performed on the ferret (Cameron and Coburn, 1981), the rabbit (Fowler and Weinreich, 1986), and the cat (Mitchell *et al.*, 1987). The rabbit (Fowler and Weinreich, 1986) contains two different types of neurons, classified on the basis of whether channels involved in generating action potentials, inactivated or not. Neurons in the cat paratracheal ganglia have been classified on the basis of *in vivo* characteristics, whether they spike during inspiration or during expiration (Mitchell *et al.*, 1987). Of two general types of neurons identified in the ferret paratracheal ganglia (Cameron and Coburn, 1981, 1984), only one appears to be similar to either of the neurons described from rabbit paratracheal ganglia.

Transmission from preganglionic to postganglionic neurons has been studied in only one airway preparation, the ferret paratracheal ganglion, termed the paratracheal parasympathetic nerve-ganglion plexus. Since the major goal of this chapter is to discuss neurotransmission, comments are limited to this species; however, neurotransmission in nonpulmonary autonomic ganglion is discussed, as studies using these preparations are the basis for our current understanding of general principles of neurotransmission in autonomic ganglia.

Only the neurons present in the superficial chains of nerve trunks and ganglia have been studied so far in the ferret paratracheal ganglia (Fig. 2). Only somal microelectrode penetrations have been used in these studies. The two general types of neurons found in this preparation have been classified on the basis of whether injections of cathodal current evoke action potentials and by the types of postsynaptic potentials that can be evoked by electrical stimulation of the major nerve trunk that connects different ganglia, the interganlionic nerve trunk (IGNT) (Fig. 3). The spiking neurons were named AH cells, because of the large action potential afterhyperpolarization, similar to that observed in AH cells

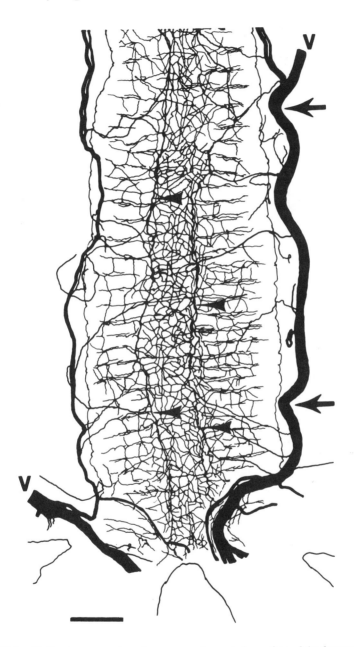

FIGURE 2. Cholinesterase mapping of neurons on the posterior surface of the ferret trachealis muscle. (Taken and modified from Baker *et al.*, 1986). Large arrows and V indicate veins; small arrowheads indicate the superficial plexus, which includes chains of ganglia and interganglionic nerve trunks. Bar: 2 μm.

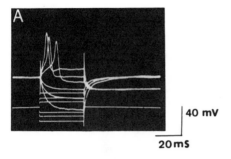

40 mV

20 mS

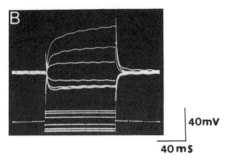

40mV

40 mS

FIGURE 3. Somal current injections into AH (A) and into type B (B) cells. Note the differences in response of cathodal current injection in the different types of ganglion neurons seen in ferret paratracheal ganglia.

in the guinea pig myenteric plexus (Wood, 1984). In these neurons, membrane depolarization, evoked by constant cathodal current injection, almost always evokes only one action potential, and it appears that inward currents are inactivated by depolarization (accommodation). (In this discussion, we use the convention wherein inward current refers to depolarizing currents carried by the cations Ca^{2+}, or Na^{+}; outward currents refers to hyperpolarizing currents carried by K^{+} ions. As expected, the evoked action potential is almost entirely inhibited by tetrodotoxin (TTX), which inhibits Na^{+} currents; TTX-resistant action potentials are likely attributable to inward Ca^{2+} current, since the rate of action potential depolarization is depressed by removing Ca^{2+} from the bathing solution, or by adding 1 mM Mn^{2+}, a Ca^{2+} antagonist ion (Coburn, 1988). The Ca^{2+} component of the action potential is important in that it appears that it can be modulated, which can cause changes of excitability of these neurons.

The input resistance of AH neurons decreases during the action potential afterhyperpolarization, and the action potential afterhyperpolarization is inhibited by hyperpolarizing the membrane potentials before evoking an action potential (Coburn, 1988). These findings indicate that afterhyperpolarization is due to an increase in K^{+} conductance. The afterhyperpolarization is not evoked by depolarizing the membrane potential to levels below the action potential threshold and is inhibited by the removal of Ca^{2+} (or increasing Mg^{2+}/Ca^{2+}) in the bathing solution or by the addition of 1 mM Mn^{2+}. These data strongly suggest that the outward K^+ currents that cause hyperpolarization are dependent on the influx of Ca^{2+} during the action potential, a mechanism established for the action potential hyperpolarization seen in other autonomic ganglia (North, 1973; Suzuki and Kusano, 1978; Pennefather et al., 1985; Gaban and Adams, 1982). Ferret paratracheal ganglion AH neurons are less excitable during the action potential afterhyperpolarization (Coburn, 1988), indicating an important role of the afterhyperpolarization in filtering neural inputs. The following section discusses evidence that altering the amplitude of the action potential afterhyperpolarization is a mechanism of altering neurotransmission.

Stimulation of preganglionic neurons results in the appearance in almost all AH cells of fast excitatory postsynaptic potentials (fast EPSPs), each of which evokes a single action potential (Cameron and Coburn, 1984). Single-pulse stimulation of the IGNT often causes multiple fast EPSPs, indicating that multiple preganglionic neurons synapse on single AH cells and that these presynaptic neurons have different conduction velocities. These potentials are inhibited in low Ca^{2+} or high Mg^{2+}/Ca^{2+}, and by hexamethonium, indicating the mechanism for generation of this fast EPSP involves binding of ACh released from presynaptic neurons to nicotinic receptors on postsynaptic neurons. AH neurons are usually present in every ganglion associated with the IGNT. We have been unable to record other fast or slow postsynaptic potentials in these neurons, as are found in neurons in other autonomic ganglia (Karczmar et al., 1986). These neurons have multiple dendrites and several axons that run for long distances in the IGNT, sometimes toward the larynx as well as the carina and apparently make contact with neurons in neighboring ganglia. However, the nature of this contact has not been determined. Axons also enter muscle. Intracellular cathodal current injection into AH soma result in excitatory junction potentials in smooth muscle cells that are completely inhibited by atropine (R. F. Coburn, unpublished data); therefore, at least some AH neurons innervate smooth muscle and contain ACh.

Type B cells have been classified on the basis that it is not possible to evoke somal action potentials with somal cathodal current injection. Electrical stimulation of the IGNT does not evoke a fast EPSP in these cells. At least some of these cells have been shown to be neurons, using HRP and lucifer yellow intracellular injection techniques (Coburn and Kalia, 1986). These cells appear not to be injured AH neurons, because of the different characteristics of postsynaptic potentials in AH and type B neurons (and because injured AH cells do not lose their fast EPSP or develop a slow postsynaptic potential). The inability to record action potentials in these cells probably means that the site of action potential generation is in dendrites or the axon hillock (Nishi and Christ, 1971) and that action potentials are not transmitted into the soma. Multiple electrical stimulation of the IGNT evokes a slow excitatory potential, which is not present in AH neurons (Cameron and Coburn, 1984). It is unlikely this potential is due to transmission of action potentials from action potential-generating sites on the axon hillock or dendrites, into the soma, because of the long duration of this potential following only a brief 2- or 4-Hz stimulation. In some type B cells, atropine inhibits this slow potential; this potential may be analogous to the muscarinic slow EPSP recorded in other parasympathetic ganglia (DeGroat and Booth, 1980; Griffith *et al.*, 1981b; Gallagher *et al.*, 1982). Type B cells project for long distances in the IGNT, apparently making contact with neurons in adjacent ganglia, as do AH cells. Although axons from type B cells are seen in smooth muscle, we have not as yet obtained direct evidence that these neurons innervate smooth muscle and the role of these neurons is not known.

Mitchell *et al.* (1987) classified neurons in the cat paratracheal ganglia, as studied *in vivo,* on the basis of neural inputs during inspiration and expiration. One class fires during inspiration and the other during expiration. It is possible that ferret AH and type B neurons may have different functions in control of smooth muscle tone during expiration and inspiration. Spontaneously firing neurons have not been observed in the *in vitro* ferret paratracheal ganglia, as occur in other parasympathetic ganglion (DeGroat and Booth, 1980).

The classification of neurons in the ferret paratracheal ganglion should be considered preliminary, in that it is likely that there will be neurons that provide axons to the epithelium and that others will provide axons to smooth muscle and other subclasses. Subclasses of neurons may provide inhibitory and excitatory inputs to various cells in the trachea.

IV. CIRCUITRY OF FERRET PARATRACHEAL GANGLION NEURONS

The anatomy of the paratracheal ganglion nerve plexus is shown in Fig. 2, which was taken from the work of Baker *et al.* (1986). Input from the central nervous system into this plexus occurs via branches of the laryngeal nerves, usually 6–8 per preparation (inlet nerves). There is usually a pair of major nerve trunks, made up of ganglia and IGNT in series, running parallel to the long axis of the trachea and perpendicular to the transversely oriented smooth muscle cells. The numbers of cell bodies in the ferret trachea have been determined using silver staining (Coburn, 1984a) or cholinesterase histochemistry (Baker *et al.*, 1986). In superficial ganglia connected to the IGNT, the number of cell bodies per trachea determined using either the silver or cholinesterase technique was found to be in the range 200–300, a value similar to that seen in mouse (Chiang and Gabella, 1986). However, cholinesterase histochemistry performed on the ferret trachealis muscle revealed about 10 times more cell bodies were present in areas not connected to the major nerve trunks, which were not detectable using silver stain. So far, studies of circuitry have considered only superficial neurons which have cell bodies in ganglia connected to IGNTs.

Mapping of transmission of signals evoked by IGNT or inlet nerve stimulation at single points in the preparation have revealed that the ganglion-nerve plexus apparently functions to coordinate neural input to smooth muscle cells (Coburn, 1984a). Single-pulse IGNT, or inlet-nerve, stimulation directed at a single point in the preparation, evokes fast EPSPs in AH cells located throughout the preparation. Similar stimulations also evoked excitatory junction potentials in smooth muscle cells located throughout the preparation. Thus, there is a marked divergence of signals initiated by single point nerve stimulation. There must be marked convergence of neural inputs to individual smooth muscle cells, arising from multiple inlet nerves. I analyzed conduction velocities of signals initiated by single-point nerve stimulation. Data suggest that rapidly conducting neurons are present in inlet nerves and that these neurons synapse with neurons in ganglia that are close to the inlet nerves. The signal is then carried by slowly conducting neurons that synapse with neurons present in other ganglia in the preparation. Electrically evoked signals are not carried over long distances via gap junctions. HRP morphological studies of single neurons in this preparation show that all the neurons studied to date send axons via the IGNT that pass through neighboring ganglia (Coburn and Kalia, 1986), as well as provide axons into smooth muscle and/or to the epithelium. No morphological evidence has been obtained for the presence of interneurons that function within an individual ganglion and do not have axons traveling long distances in the tissue and the IGNT. The ferret ganglion–nerve plexus microcircuitry has also been studied by determining the distribution of filled cell bodies after placing a cut end of the IGNT in lucifer yellow or HRP for long periods (Greene and Coburn, 1988). These studies, which show that cell bodies of axons running at a given point in the IGNT, are distributed in several different ganglia in the preparation, are consistent with the above-described circuit. In this study, HRP-filled neurons could be visualized intermingling in close aposition to cell bodies within the same ganglia (Fig. 4). This circuit shows major differences from circuits that have been studied in other autonomic ganglia. Using the guinea pig superior cervical ganglion and the submandibubular parasympathetic ganglion, studies by Purves and by Lichtman indicate that single presynaptic neurons synapse with multiple postsynaptic neurons, independent of topography (Lichtman *et al.*, 1979; Lichtman, 1980; Purves and Wigstron, 1983). Neurons innervated by single presynaptic neurons often are surrounded by neurons not innervated by this presynaptic neuron. However, to our knowledge, it is unknown in these structures whether postsynaptic neurons provide axons that synapse on other postsynaptic neurons, such as seems to occur in the ferret paratracheal ganglion-nerve plexus.

It is known that there is a rich afferent network of neurons in airway smooth muscle (see Chapter 3, this volume). Whether afferent neurons synapse with motor neurons in peripheral airway ganglia is unknown.

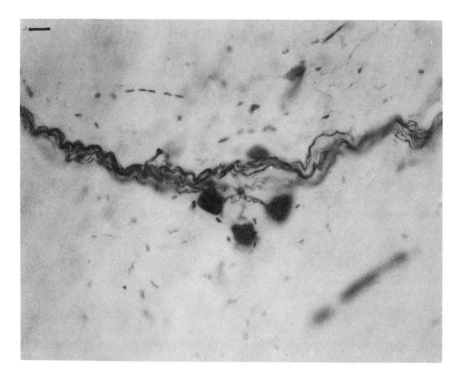

FIGURE 4. Wiring in a ferret paratracheal ganglion. The cut end of the interganglionic nerve trunk (IGNT) was placed in a well of horseradish peroxidase (HRP) for several hours and the specimen developed. This photograph shows axons filled with HRP, some of which had cell bodies in this ganglion; many axons passed through the ganglion. (From Greene and Coburn, unpublished data) Bar: 25 μm.

V. INTEGRATION AND MODULATION OF NEURAL INPUTS

A. Peripheral Airway Ganglia

The basis of our understanding of neurointegration in autonomic ganglia comes from the results of years of research in sympathetic, enteric, and other parasympathetic ganglia (Karczman *et al.*, 1986). It is known that ganglionic transmission in sympathetic and parasympathetic ganglia is more complex than simply a 1 : 1 monosynaptic transmission of pre- to postsynaptic neurons. Mechanisms of modulation of neurotransmission are present in various autonomic ganglia. These mechanisms can be classified as presynaptic or postsynaptic.

Presynaptic modulation of neurotransmission occurs via control of release of a neurotransmitter from preganglionic cholinergic, nicotinic neurons that provide signals generated in the CNS. In this discussion, we use the term command neurons to indicate neurons which are providing cholinergic nicotinic inputs to postsynaptic neurons, and the term modulator to indicate an agent that alters nicotinic synaptic function. In the parasympathetic neurons studied to date, the only known neurotransmitter released from command neurons, is ACh, which acts on postsynaptic neurons via a nicotinic mechanism producing a fast EPSP. The response of postsynaptic neurons to ACh released from command neurons in some parasympathetic ganglia includes an effect of ACh binding to muscarinic receptors, which evokes a slow excitatory postsynaptic potential (slow EPSP) (Nishi and Christ, 1971; DeGroat and Booth, 1980; Griffith et al., 1981b; Gallagher et al., 1982). Presynaptic modulation either increases or decreases the quanta of neurotransmitter released from a presynaptic command neuron terminal, per action potential. Presynaptic release of ACh can be altered by effects of modulators that bind to receptors on presynaptic nerve terminals and alter cellular signals which control ACh release. In nonairway autonomic ganglia, presynaptic effects on ACh release from command neurons are seen during administration of opiates (Simonds et al., 1983), norepinephrine (Christ and Nishi, 1969; DeGroat and Booth, 1980; Koketsu, 1981; Koketsu and Minota, 1975), seratonin (Akasu et al., 1981a), dopamine (Ashe and Libet, 1981), VIP (Kawatani et al., 1986), purinergic agonists (Akasu et al., 1981b), and histamine (Tamura et al., 1988). The synaptic ACh concentration may modulate ACh release via a presynaptic mechanism (Gallagher et al., 1982).

Facilitation is a phenomenon whereby there is a progressive augmentation of fast EPSP amplitude during multiple presynaptic stimulation. This phenomenon probably is due to a presynaptic mechanism; however, in some parasympathetic ganglia there may also be a mechanism for facilitation that operates postsynaptically, via generation of a muscarinic slow EPSP (DeGroat and Booth, 1980).

Postsynaptic mechanisms for modulating neurotransmission are more diverse. One mechanism is that the basic train of information from preganglionic command neurons that release ACh and activate neurons via nicotinic receptors, is modulated by the activation of peptidergic neurons that synapse on the same postsynaptic neuron. Modulators released from presynaptic neurons into the synaptic cleft alter the properties of postsynaptic neurons and their response to ACh released from command neurons, by evoking slow potentials, either excitatory or inhibitory. Modulators can also alter the properties of slow postsynaptic potentials (Ashe and Libet, 1981). Modulators can control neurotransmission by evoking changes in sensitivity or number of postsynaptic ACh receptors (Akasi et al., 1981a; Ohta et al., 1984) or by evoking changes in properties of channels controlling currents involved in the generation of action potentials or channels involved in controlling the action potential afterhyperpolarization (Koketsu and Minota, 1975, Koketsu, 1981; Gaban and Adams, 1982; Pennefather et al., 1985).

Of the many modulators or neurotransmitters that influence neurotransmission in various autonomic ganglia, norepinephrine has been the best studied to date. This agent has complex presynaptic and postsynaptic effects on neurotransmission. Some autonomic ganglia contain interneurons which release peptidergic neurotransmitters or modulators (Karczman et al., 1986). The SIF cell (Eranko and Harkonen, 1963) appears to function as an interneuron in both sympathetic and parasympathetic ganglia. This neuron contains multiple transmitters including norepinephrine, serotonin, and other peptidergic compounds.

The study of neurotransmission in peripheral airway ganglia is in its infancy, but there is evidence of several different mechanisms that modulate neurotransmission. The only reported example of presynaptic modulation is norepinephrine-induced inhibition of the amplitude of fast EPSPs in ferret AH cells, which is probably due to a norepinephrine-induced decrease in release of ACh/per action potention from command neurons (Baker et al., 1983; Coburn, 1988). This effect of norepinephrine can be seen at concentrations as low as 10^{-7} M, is completely reversible, is inhibited by phentolamine, and has a major inhibitory effect on transmission of presynaptic to postsynaptic AH neurons during stimulation of the IGNT (Coburn, 1988). Although a physiological role of norepinephrine in controlling neurotransmission has not been proved, the finding that adrenergic neurons synapse on neurons in peripheral airway ganglia in some species (Jacobowitz et al., 1973; Knight et al., 1981) supports this concept. Skoog (1986) showed a large inhibitory effect of norepinephrine on neurotransmission in ferret paratracheal ganglion, using an in vivo approach. Excitatory modulation, i.e., facilitation, has not been studied in any peripheral airway ganglion.

Norepinephrine has multiple postsynaptic effects on the ferret paratracheal AH cell (Coburn, 1988). This transmitter depolarizes the plasma membrane potential, decreases the rate of action potential upsweep, and inhibits the amplitude and duration of the action potential afterhyperpolarization. The norepinephrine-evoked decrease in the rate of action potential depolarization (Coburn, 1988) is similar to that seen after the muscle is placed in zero Ca^{2+}; thus, norepinephrine very likely acts by inhibiting the Ca^{2+} component of the action potential with resultant decrease in Ca^{2+} influx across the plasma membrane. The mechanism for norepinephrine-induced inhibition of action potential afterhyperpolarization amplitude also appears to be related to inhibition of the Ca^{2+} component of the action potential and consequent less activation of Ca^{2+}-induced late K^+ currents that appear to cause action potential afterhyperpolarization. In runs in which the IGNT is stimulated with trains at 10–20 Hz, it can be shown that whether an antidromic or orthodromic-evoked action potential is activated is a function of the afterhyperpolarization amplitude. During treatment with norepinephrine, the efficiency of transmission of presynaptic signals is decreased, despite a norepinephrine-evoked decrease in afterhyperpolarization amplitude. This finding serves to point out the complexity of modulation of neurotransmission, whereby the same agent has a major presynaptic effect, as well as several different postsynaptic effects.

Peripheral airway neurons contain multiple neuropeptides that may function as modulators (the ferret paratracheal ganglion so far has been shown to contain VIP and substance P (R. Dey and R. F. Coburn, unpublished data), and other neuropeptides have been seen in peripheral airway ganglia in other species (see Chapter 4, this volume). Airway ganglia are often in close proximity to mast cells that contain histamine, or seratonin (see Chapter 1, this volume). Histamine has been shown to exert an inhibitory effect on nicotinic synapses in enteric ganglia (Tamura et al., 1988), and ACh inhibits presynaptic ACh release in nicotinic synapses in the cat vesicular parasympathetic ganglia (Gallagher et al., 1982). VIP enhances presynaptic ACh release in the cat vesicular parasympathetic ganglion (Kawatani et al., 1986). It is unknown whether SIF cells are present in peripheral airway ganglia, and the area of delivery to various modulators to neurons in peripheral airway ganglia is not developed well enough to be able to determine which modulators function during normal or diseased conditions.

B. Neuromuscular Junction

The area of local control of release of neurotransmitters from postsynaptic neurons innervating end organs (prejunctional neurons) is well developed for adrenergic innervation of vascular tissue. This information is summarized by Vanhoutte et al. (1981). Similar mechanisms by which a given neurotransmitter or mediator can alter effects of another neurotransmitter appear to exist in airway smooth muscle. Most information relates to prejunctional mechanisms, but post-junctional effects are also operative. Mechanisms for prejunctional modulation are likely to be similar to mechanisms of presynaptic modulation of neurotransmission in ganglia. At the nerve–muscle junction, this mechanism can either result in augmentation or inhibition of the quanta of neurotransmitter released per action potential. Evidence supports a mechanism by which a plasma membrane receptor on the prejunctional nerve terminal can activate coupling mechanisms that alter the properties of voltage-dependent Ca^{2+} channels. This results in changes in the rate of influx of Ca^{2+} into the cell, and the cytosolic free $[Ca^{2+}]$, which is thought to be involved in fusion of vesicles (containing neurotransmitters) to the plasma membrane, an initial step in the process of neurotransmitter secretion (Vanhoutte et al., 1981).

The data base for neuromuscular modulation in airway smooth muscle can be divided into modulation via known neurotransmitters (which infers these mechanisms may operate during the normal state) and modulation occurring under pathological conditions. For the first case, evidence is available that ACh inhibits prejunctional release of norepinephrine (Russell and Bartlett, 1981). The evidence is getting stronger for an ACh-mediated inhibitory effect on prejunctional ACh release in airway smooth muscle (Fryer and Maclagan, 1984; Minette and Barnes, 1988). Prostaglandin E_2 (PGE$_2$) can decrease prejunctional ACh release in airway smooth muscle (Walters *et al.*, 1984; Inouet *et al.*, 1984; Inouet and Ito, 1985); however, it is unknown whether this occurs under normal conditions, such as has been shown to occur for the effect of this prostaglandin on norepinephrine release from adrenergic nerve terminales. Substance P, seratonin, leukotriene C_4, enkephalin, and thromboxane A_2 appear to increase prejunctional ACh release in various airway smooth muscle preparations (Tanaka and Grunstein, 1984; Inoue and Ito, 1985; Sheller *et al.*, 1982; Chung *et al.*, 1985; Seria and Daniel, 1988). Enkephalin appears to inhibit prejunctional ACh release in canine trachealis muscle (Russell and Simon, 1985). The importance of these findings needs to be studied.

ACKNOWLEDGMENTS. This work was supported by grant 1 R37 HL 37498 from the National Heart, Lung and Blood Institute, National Institutes of Health, Bethesda.

VI. REFERENCES

Akasu, T., Hirai, K., and Koketsu, K., 1981a, 5-Hydroxytryptamine controls ACh-receptor sensitivity of bullfrog sympathetic ganglion cells, *Brain Res.* **212:**217–220.

Akasu, T., Hirai, K., and Koketsu, K., 1981b, Increase of acetylcholine-receptor sensitivity by adnosine triphosphate: A novel action of ATP on ACh-sensitivity, *Br. J. Pharmacol.* **74:**505–507.

Altier, R. J., and Diamond, L., 1985, Effect of alpha chymotrypsin on the non-adrenergic, noncholinergic inhibitory system in cat airways, *Eur. J. Pharmacol.* **114:**75–78.

Altier, R. J., and Diamond, L., 1986, Role of vagal sensory fibers in nonadrenergic noncholinergic inhibitory responses in cat airways, *Am. Rev. Respir. Dis.* **133:**1159–1162.

Altier, R. J., Szarek, J. L., and Diamond, L., 1984, Neural control of relaxations in cat airway smooth muscle, *J. Appl. Physiol.* **57:**1536–1548.

Andersson, R. G., and Grundstrom, N., 1983, The excitatory non-cholinergic non-adrenergic nervous system of the guinea-pig airways, *Eur. J. Respir. Dis.* **131**(Suppl.):141–157.

Ashe, J. H., and Libit, B., 1981, Modulation of slow postsynaptic potentials by dopamine in rabbit sympathetic ganglion, *Brain Res.* **217:**93–106.

Baker, D. G., Basbaum, C. B., Herbert, D. A., and Mitchell, R. A., 1983, Transmission in airway ganglia of ferrets: Inhibition by norepinephrine, *Neurosci. Lett.* **41:**139–143.

Baker, D. G., McDonald, D. M., Basbaum, C. B., and Mitchell, R. A., 1986, The architecture of nerves and ganglia of the ferret trachea as revealed by acetylcholinesterase histochemistry, *J. Comp. Neurol.* **246:**513–526.

Baluk, P., Fumiwara, T., and Martsuda, S., 1985, The fine structure of the ganglia of the guinea pig trachea, *Cell Tissue Res.* **239:**51–60.

Blackman, J. G., Crawcroft, P. J., Devine, C. E., Holman, M. E., and Yonemura, K., 1969, Transmission from preganglionic fibres in the hypogastric nerve to peripheral ganglia of male guinea pigs, *J. Physiol. (Lond.)* **201:**723–743.

Bloomquist, E. I., and Kream, R. M., 1988, Leukotriene D4 acts in part to control guinea pig ileum smooth muscle by releasing substance P, *J. Pharmacol. Exp. Ther.* **240:**523–528.

Boyle, J. P., Davies, J. M., Foster, R. W., Morgan, G. W., and Small, R. C., 1987, Inhibitory responses to nicotine and transmural stimulation in hyoscine-treated guinea-pig isolated trachealis muscle, *Br. J. Pharmacol.* **90:**733–744.

Cabezas, G. A., Graf, P. D., and Nadel, J., 1971, Sympathetic versus parasympathetic nervous regulation of airways in dogs, *J. Appl. Physiol.* **31:**651–665.

Cameron, A. R., and Coburn, R. F., 1981, Electrical properties of the cells of the ferret paratracheal ganglion, *Physiologist* **24:**84.

Cameron, A. R., and Coburn, R. F., 1984, Electrical and anatomic characteristics of cells of the ferret paratracheal ganglion, *Am. J. Physiol.* **246:**C450–458.

Cameron, A. R., and Kirkpatrick, C. T., 1977, A study of excitatory neuromuscular transmission in the bovine trachea, *J. Physiol. (Lond.)* **270:**733–745.

Cameron, A. R., Bullock, C. G., and Kirkpatrick, C. T. 1982, The ultrastructure of bovine tracheal smooth muscle, *J. Ultrastruct. Res.* **81:**290–305.

Cameron, A. R., Johnston, C. F., Kirkpatrick, C. T., and Kirkpatrick, M. C. A., 1983, The quest for the inhibitory neurotransmitter in bovine tracheal smooth muscle, *Q. J. Exp. Physiol.* **68:** 413–426.

Chiang, C. H., and Gabella, G., 1986, Quantitative study of the ganglion neurons of the mouse trachea, *Cell Tissue* **100:**243–252.

Chesrown, S. E., Venugopalan, C. S., Gold, W. M., and Drazen, J. M., 1980, In vivo demonstration of nonadrenergic inhibitory innervation of the guinea pig trachea, *J. Clin. Invest.* **654:**315–320.

Christ, D. D., and Nishi, S., 1969, Presynaptic action of epinephrine on sympathetic ganglia, *Life Sci.* **8:**1235–1238.

Chung, K. F., Evans, T. W., Graf, P. D., and Nadel, J. A., 1985, Modulation of cholinergic neurotransmission in canine trachealis muscle by thromboxane mimetic U46619, *Br. J. Pharmacol.* **117:**373–375.

Coburn, R. F., 1984a, Neural coordination of excitation of ferret trachealis muscle, *Am. J. Physiol.* **246:**C459–466.

Coburn, R. F., 1984b, The anatomy of the ferret paratracheal parasympathetic nerve–ganglion plexus, *Exp. Lung Res.* **7:**1–9.

Coburn, R. F., 1987, Colinergic neuroeffector mechanisms in airway smooth muscle, in: *The Airways, Neural Control in Health and Disease,* Vol. 33: *Lung Biology in Health and Disease,* M. A. Kaliner and P. T. Barnes, eds.), pp. 159–186, Marcel Dekker, New York.

Coburn, R. F., and Kalia, M. P., 1986, Morphological features of spiking and nonspiking cells in the paratracheal ganglion of the ferret, *J. Comp. Neurol.* **254:**341–351.

Coburn, R. F., and Tomita, T., 1973, Evidence for nonadrenergic inhibitory nerves in the guinea pig trachealis muscle, *Am. J. Physiol.* **224:**1072–1080.

Coburn, R. F., 1987, Effect of norepinephrine on neurotransmission in the AH cell of the ferret paratracheal ganglion, *Fed. Proc.* **46:**704.

Coleman, R. A., and Levy, G. P., 1974, A non-adrenergic inhibitory nervous pathway in the guinea pig trachea, *Br. J. Pharmacol.* **52:**167–174.

Dahlstrom, A., Fuxe, K., Hokfelt, T., and Norberg, K. A., 1966, Adrenergic innervation of the bronchial muscle of the cat, *Acta Physiol. Scand.* **66:**507–508.

DeGroat, W. C., and Booth, A. M., 1980, Inhibition and facilitation in parasympathetic ganglia of the urinary bladder, *Fed. Proc.* **39:**2990–2996.

DeGroat, W. C., and Saum, W. A., 1971, Adrenergic inhibition in mammalian parasympathetic ganglia, *Nature New Biol.* **231:**188–189.

Dey, R. D., Shannon, J.W.A., and Said, S. A., 1981, Localization of VIP immunoreactive nerves in airway and pulmonary vessels of dogs, cats and human subjects, *Cell Tissue Res.* **222:**231–239.

Diamond, L., and O'Donnell, M., 1980, A nonadrenergic vagal inhibitory pathway in feline airways, *Science* **208:**185–188.

Eranko, O., and Harkonen, M., 1963, Histochemical demonstration of fluorgenic amines in the cytoplasm of sympathetic ganglion cells of the rat, *Acta Physiol.* **58**:285–286.

Fowler, J. C., and Weinreich, D., 1986, Electrophysiological membrane properties of paratracheal ganglion neurons of the rabbit, *Neurosci. Abstr.* **11**:1182.

Fryer, A. D., and Maclagan, J., 1984, Muscarinic inhibitory receptors in pulmonary parasympathetic nerves in the guinea pig, *Br. J. Pharmacol.* **83**:973–978.

Gaban, M., and Adams, P. R., 1982, Control of calcium current in rat sympathetic neurons by norepinephrine, *Brain Res.* **244**:155–144.

Gallagher, J. P., Griffith, W. J., III, and Shinnick-Gallagher, P., 1982, Cholinergic transmission in cat parasympathetic neurons, *J. Physiol. (Lond.)* **332**:473–486.

Goyal, R. K., Rattan, S., and Said, S. I., 1980, VIP as a possible neurotransmitter of non-cholinergic, non-adrenergic inhibitory neurones, *Nature (Lond.)* **288**:378–380.

Greene, J. H., and Coburn, R. F., 1988, VIP and the inhibitory neurotransmitter in the guinea pig trachealis muscle (submitted).

Griffith, W. H. III, Gallagher, J. P., and Shinnick-Gallagher, P., 1981a, Sucrose-gap recordings of nerve-evoked potentials in mammalian parasympathetic ganglia, *Brain Res.* **209**:446–451.

Griffith, W. H. III, Gallagher, J. P., and Shinnick-Gallagher, P., 1981b, Mammalian parasympathetic ganglia fire spontaneous action potentials and transmit slow potentials, in: *Advances in Physiological Sciences Vol 4. Physiology of Excitable Membranes* (J. Salanki, ed.), pp. 347–350, Adademial Kiado, Budapest.

Inoue, T., and Ito, Y., 1985, Pre- and post-junctional effects of prostaglandin I_2 and leukotriene C_4 in dog tracheal tissue. *Br. J. Pharmacol.* **84**:289–298.

Inoue, T., Ito, Y., and Takeda, K., 1984, Prostaglandin induced inhibition of acetylcholine release from neuronal elements of dog tracheal tissue, *J. Physiol. (Lond.)* **349**:553–570.

Jacobowitz, D., Kent, K. M., Fleisch, J. H., and Cooper, T., 1973, Histofluorescent study of catecholamine-containing elements in cholinergic ganglion from the calf and dog lung, *Proc. Soc. Exp. Biol. Med.* **144**:464–466.

Karczmar, A. G., Koketsu, K., and Nishi, S., 1986, *Autonomic and Enteric Ganglia*, Plenum, New York.

Kawatani, M., Rutigliano, M., and DeGroat, W. C., 1986, Selective facilitatory effects of vasoactive intestinal polypeptide on muscarinic mechanisms in sympathetic and parasympathetic ganglia of the cat, in: *Dynamics of Cholinergic Function*, (I. Hanin, ed.), Plenum, New York.

Knight, D. S., 1980, A light and electron microscopic study of feline intrapulmonary ganglia, *J. Anat. (Lond.)* **131**:413–428.

Knight, D. S., Hyman, A. L., and Kadowitz, P. J., 1981, Innervation of intrapulmonary airway smooth muscle of the dog, monkey and baboon, *J. Auton. Nerv. Syst.* **3**:31–43.

Koketsu, K., 1981, Electropharmacological actions of catecholamine in sympathetic ganglia: Multiple modes of actions to modulate the nicotinic transmission, *Jpn. J. Pharmacol.* **31**(Suppl.): 27P–28P.

Koketsu, K., and Minota, S., 1975, The direct action of adrenaline on the action potentials of bullfrog's sympathetic ganglion cells, *Experientia* **31**:822–823.

Laitinen, A., Partanen, M., Hervonen, A., and Laitinen, L. A., 1985, Electron microscopic study on the innervation of the human lower respiratory tract: Evidence of adrenergic nerves, *Eur. J. Respir. Dis.* **67**:209–215.

Leff, A. R., Munoz, N. M., and Hendrix, S. G., 1983, Parasympathetic and adrenergic contractile responses in canine trachea and bronchus, *J. Appl. Physiol.* **55**:113–120.

Lichtman, J. W., 1980, On the predominantly single innervation of submandibular ganglion cells in the rat, *J. Physiol. (Lond.)* **302**:121–130.

Lichtman, J. W., Purves, D., and Yip, J. W., 1979, On the purpose of selective innervation of guinea-pig superior cervical ganglion cells, *J. Physiol. (Lond.)* **292**:69–84.

Lundberg, J. M., and Hokfelt, T., 1983, Coexistence of peptides and classification of neurotransmitters, *Trends Neurosci.* **6**:325–333.

Lundberg, J. M., and Saria, A., 1982, Bronchial smooth muscle contraction induced by stimulation of capsaicin-sensitive sensory neurons, *Acta Physiol. Scand.* **116**:473–476.

Lundberg, J. M., Saria, A., Brodin, E., Rosell, S., and Folkers, K., 1983, A substance P antagonist inhibits vagally induced increase in vascular permeability and bronchial smooth muscle contraction of the guinea pig, *Proc. Natl. Acad. Sci. USA* **80**:1120–1124.

Mann, S., 1971, The innervation of mammalian bronchial smooth muscle: The localization of catecholamines and cholinesterases, *Histochem. J.* **3**:319–331.

Matsuzaki, Y., Hamasaki, Y., and Said, S. I., 1980, Vasoactive intestinal peptide: A possible transmitter of nonadrenergic relaxation of guinea pig airways, *Science* **210**:1252–1253.

Middendorf, W. F., and Russell, J. A., 1980, Innervation of airway smooth muscle in the baboon. Evidence for a nonadrenergic inhibitory system, *J. Appl. Physiol.* **48**:947–956.

Minette, P. A., and Barnes, P. J., 1988, Prejunctional inhibitory muscarinic receptors on cholinergic nerves in human and guinea pig airways, *J. Appl. Physiol.* **64**:2532–2537.

Mitchell, R. A., Herbert, D. A., Baker, D. G., and Basbaum, C. B., 1987, In vivo activity of tracheal parasympathetic ganglion cells innervating tracheal smooth muscle, *Brain Res.* **437**:157–160.

Nishi, S., and Christ, D. D., 1971, Electrophysiological and anatomical properties of mammalian parasympathetic ganglion cells, in: *Proceedings of the International Union of Physiological Sciences,* Vol. 9, pp. 421–431, German Physiological Society, Munich.

North, R. A., 1973, The calcium-dependent slow after hyperpolarization in myenteric neurons with TTX-resistant action potentials, *Br. J. Pharmacol.* **49**:709–711.

O'Donnell, S. R., and Saar, N., 1973, Histochemical localization of adrenergic nerves in the guinea pig trachea, *Br. J. Pharmacol.* **47**:707–710.

Ohta, Y., and Koketsu, K., 1984, Histamine as an endogenous antagonist of nicotinic Ach-receptor, *Brain Res.* **306**:370–373.

Pennefather, P., Lancaster, B., Adams, P. R., and Nicoll, R. A., 1985, Two distinct Ca-dependent K currents in bullfrog sympathetic ganglion cell, *Proc. Nat. Acad. Sci. USA* **82**:3040–3042.

Purves, D., and Wigston, D. J., 1983, Neural units in the superior cervical ganglion of the guinea-pig, *J. Physiol. (Lond.)* **334**:169–178.

Richardson, J., and Beland, J., 1976, Nonadrenergic inhibitory nerves in human airways, *J. Appl. Physiol.* **41**:764–771.

Russell, J. A., 1978, Responses of isolated canine airways to electrical stimulation and acetylcholine, *J. Appl. Physiol.* **45**:690–698.

Russell, J. A., and Bartlett, J., 1981, Adrenergic neurotransmission in airways: Inhibition by acetylcholine, *J. Appl. Physiol.* **52**:376–383.

Russell, J. A., and Simon, E. J., 1985, Modulation of cholinergic neurotransmission in airways by enkephalin, *J. Appl. Physiol.* **58**:853–858.

Seria, R., and Daniel, E. E., 1988, Thromboxane effect on canine trachealis neuromuscular function, *J. Appl. Physiol.* **64**:1979–1988.

Sheller, J. R., Holtzman, J. J., Skoogh, B. E., and Nadel, J. A., 1982, Interaction of serotonin with vagal and acetylcholine induced bronchoconstriction in canine lungs, *J. Appl. Physiol.* **52**:964–966.

Simonds, W. F., Booth, A. M., Thor, K. B., Ostrowski, N. L., Nagel, J. R., and DeGroat, W. C., 1983, Parasympathetic ganglia: Naloxone antagonizes inhibition by leucine-enkephalin and GABA, *Brain Res.* **271**:365–370.

Skoogh, B. E., 1986, Transmission through airways ganglia, *Eur. J. Respir. Dis.* **131**(Suppl.):159–170.

Suzuki, T., and Kusano, K., 1978, Hyperpolarizing potentials induced by Ca-mediated K conduction increase in hamster submandibular ganglion cells, *J. Neurobiol.* **9**:367–392.

Tamura, K., Palmer, J. M., and Wood, J. D., 1988, Presynaptic inhibition produced by histamine at nicotinic synapses in enteric ganglia, *Neurosci.* **25:**171–179.

Tanaka, D. T., and Grunstein, M. D., 1984, Mechanisms of substance P-induced contraction of rabbit airway smooth muscle, *J. Appl. Physiol.* **57:**1551–1557.

Uddman, R. J., Aluments, J., Senset, O., Hakanson, R., and Sundler, F., 1978, Occurrence and distribution of VIP nerves in the nasal mucosa and tracheal-bronchial wall, *Acta Otolaryngol. (Stockh.)* **86:**433–448.

Vanhoutte, P. M., Verbeurent, J., and Webb, R. C., 1981, Local modulation of adrenergic neuroeffector interactions in the blood vessel wall, *Physiol. Rev.* **61:**151–247.

Walters, E. H., O'Bryne, P. M., Fabbri, L. M., Graf, P. D., Holtzman, M. J., and Nadel, J. A., 1984, Control of neurotransmission by prostaglandins in canine trachealis smooth muscle, *J. Appl. Physiol.* **57:**128–134.

Wood, J. D., 1984, Enteric neurophysiology, *Am. J. Physiol.* **247:**G585–598.

Yip, P., Palombini, B., and Coburn, R. F., 1981, Inhibitory innervation to the guinea pig trachealis muscle, *J. Appl. Physiol.* **50:**374–382.

Chapter 3

Nervous Receptors in the Tracheobronchial Tree
Airway Smooth Muscle Reflexes

John G. Widdicombe

Department of Physiology
St. George's Hospital Medical School
London SW17 0RE, England

I. INTRODUCTION

Activity in motor nerves to airway smooth muscle can be influenced by a very large number of reflex inputs; in this respect, the control of tracheobronchial smooth muscle tone resembles other motor components of the autonomic nervous system. This chapter concentrates on the reflexes from the lower airways, where the afferent sense organs are already close to the motor tissues we are concerned with. However, activity originating in these end organs will also affect reflexly many other physiological variables, including some in the respiratory tract. Some of the reflex changes may have secondary influences on airway smooth muscle.

There seems to be a consensus that there are four groups of afferent end organ in the walls of the trachea and bronchi, although some of these groups may be subdivided (Barnes, 1986; Coleridge and Coleridge, 1986; Sant'Ambrogio, 1982; Widdicombe, 1986a).

II. PULMONARY STRETCH RECEPTORS

It has been known for many years that slowly adapting stretch receptors signal distention of the lungs and airways, and reflexly evoke the Breuer–Hering reflex; this consists of an inhibition of inspiration and a prolongation of expiration. Indirect evidence suggests that these receptors are found in the smooth muscle of the larger airways. Recently, this was convincingly confirmed by Krauhs (1984), who used three-dimensional electron microscopy to show that the receptors were small but complex structures entirely within the smooth muscle (Figs. 1 and 2). Physiological studies show that most of the endings are in the trachea and larger bronchi.

Although the structural connections between airway smooth muscle and collagen and the nervous receptors have not been fully clarified, physiological studies show that the receptors can be stimulated or sensitized by contraction of airway smooth muscle surrounding them (Sant'Ambrogio, 1982). Thus, they will be subject to a negative motor feedback control, since their reflex action is to relax smooth muscle.

The receptors have been extensively studied using the classic method of recording nerve impulses from single fibers from the receptors (Adrian, 1933; Knowlton and Larrabee, 1946; Widdicombe, 1954b). The nerves are myelinated with high conduction velocities and run almost exclusively in the vagus nerves. The receptors are slowly adapting, many spontaneously discharging at FRC, and the others have low-volume thresholds. They have a regular pattern of discharge. They seem to be insensitive to most chemical changes; inflammatory mediators

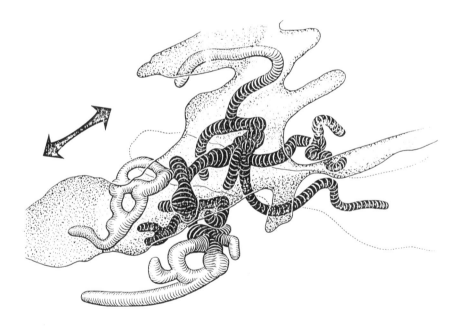

FIGURE 1.   Reconstruction from electron micrographs of a slowly adapting stretch receptor from the trachea of a dog. The stippled area represents smooth muscle and the tubular structure the receptor complex. The arrows show the longitudinal direction of contraction of the muscle. (From Krauhs, 1984.)

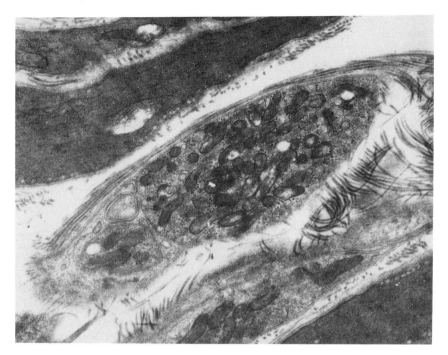

FIGURE 2.   Group of nerve fibers within smooth muscle of a mouse. This group includes one large fiber packed with mitochondria. The collagen surrounding the nerve is in particularly close association with the neurite. The nerve has the characteristics of an afferent fiber, possibly a pulmonary stretch receptor. (From Pack *et al.*, 1984.)

and drugs probably affect them mainly by a secondary mechanical effect due to contraction of airway smooth muscle. The important exception to this chemical insensitivity is the fact that many of them are inhibited by an increase in CO_2 concentration in the airway luminal gas (Bartlett and Sant'Ambrogio, 1976; Sant'Ambrogio, 1982). There has been much speculation as to whether this phenomenon is of physiological importance.

A. Reflex Actions on Airway Smooth Muscle

It was shown more than 80 years ago that, when the lungs were distended, as well as Breuer-Hering inhibition of breathing, there was a relaxation of airway smooth muscle (Kahn, 1907). This observation has been repeatedly confirmed (e.g., Loofbourrow *et al.*, 1957; Widdicombe and Nadel, 1963a) (Fig. 3). It may account for the transient (about 1 min) bronchodilation often seen in healthy humans who have performed a large inspiration (Orehek *et al.*, 1975). In experimental animals, this reflex airway dilation is predominantly attributable to inhi-

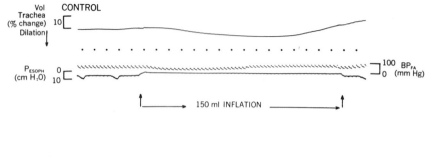

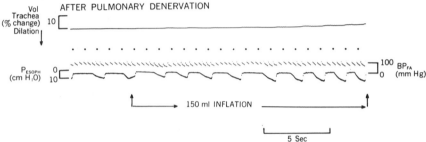

FIGURE 3. Effect of lung inflation (150 ml) on tracheal volume (vol trachea) in a spontaneously breathing dog before and after pulmonary denervation. P_{ESOPH}, esophageal pressure; BP_{FA}, femoral arterial blood pressure. Arrows indicate period of maintained inflation. Two sec after the onset of lung inflation, the trachea began to dilate. After ~10 sec of maintained inflation, the trachea began to constrict due to asphyxia. During lung inflation, the dog made no respiratory efforts (see P_{ESOPH}). After pulmonary denervation, maintained lung inflation caused no tracheal dilatation or inhibition of respiratory efforts but still caused late tracheal constriction. (From Widdicombe and Nadel, 1963b.)

bition of parasympathetic cholinergic motor tone; however, the roles of sympathetic and of nonadrenergic noncholinergic (NANC) nerves in this reflex do not seem to have been studied. At least in health, the latter two systems may not be important in this respect (Barnes, 1986); the lung inflation/bronchodilator reflex is prevented by cholinoceptor antagonists such as atropine, indicating that it works by decreasing cholinergic constrictor tone, although such a fully dilated airway would not be suitable for studying the two dilator nervous motor systems.

The functional advantage of the reflex is not very clear. Increased FRC would lead to greater distention of the airways (since the smooth muscle relaxes) than that due to mechanical factors alone, and this might be of benefit to patients with bronchoconstriction. The reflex might modulate motor activity to the airway smooth muscle, which is known to have phasic respiratory discharge. It could adjust airway geometry in different patterns of breathing to optimize the relationship between airways resistance and deadspace (Widdicombe and Nadel, 1963b).

B. Other Reflex Actions

Other reflex actions of slowly adapting pulmonary stretch receptors on the airways include an inhibition of the accessory skeletal muscles of respiration of the upper airways, corresponding to the inhibition of the diaphragm (Bartlett, 1986). The abductor (dilator) muscles of the larynx are also inhibited, so the laryngeal movements are arrested with the larynx held partly open. The receptors do not seem to have an appreciable reflex action on the vascular beds of the nose (Lung and Widdicombe, 1987) and tracheobronchial tree (Sahin *et al.*, 1987), and any reflex action they may have on airway mucus secretion has not been established.

III. RAPIDLY ADAPTING (IRRITANT) RECEPTORS

Rapidly adapting receptors also have vagal myelinated fibers but are of thinner diameter than those of pulmonary stretch receptors (Coleridge and Coleridge, 1986; Sant'Ambrogio, 1982). As their name implies, these receptors adapt rapidly to maintained lung inflations and also are activated by lung deflation. Their discharge has a characteristic irregular pattern. Furthermore, unlike the slowly adapting receptors, they are stimulated by a number of chemical irritants and inflammatory mediators such as histamine, serotonin, and prostaglandin $F_{2\alpha}$ ($PGF_{2\alpha}$) (Sampson and Vidruk, 1975). Cigarette smoke is one of the effective irritants. The extent to which the irritants and mediators act directly on the receptors, or indirectly by causing bronchoconstriction with a mechanical action on the receptor, is still being debated.

There is considerable evidence that the rapidly adapting receptors are the nonmyelinated fibers seen in, and possibly under, the epithelium of the large airways in many species (Widdicombe, 1986; Das *et al.*, 1978). The distribution of these fibers in the cat corresponds closely to that which can be mapped during recording activity in the single vagal myelinated fibers from rapidly adapting receptors (Widdicombe, 1954b; Das *et al.*, 1978). The nerves are established as afferent by degeneration experiments (Das *et al.*, 1979). Many lie close to the airway lumen (Fig. 4), where they would be well-positioned to respond to intraluminal irritants.

The rapidly adapting receptors are stimulated or sensitized during a number of lung pathological conditions, including microembolism, anaphylactic reactions, pneumothorax, pulmonary edema, and drug-induced bronchoconstriction (Coleridge and Coleridge, 1986). As with the action of mediators described above, it is not clear whether these responses are caused by the direct action of mediators released in lung pathology or are secondary to mechanical changes in the airways; in the case of pneumothorax and atelectasis, the latter explanation seems more likely.

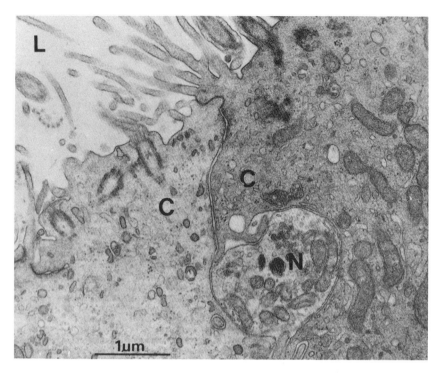

FIGURE 4.   Axon profile (N) with mitochondria and vesicles located closely to the airway lumen (L) between two ciliated cells (C), human tissue. (From Laitinen, 1985.) (×24,600)

A. Reflex Actions on Airway Smooth Muscle

Early evidence that rapidly adapting receptors caused reflex contraction of airway smooth muscle was largely indirect and was based on the correlation of properties of the receptors with those of the reflex (Mills *et al.*, 1970; Widdicombe, 1986a). Recently more direct experiments have been performed that seem to establish the tracheobronchoconstrictor reflex (Jammes *et al.*, 1985) (Fig. 5). The fact that the receptors are themselves stimulated or sensitized by contraction of the underlying smooth muscle points to the possibility of a vicious circle being set up and maintaining a bronchoconstriction. Such a mechanism could be of considerable importance in human conditions like asthma.

B. Other Reflex Actions

The rapidly adapting receptors in the trachea, especially at the carina, cause coughing when stimulated (Widdicombe, 1954a). Those deeper in the lungs seem

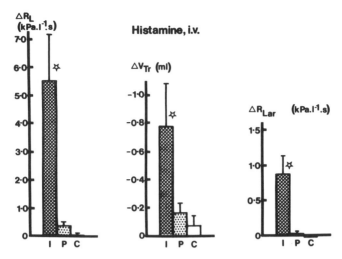

FIGURE 5. Effects of intravenous histamine to stimulate lung rapidly adapting irritant receptors on total lung resistance (R_L), tracheal volume (V_{tr}), and laryngeal resistance (R_{lar}) in dogs vagally intact (I), with partial vagal block (P), and with complete vagal blockade (C). Results are given as changes from control. $☆p < 0.05$. Note the changes in all three variables due to histamine, abolished by vagal blockade. (From Jammes *et al.*, 1985.)

more likely to cause hyperpnea. More than 40 years ago, it was suggested that the rapidly adapting receptors are responsible for augmented breaths or sighs (Larrabee and Knowlton, 1946), the occasional deep breaths taken by most mammals including humans. This view has recently been supported (Davies and Roumy, 1982). In addition, it has been shown that the receptors shorten expiration, thereby causing accelerated breathing (Davies and Roumy, 1986). Their effect on tidal volume (apart from augmented breaths) is less clear. It is obvious that the receptors can have complex and varied effects on the pattern of breathing, and the dominant response may depend on the localization of the receptors stimulated and possibly the number of receptors and their degree of activation.

Stimulation of rapidly adapting receptors causes laryngeal constriction (Stransky *et al.*, 1973; Jammes *et al.*, 1985) (Fig. 5), especially in the expiratory phase, whether or not coughing takes place. In the case of cough, the laryngeal closure is interrupted during the expulsive phase.

Reflex activation of the receptors evokes secretion of mucus from the trachea (Phipps and Richardson, 1976) and possibly other parts of the respiratory tract. The reflex role of the receptors on airway function cranial to the larynx does not seem to have been studied. Their reflex action on vascular beds, including those in the respiratory tract and lungs, has not been reported and may be weak or absent.

IV. C-FIBER RECEPTORS

Nonmyelinated fibers are seen in the epithelium of the larger airways, and these seem to be connected to myelinated fibers in the vagal trunk. However, a further group of afferent endings with nonmyelinated vagal fibers has been described (Coleridge and Coleridge, 1986; Sant'Ambrogio, 1982). The receptors have not been unequivocally identified, although nonmyelinated fibers occur in the alveolar wall and may be sensory (Meyrick and Reid, 1974; Fox *et al.*, 1980) (Fig. 6); degeneration studies would have to be done for this, and there might be difficulty in distinguishing between C-fiber endings and the branches of rapidly adapting receptors. Ultrastructural appearances of the nerves alone are not convincing evidence. However, in the vagus of the cat, nonmyelinated afferent fibers outnumber the myelinated by more than 3 : 1 (Coleridge and Coleridge, 1986).

It follows that most of the studies on C-fiber receptors have been physiological, in particular by recording afferent discharges from single vagal fibers (Paintal, 1973). These experiments show that the endings are stimulated by a large number of inflammatory mediators, including histamine, prostaglandins, serotonin, and bradykinin (Coleridge and Coleridge, 1984, 1986). They are also excited by exogenous agents, such as capsaicin and phenylbiguanide. Inhaled irritant gases stimulate the C-fiber receptors, and the endings increase their activity in a number of lung diseases, such as microembolism and pneumonia. In general, the patterns of response of the lung C-fiber receptors are similar to those found for C-fiber receptors in many viscera, the receptors seeming to respond nonspecifically to a wide variety of forms of tissue damage and pathology.

The C-fiber receptors can be subdivided into two groups: pulmonary and bronchial (Coleridge and Coleridge, 1984, 1986). The original distinction was made because the former were responsive to drugs injected into the right heart and the latter to those in the left heart, presumably reaching the receptors via the bronchial arteries. Although nonmyelinated fibers of presumed sensory receptors have been seen in the alveolar wall (Fig. 6), they have not been identified in the airway wall apart from the branches of rapidly adapting receptors in the large airway epithelium. The two groups of C-fiber receptor can also be distinguished by their responses to inflammatory mediators, especially bradykinin and prostaglandins.

A. Reflex Actions on Airway Smooth Muscle

Both the pulmonary and bronchial C-fiber receptors cause a reflex tracheobronchial constriction when activated (Coleridge *et al.*, 1982) (Figs. 7 and 8). Unlike the rapidly adapting receptors, there is no evidence that the C-fiber endings are sensitized by smooth muscle contraction. The reflex response seems to be predominantly due to an increase in vagal cholinergic tone to the smooth

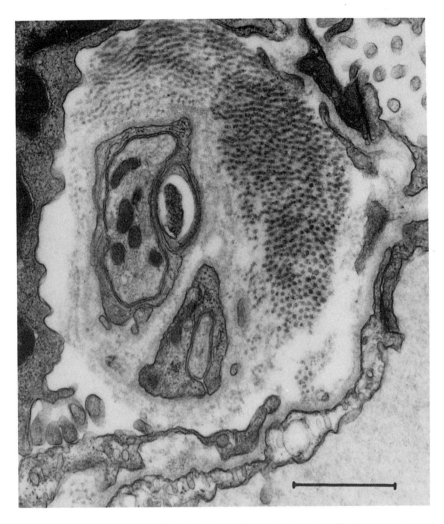

FIGURE 6. Two separate nerve fibers in the interstitium of the alveolar wall of humans. Air space (right) with surface of type II pneumocyte (top right) and fibroblast (left). Bar: 1 μm. (From Fox *et al.*, 1980.)

muscle. Recently it has been shown that if this cholinergic bronchoconstriction is prevented by drugs such as atropine, stimulation of C-fiber receptors can cause bronchodilation (Ichinose *et al.*, 1987); since this bronchodilation is not prevented by β-adrenoceptor antagonists, it is believed that it could be due to release of a neuropeptide such as VIP. The significance of such a phenomenon in relation to co-transmission in airway motor nerves is mentioned in Section 8.

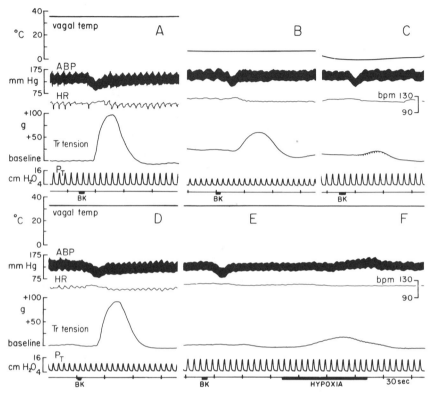

FIGURE 7. Effect of injecting bradykinin into right bronchial artery on smooth muscle tone in upper tracheal segment of dog. Recurrent laryngeal nerves are cut such that segment is innervated only by superior laryngeal nerves. (A–E) bradykinin (1.5 μg) is injected into bronchial artery at signal. (A) Both vagi at 39°C. (B) Vagi cooled to 7°C. Note increase in baseline tension. (C) Vagi cooled to 0°–1°C. Note decrease in baseline tension. (D) Vagi rewarmed. (D,E) Both cervical nerves are cut. (E) Response to bradykinin is abolished by vagotomy. (F) Tracheal contraction can still be evoked reflexly when carotid bodies are stimulated by ventilating lungs with 5% O_2 in N_2. At all other times (A–F), lungs are ventilated with 50% O_2 in air. (A,D) Initial baseline tracheal tension is set at 50 g; it is not adjusted in (B), (C), or (E). Note changes in tracheal pressure; lungs briefly hyperinflated between (A) and (B) and between (C) and (D). Reduced response after vagal cooling to 7°C (A,B) is not caused by blockade of myelinated vagal afferents but by temperature-dependent decrease in firing frequency in afferent C fibers. Vagal temp, temperature of cervical vagus nerves; ABP, arterial blood pressure; HR, heart rate; Tr, tension, tracheal tension in grams above resting baseline tension; P_T, tracheal pressure. (From Coleridge and Coleridge, 1986.)

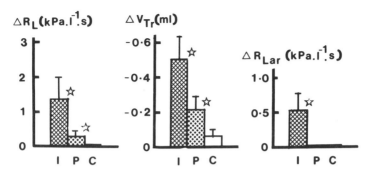

FIGURE 8. Effects of intravenous capsaicin to stimulate lung C-fiber receptors on total lung resistance (R_L), tracheal volume (V_{tr}), and laryngeal resistance (R_{lar}) in dogs vagally intact (I), with partial vagal block (P) and with complete vagal blockade (C). Results are given as changes from control. ☆$p < 0.05$. Note the changes in all three variables due to capsaicin, abolished by vagal blockade. (From Jammes et al., 1985.)

B. Other Reflex Actions

C-fiber receptors have manifold actions, including that on airway smooth muscle. On respiration they cause apnea, followed by rapid shallow breathing. On the other motor systems of the airways they cause secretion of mucus (Phipps and Richardson, 1976), laryngeal constriction, or even closure (Stransky et al., 1973; Jammes et al., 1985) (Fig. 8), as well as dilation of the nasal (Lung and Widdicombe, 1987) and tracheal (Sahin et al., 1987) vascular beds. On the systemic circulation, they cause bradycardia and hypotension due to dilation of many vascular beds. They can produce long-lasting inhibition of skeletal muscle, their injection into unanesthetized animals causing a kind of defensive paralysis (Ginzel and Eldred, 1977; Deshpande and Devanandan, 1970).

It has been suggested that C-fiber receptor stimulation can cause cough (Paintal, 1986); the main evidence is that inhaled capsaicin can cause cough and that capsaicin is known to stimulate C-fiber endings. However, rapidly adapting receptors also have nonmyelinated terminals and could be stimulated by capsaicin. Furthermore, in all the extensive literature on injection of "specific" C-fiber stimulants into anesthetized and unanesthetized animals, there does not seem to be a single report of cough.

V. LOCAL (AXON) REFLEX ACTIONS

It has been known for some years that noncholinergic nerves can be found in the airways, which cause smooth muscle contraction; recently, evidence has accumulated that C-fiber receptors can cause axon reflex responses due to release of neuropeptides (Barnes, 1986; McDonald, 1987; Lundberg and Saria, 1983) (Fig. 9). The receptors branch out over a wide but undefined area of tissue, and activation of one part of the receptor complex can lead to spread throughout the entire complex with motor actions at its terminals. The response can be mimicked by antidromic stimulation down sensory nerves (McDonald, 1987) and by appropriate field-stimulation experiments.

This important concept has been shown to apply especially to axon reflex vasodilation in the airways, including the nose and lower airways (Lundblad, 1984). Some experiments suggest that the same mechanism applies to smooth muscle contraction, especially near the hilum of the lungs (Barnes, 1986; Karlsson and Persson, 1983). Thus, activation of C-fiber nerves in the epithelium

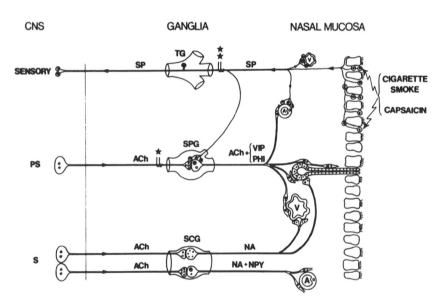

FIGURE 9. Schematic illustration of nasal autonomic innervation. Preganglionic sympathetic (S) neurons contain acetylcholine (ACh) and originate from thoracic region of spinal cord. These neurons relay in superior cervical ganglion (SCG). Most postganglionic noradrenergic (NA) fibers run together (not shown) with preganglionic parasympathetic (PS) nerve to form so-called Vidian nerve (*), which enters sphenopalatine ganglion (SPG). Sympathetic nerves predominantly innervate blood vessels (A, arterioles; V, venous sinusoids and venules), whereas adrenergic innervation of glands is sparse. Neuropeptide Y (NPY) occurs in noradrenergic periarterial nerves. Postganglionic cholinergic fibers innervate both blood vessels and exocrine glands. Vasoactive intestinal peptide (VIP) and peptide histidine isoleucine (PHI) are present in sphenopalatine ganglion cells, presumably together with ACh. Sensory supply from trigeminal ganglion (TG) travels in maxillary portion (**) and joins posterior nasal nerves. Substance P (SP) neurons constitute population of trigeminal afferents with peripheral branches within respiratory epithelium as well as around arterioles, venules, and sphenopalatine ganglion cells. Local axon connections of sensory nerves are shown. (From Lundblad, 1984.)

would cause local contraction of smooth muscle underlying this zone. The same nervous activation would presumably lead to central nervous reflexes as impulses pass up the vagus nerves to the brain.

A number of neuropeptides have been suggested to be responsible for this phenomenon. Originally substance P was the most popular, but more recently neurokinins A and B and calcitonin gene-related peptide have been shown to occur in airway nerves (Uddman and Sundler, 1987). The relative importance of these peptides has not been established, and there may be species differences. However, the concept of a local axon reflex mechanism, including bronchoconstriction, is important and seems to be accepted.

VI. NEUROEPITHELIAL BODIES

Neuroendocrine cells containing bioactive peptides can be seen in the epithelium of the airways of many species and are sometimes collected into clusters called neuroepithelial bodies (Lauweryns *et al.*, 1982; Pack and Widdicombe, 1984) (Fig. 10). These are especially common in fetal and neonatal airways and

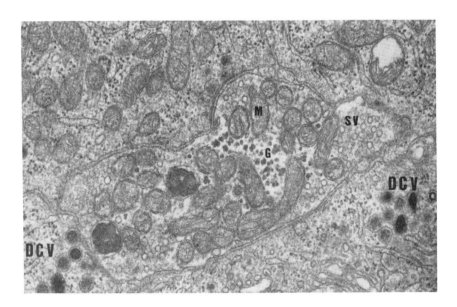

FIGURE 10.   An intracorpuscular nerve ending in a neuroepithelial body (NEB) in a rabbit. Type Ia nerve endings are morphologically afferent, containing numerous mitochondria (M) and relatively few synaptic vessels (SV). This nerve ending also contains glycogen granules (G). In the surrounding corpuscular cells, dense-cored vesicles (DCV) are seen. (From Lauweryns *et al.*, 1985.) (×45,000)

in some species such as the rabbit. They have been described in humans. The neuroepithelial bodies are innervated. Early controversy as to whether the innervation was motor or sensory or both has not been dispelled, but recent evidence shows that at least some of the nerves are certainly afferent (Lauweryns *et al.*, 1985). Indirect (morphological) evidence suggests that the stimulus for nervous discharge is hypoxia, which may also release peptides such as bombesin and calcitonin gene-related peptide from the cells. The peptides could act as neurotransmitters to cause nerve discharge and/or have actions on motor tissues such as vascular beds. The scarcity of physiological compared with morphological studies of the neuroepithelial bodies limits our understanding of their function at present.

VII. OTHER REFLEX ACTIONS ON AIRWAY SMOOTH MUSCLE

A. Reflexes from the Upper Respiratory Tract

The nose and pharynx are powerful sites for reflex changes in airway smooth muscle tone (Widdicombe, 1963, 1986b; Nadel, 1980). They are clearly important in this respect both in airway pathophysiology and in changes in bronchomotor tone induced by inhalation of gases, aerosols, and agents.

From the nose, both bronchoconstriction and bronchodilation have been described, and it seems probable that more than one type of afferent end organ exists in the nose and that more than one type of response can be obtained (Widdicombe, 1986b, 1987). However, we know little about the structure and physiology of nasal afferent endings.

For the larynx, irritation and mechanical stimulations consistently cause a tracheobronchial constriction (Widdicombe, 1986b; Mathews and Sant'Ambrogio, 1987). Laryngeal receptors have been extensively studied by physiological means, and at least four types of receptor have been identified (Widdicombe *et al.*, 1987). Of these four, those that respond to irritants and chemical changes seem most likely to cause bronchoconstriction. The larynx also has C-fiber receptors, which have been little studied, and the role of these on airway smooth muscle tone has not been defined.

In addition to the action of nasal and laryngeal receptors on airway smooth muscle, they have manifold effects on other motor systems in the body. Some of these are listed in Table I.

B. Bronchomotor Reflexes from Other Sites

Bronchomotor tone is affected by reflexes from peripheral and central chemoreceptors, arterial baroreceptors, skeletal muscle receptors, pain endings in many parts of the body, temperature receptors, and probably other groups of

TABLE I
Reflex Control of Airway Motor Functions[a]

	Airway responses				
Afferent input	Smooth muscle	Mucus glands	Larynx	Nasal vessels	Tracheal vessels
Nose	+/−	+	Const	Dilat	?
Larynx	+	+	Const	?	Dilat
Lung, SAR	−	0	Dilat	0	0
Lung, RAR	+	+	Const	?	?
Lung, C fiber	+	+	Const	Dilat	Dilat
Peripheral chemoreceptor	+	+	Dilat	Const	(Const)
Central chemoreceptor	+	+	Dilat	Const	Const
Arterial baroreceptor	−	?	?	?	0

[a] +, smooth muscle contraction or gland secretion; −, smooth muscle relaxation; 0, no effect; Const, constriction; Dilat, dilatation; ? unknown; SAR, slowly adapting receptor; RAR, rapidly adapting receptor.

afferent end organ (Widdicombe, 1963; Nadel, 1980). Table I lists some of these. It is clear not only that bronchomotor tone represents the cumulative interaction of large numbers of inputs, constrictor and dilator, but that the actions of these reflexes on other physiological variables may produce secondary changes in bronchomotor tone. For example, any reflex that changes blood pressure or respiration will affect bronchomotor tone by both primary and secondary mechanisms. Quantitation of some of these interactions has been successfully attempted.

VIII. MOTOR PATHWAYS TO TRACHEOBRONCHIAL SMOOTH MUSCLE

Most studies indicate that, at least in experimental animals, the dominant motor control of tracheobronchial smooth muscle is vagal and cholinergic and that the reflexes described above act by modulation of this tone. However, the airways also receive an innervation via sympathetic nerves, at least in some species; even if this pathway is only weakly dilated, changes in its activity could modify the reflex responses described (Barnes, 1984, 1986).

Of greater current interest and probably importance is the role of the NANC systems to the airways, discussed elsewhere (Barnes, 1984, 1986; Said, 1987). At least two recent studies have shown that two clear and dominant bronchoconstrictor reflexes, from the larynx (Fig. 11) and from lung C receptors, respectively, can be reversed to bronchodilations if the cholinergic component of the motor side is pharmacologically blocked (Szarek et al., 1986; Ichinose et al.,

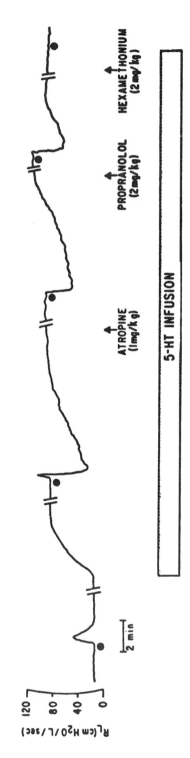

FIGURE 11. Changes in lung resistance (R_L) caused by mechanical stimulation of the larynx (filled circles) in an anesthetized artificially ventilated cat. Initially, the larynx was stimulated under basal conditions (far left). Thereafter, baseline airway muscle tone was enhanced by a continuous infusion of 5-hydroxytryptamine (5-HT) in order to observe bronchodilator responses. (From Szarek *et al.*, 1986.)

1987). If there is co-transmission in motor nerves to the smooth muscle, the two transmitters being constrictor and dilator, respectively, we need to re-evaluate our concept of bronchomotor reflexes. The result of such re-evaluation, together with consideration of the role of axon reflexes described earlier, is much fruitful research and interest in the nervous control of tracheobronchial muscle in health and disease.

IX. REFERENCES

Adrian, E. D., 1933, Afferent impulses in the vagus and their effect on respiration, *J. Physiol. (Lond.)* **79**:332–358.

Barnes, P. J., 1984, The third nervous system in the lung: Physiology and clinical perspectives, *Thorax* **39**:561–567.

Barnes, P. J., 1986, Neural control of human airways in health and disease, *Am. Rev. Respir. Dis.* **134**:1289–1314.

Bartlett, D., Jr., 1986, Upper airway motor systems, in: *Handbook of Physiology, Part 3: The Respiratory System,* Vol. II: *Control of Breathing* (N. S. Cherniack and J. G. Widdicombe, eds.), pp. 223–246, American Physiological Society, Bethesda, Maryland.

Bartlett, D., Jr., and Sant'Ambrogio, G., 1976, Effects of local and systemic hypercapnia on the discharge of stretch receptors in the airways of the dog, *Respir. Physiol.* **26**:91–99.

Coleridge, H. M., and Coleridge, J. C. G., 1986, Reflexes evoked from the tracheobronchial tree and lungs, in: *Handbook of Physiology, Part 3: The Respiratory System,* Vol. II: *Control of Breathing* (N. S. Cherniack and J. G. Widdicombe, eds.), pp. 395–430, American Physiological Society, Bethesda, Maryland.

Coleridge, J. C. G., and Coleridge, H. M., 1984, Afferent vagal C-fiber innervation of the lungs and airways and its functional significance, *Rev. Physiol. Biochem. Pharmacol.* **99**:1–110.

Coleridge, J. C. G., Coleridge, H. M., Roberts, A. M., Kaufman, P., and Baker, D. G., 1984, Tracheal contraction and relaxation initiated by lung and somatic afferents in dogs, *J. Appl. Physiol.* **52**:984–990.

Das, R. M., Jeffery, P. K., and Widdicombe, J. G., 1978, The epithelial innervation of the lower respiratory tract of the cat, *J. Anat.* **126**:123–131.

Das, R. M., Jeffery, P. K., and Widdicombe, J. G., 1979, Experimental degeneration of intra-epithelial nerve fibers in cat airways, *J. Anat.* **128**:259–263.

Davies, A., and Roumy, M., 1982, The effect of transient stimulation of lung irritant receptors on the pattern of breathing in rabbits, *J. Physiol. (Lond.)* **324**:389–401.

Davies, A., and Roumy, M., 1986, A role of pulmonary rapidly adapting receptors in control of breathing, *Aust. J. Exp. Biol. Med. Sci.* **64**:67–78.

Deshpande, S. S., and Devanandan, M. S., 1970, Reflex inhibition of monosynaptic reflexes by stimulation of type J pulmonary endings, *J. Physiol. (Lond.)* **206**:345–357.

Fox, B., Bull, T. B., and Guz, A., 1980, Innervation of alveolar walls in the human lung: An electron microscopic study, *J. Anat.* **131**:683–692.

Ginzel, H. K., and Eldred, E., 1977, Reflex depression of somatic motor activity from heart, lungs and carotid sinus, in: *Krogh Centenary Symposium on Respiratory Adaptations, Capillary Exchange and Reflex Mechanisms* (A. S. Paintal and P. Gill-Kumar, eds.), pp. 358–394, Vallabhbhai Patel Chest Institute, Delhi.

Ichinose, M., Inoue, H., Miura, M., Yafuso, N., Nogami, H., and Takishima, T., 1987, Possible sensory receptor of nonadrenergic inhibitory nervous system, *J. Appl. Physiol.* **63**:923–929.

Jammes, Y., Davies, A., and Widdicombe, J. G., 1985, Tracheobronchial and laryngeal responses to hypercapnia, histamine and capsaicin in dogs, *Clin. Respir. Physiol.* **21**:515–520.

Kahn, R. H., 1907, Zur Physiologie der Trachea, *Arch. Anat. Physiol.* pp. 398–426.

Karlsson, J-A., and Persson, C. G. A., 1983, Evidence against vasoactive polypeptide (VIP) as a dilator and in favour of substance P as a constrictor in airway neurogenic responses, *Br. J. Pharmacol.* **79:**634–636.

Knowlton, G. C., and Larrabee, M. G., 1946, A unitary analysis of pulmonary volume receptors, *Am. J. Physiol.* **147:**100–114.

Krauhs, J. M., 1984, Morphology of presumptive slowly adapting receptors in dog trachea, *Anat. Res.* **210:**73–85.

Laitinen, A., 1985, Ultrastructural organisation of intraepithelial nerves in the human airway tract, *Thorax* **40:**488–492.

Larrabee, M. G., and Knowlton, G. C., 1946, Excitation and inhibition of phrenic motoneurones by inflation of the lungs, *Am. J. Physiol.* **147:**90–99.

Lauweryns, J. M., Cokelaere, M., and Theunynck, P., 1972, Neuro-epithelial bodies in the respiratory mucosa of various mammals, *Z. Zellforsch.* **135:**569–592.

Lauweryns, J. M., Van Lommel, A. T., and Dom, R., 1985, Innervation of rabbit intrapulmonary neuroepithelial bodies, *J. Neurol. Sci.* **67:**81–92.

Loofbourrow, G. N., Wood, W. B., and Baird, I. L., 1957, Tracheal constriction in the dog, *Am. J. Physiol.* **191:**411–415.

Lundberg, J. M., and Saria, A., 1983, Capsaicin-induced desensitization of the airway mucosa to cigarette smoke, mechanical and chemical irritants, *Nature (Lond.)* **302:**251–253.

Lundblad, L., 1984, Protective reflexes and vascular effects in the nasal mucosa elicited by activation of capsaicin-sensitive substance P-immunoreactive trigeminal neurones, *Acta Physiol. Scand.* **529**(Suppl.): 1–42.

Lung, M. A., and Widdicombe, J. G., 1987, Lung reflexes and nasal vascular resistance in the anesthetized dog, *J. Physiol. (Lond.)* **386:**465–474.

Mathew, O. P., and Sant'Ambrogio, F., 1987, Laryngeal reflexes, in: *Respiratory Function of the Upper Airway* (G. Sant'Ambrogio and O. P. Mathew, eds.), pp. 259–302, Dekker, New York.

McDonald, D. M., 1987, Neurogenic inflammation in the respiratory tract: Actions of sensory nerve mediators on blood vessels and epithelium of the airway mucosa, *Am. Rev. Respir. Dis.* **136:** S65–S71.

Meyrick, B., and Reid, L., 1971, Nerves in rat intra-acinar alveoli: An electron microscopic study, *Respir. Physiol.* **11:**367–377.

Mills, J. E., Sellick, H., and Widdicombe, J. G., 1970, Epithelial irritant receptors in the lungs, in: *Breathing: Hering–Breuer Centenary Symposium* (R. Porter, ed.), pp. 77–92, Churchill Livingstone, London.

Nadel, J. A., 1980, Autonomic regulation of airway smooth muscle, in: *Physiology and Pharmacology of the Airways* (J. A. Nadel, ed.), pp. 217–257, Dekker, New York.

Orehek, J., Gayrard, P., Grimaud, C., and Charpin, J., 1975, Effect of maximal respiratory manoeuvres on bronchial sensitivity of asthmatic patients as compared to normal people, *Br. Med. J.* **1:**123–125.

Pack, R. J., and Widdicombe, J. G., 1984, Amine-containing cells of the lung, *Eur. J. Respir. Dis.* **65:**559–578.

Pack, R. J., Al-Ugaily, L. H., and Widdicombe, J. G., 1984, The innervation of the trachea and extrapulmonary bronchi of the mouse, *Cell Tissue Res.* **238:**61–68.

Paintal, A. S., 1973, Vagal sensory receptors and their reflex effects, *Physiol. Rev.* **53:**159–227.

Paintal, A. S., 1986, The visceral sensations—Some basic mechanisms, in: *Visceral Sensation* (F. Cervero and J. F. B. Morrisson, eds.), pp. 3–20, Elsevier, Amsterdam.

Phipps, R. J., and Richardson, P. S., 1976, The effects of irritation at various levels of the airway upon tracheal mucus secretion in the cat, *J. Physiol. (Lond.)* **261:**563–581.

Sahin, G., Webber, S. E., and Widdicombe, J. G., 1987, Lung and cardiac reflex actions on the tracheal vasculature in anaesthetised dogs, *J. Physiol. (Lond.)* **387:**47–57.

Said, S. I., 1987, Influence of neuropeptides on airway smooth muscle, *Am. Rev. Respir. Dis.* **136:** S52–S58.

Sampson, S. R., and Vidruk, E. H., 1975, Properties of "irritant" receptors in canine lung, *Respir. Physiol.* **25:**9–22.

Sant'Ambrogio, G., 1982, Information arising from the tracheobronchial tree of mammals, *Physiol. Rev.* **62:**531–569.

Stransky, A., Szereda-Przestaszewska, M., and Widdicombe, J. G., 1973, The effect of lung reflexes on laryngeal resistance and motoneurone discharge, *J. Physiol. (Lond.)* **231:**417–438.

Szarek, J. L., Gillespie, M. N., Altiere, R. J., and Diamond, L., 1986, Reflex activation of the noradrenergic noncholinergic inhibitory nervous system in feline airways, *Am. Rev. Respir. Dis.* **133:**1159–1162.

Uddman, R., and Sundler, F., 1987, Neuropeptides in the airways: A review, *Am. Rev. Respir. Dis.* **136:**S3–S8.

Widdicombe, J. G., 1954a, Respiratory reflexes from the trachea and bronchi of the cat, *J. Physiol. (Lond.)* **123:**55–70.

Widdicombe, J. G., 1954b, Receptors in the trachea and bronchi of the cat, *J. Physiol. (Lond.)* **123:** 71–104.

Widdicombe, J. G., 1963, Regulation of tracheobronchial smooth muscle, *Physiol. Rev.* **43:**1–37.

Widdicombe, J. G., 1986a, Sensory innervation of the lungs and airways, in: *Visceral Sensation* (F. Cervero and J. F. B. Morrisson, eds.), pp. 49–64, Elsevier, Amsterdam.

Widdicombe, J. G., 1986b, Reflexes from the upper respiratory tract, in: *Handbook of Physiology, Section 3: The Respiratory System,* Vol. II: *Control of Breathing* (N. S. Cherniack and J. G. Widdicombe, eds.), pp. 363–394, American Physiological Society, Bethesda, Maryland.

Widdicombe, J. G., 1987, Nasal and pharyngeal reflexes: Protective and respiratory functions, in: *Respiratory Functions of the Upper Airway* (G. Sant'Ambrogio and O. P. Mathew, eds.), pp. 233–258, Dekker, New York.

Widdicombe, J. G., and Nadel, J. A., 1963a, Reflex effects of lung inflation on tracheal volume, *J. Appl. Physiol.* **18:**681–686.

Widdicombe, J. G., and Nadel, J. A., 1963b, Airway volume, airway resistance, and work and force of breathing: Theory, *J. Appl. Physiol.* **18:**863–868.

Widdicombe, J. G., Sant'Ambrogio, G., and Mathew, O. P., 1987, Nerve receptors of the upper airway, in: *Respiratory Function of the Upper Airway* (G. Sant'Ambrogio and O. P. Mathew, eds.), pp. 193–232, Dekker, New York.

Chapter 4

Polypeptide-Containing Neurons and Their Function in Airway Smooth Muscle

Sami I. Said

Department of Medicine
University of Illinois College of Medicine
Chicago, Illinois 60612

I. NEUROPEPTIDES AS PHYSIOLOGICAL REGULATORS

A large number of biologically active peptides have recently been identified and characterized in the central and peripheral nervous systems of mammals as well as those of primitive invertebrates (Said, 1980; Dockray, 1979). These neuropeptides, acting centrally or locally, are now recognized as one of the principal means of neurohormonal regulation of various body functions. The occurrence of bronchoactive and vasoactive peptides in the lung was first documented 20 years ago (Said, 1967). An increasing number of peptides in the lung have since been detected, localized, and chemically and biologically characterized by radioimmunoassay (RIA), immunofluorescence, and biochemical and bioassay techniques (Said, 1982, 1984) (Table I).

II. AUTONOMIC INNERVATION OF AIRWAYS: PEPTIDERGIC NEURONS

The innervation of the airways with adrenergic and cholinergic nerves has been known and described for some time (Nadel and Barnes, 1984; Richardson, 1979), but our knowledge of peptide-containing nerves supplying the airways is more recent and less complete. It is already apparent, however, that neuropeptides have a major influence on airway smooth muscle function and that the

TABLE I
Some Neuropeptides That May Influence Airway
Smooth Muscle Function

ACTH	Neuropeptide Y
Angiotensin II	Neurotensin
Atrial natriuretic peptide (atriopeptin)	Opioids
Bombesin (or Gastrin-Releasing Peptide)	Oxytocin
Bradykinin	Peptide histidine methionine (PHM)
Calcitonin gene-related peptide	Sauvagine
Cholecystokinin	Somatostatin
Corticotropin-releasing factor	Substance P
Galanin	Thyrotropin-releasing hormone
Leukotrienes C4 and D	Vasoactive intestinal peptide (VIP)
Neurokinins A and B	Vasopressin

classic descriptions of autonomic and sensory innervation of the lung must be revised to include the peptidergic, as well as adrenergic and cholinergic, components (Dey and Said, 1985).

The airways are supplied by motor nerves, comprising the sympathetic and parasympathetic autonomic nervous system, and by sensory nerves that originate from sensory ganglia of the vagus nerve or from dorsal root ganglia (Dey and Said, 1985). The sensory nerves, with nerve cell bodies located mainly in the vagal ganglia, project axons into the lung that terminate on bronchi, blood vessels, and other structures. These fibers are in part responsible for the afferent limb of reflexes that carry impulses from irritant, stretch, and chemoreceptors in the lung to the central nervous system (CNS) (Coleridge et al., 1973; Widdicombe, 1982). Parasympathetic preganglionic nerve fibers project axons to the lung along the vagus nerve. Postganglionic neurons, arising from intrapulmonary ganglia in the adventitia and submucosa of the airways (Coburn, 1984), innervate vascular and bronchial smooth muscle and bronchial smooth muscle and bronchial glands (Dey and Said, 1985). Some peptides, such as vasoactive intestinal peptide (VIP), occur in sympathetic ganglia, as well as in preganglionic and postganglionic parasympathetic nerve fibers, and in intrapulmonary ganglia (Dey et al., 1981; Lundberg et al., 1978). Other peptides, notably the neurokinins and calcitonin gene-related peptide (CGRP), are found mainly in the peripheral projections of sensory nerves (Lundberg and Saria, 1987).

Despite recent advances, relatively little is known about peptidergic and other autonomic innervation of the lung and their interactions, compared with what is known about the corresponding innervation of the gastrointestinal (GI) tract (Wood, 1984).

III. IDENTIFICATION, LOCALIZATION, AND DISTRIBUTION OF PEPTIDERGIC NEURONS

The demonstration by RIA or immunohistochemistry of a given peptide immunoreactivity does not establish the presence of that particular peptide in that location. Thus, structurally related peptides, such as gastrin and cholecystokinin (CCK) or bombesin and gastrin-releasing peptide (GRP), exhibit sufficiently similar immunoreactivity as to be indistinguishable except by the most specific and well-characterized antisera. Full confirmation, therefore, requires either chemical characterization of the peptide following its isolation (Mutt, 1983) or its identification by newer biochemical techniques (Eipper et al., 1986).

Large variations in the content of certain peptides have been noted in the fetal, neonatal, and adult lungs of the same species (Ghatei et al., 1983), as well as in different species (Dockray, 1979). Among these variations are the higher content of bombesin in fetal versus adult lungs (Ghatei et al., 1983) and the lower density of tachykinin- and CGRP-immunoreactive nerves in human than in guinea pig airways (Lundberg and Saria, 1987).

IV. FAMILIES OF NEUROPEPTIDES

On examining the structure and biological effects of neuropeptides, several distinct groups, or families, can be identified (Table II). Each includes peptides that are not only structurally related but that also exhibit similar or comparable biological actions, although frequently differing in potency or effectiveness in producing a given effect.

TABLE II
Some Families of Neuropeptides
and Their Principal Members[a]

Bombesin, gastrin-releasing peptide
CCK, gastrin, cerulein
Neurokinins (tachykinins): neurokinin A, neurokinin B, substance P
Neuropeptide Y, pancreatic polypeptide
Opioid peptides: β-endorphin, enkephalins, dynorphins
VIP, seretin, glucagon, PHI (or PHM), helodermin, CRF, GRF,
  Sauvagine, urotensin, gastric inhibitory peptide

[a]CCK, cholecystokinin; PHI, peptide histidine isoleucine; PHM, peptide histidine methionine; CRF, corticotropin-releasing factor; GRF, growth hormone-releasing factor.

V. PEPTIDES AS NEUROTRANSMITTERS AND NEUROMODULATORS

A. When Are Neuropeptides Neurotransmitters?

In order for a neuropeptide, or other candidate neurotransmitter, to qualify for this function, it must fulfill a set of basic criteria, including (1) localization and biosynthesis in neurons; (2) concentration in nerve terminals or their biochemical equivalent, synaptosomal fractions; (3) release by depolarizing stimuli in a calcium-dependent manner; (4) similarity of the effects of exogenous application to those of endogenous material released; (5) binding to specific receptors on target cells and tissues; and (6) degradation, inactivation, or reuptake at these sites. At least some of these criteria have been satisfied for most neuropeptides occurring in the lung, and a few peptides, e.g., VIP, have already been shown to meet all these requirements (Giachetti et al., 1977; Said, 1980, 1986).

B. Neurotransmitter versus Endocrine and Paracrine Secretions

The nervous system and the endocrine system employ a variety of secretory products to accomplish the same physiological purpose: that of mediating, modulating, and regulating neuroendocrine function. These secretory products provide a spectrum of chemical messengers designed for communication at short range (neurotransmitters and paracrine secretions) or at long range (blood-borne hormones) (Scharrer, 1978).

Despite the fundamental similarity in function of circulating hormones and paracrine or neurocrine secretions, the distinction between them is important, as it determines the experimental approach to these peptides and defines the criteria required to establish their physiological status (Said and Zfass, 1978). Thus, to establish the hormonal relevance of a peptide, it is necessary first to demonstrate the existence of a physiological mechanism capable of releasing the peptide into the circulation, and then to show that the circulating levels of the peptide are capable of reproducing its postulated action. Blood levels are irrelevant to paracrine or neurocrine effects, which can result from locally high concentrations not reflected in peripheral blood and mimicked only by large amounts of injected peptide (Laitinen et al., 1985).

The same peptide may serve more than one messenger role in different locations: VIP released by the adenohypophysis into the portal hypophyseal vessels is a neuroendocrine secretion, but elsewhere it acts mainly as a neurotransmitter. Bombesin, or GRP, a neuropeptide with neurotransmitter properties, is also present in neuroendocrine cells in the airways, where it is believed to serve a paracrine function (Ghatei et al., 1982).

C. Coexistence of Peptides and Other Neurotransmitters

There is increasing evidence that two or more peptides frequently coexist in the same neuron. In many instances, a classic neurotransmitter, such as acetylcholine (ACh), norepinephrine, or dopamine, is also present with the neuropeptides (Lundberg and Hökfelt, 1983) (Table III). The coexistence in the same neuron of these chemical messengers multiplies the opportunities for interactions among these transmitters and adds to the complexity of the regulatory mechanisms.

D. How Neuropeptides May Influence Respiratory Function

In general, neuropeptides may exert their regulatory influence on particular organ systems in one or more of several ways (Scharrer, 1978):

1. As neurotransmitters in the CNS, acting on central regulatory neurons
2. As neurotransmitters in the peripheral (autonomic) nervous system, acting locally at receptor sites on different cells and tissues
3. As neurohormones, reaching their target organs via a specialized circulation (e.g., the hypothalamic releasing-factors, transported by the hypophyseal portal vessels) or via the general circulation (e.g., oxytocin and vasopressin)

In the regulation of airway and other respiratory function, neuropeptides most commonly act through their presence and release near specific receptors in the airways, pulmonary vessels and alveoli. Certain peptides, e.g., thyrotropin-releasing hormone, may also influence function through their release near key

TABLE III
Examples of Coexistence of Neuropeptides
and Classic Neurotransmitters in Some Neurons

Peptide	Other peptide or "classic" neurotransmitter
CCK	Dopamine
CGRP	Substance P, acetylcholine
Enkephalin	Dompamine, norepinephrine, serotonin
Neuropeptide Y	Epinephrine, norepinephrine
Neurotensin	Dopamine, epinephrine
Somatostatin	Norepinephrine, γ-aminobutyric acid (GABA)
Substance P	Neurokinin A, CGRP, serotonin, acetylcholine
TRH	Serotonin
VIP	Acetylcholine, PHI(PHM), vasopressin, substance P

regulatory areas in the brain with major influence on respiratory function (Holtman *et al.*, 1986). In both cases, peptides elicit their action either directly or by modifying or modulating the actions of other hormones or neurotransmitters.

E. Bronchoactive Peptides

The majority of peptides with activity on airway smooth muscle are bronchoconstrictors:

1. *Bombesin:* Both bombesin, found in nonmammalian tissues, and its mammalian counterpart, GRP (McDonald *et al.*, 1979), elicit contraction of many smooth muscle organs, including larger airways. GRP-immunoreactive nerve fibers have been localized in the respiratory tract (Uddman *et al.*, 1984).

2. *Activated complement fragment C5A anaphylatoxin:* This complement can be generated in the lung in inflammatory conditions (Shaw *et al.*, 1985), constricts airways and alveolar ducts (Stimler *et al.*, 1983), and induces other changes associated with acute lung injury (Till and Ward, 1985).

3. *CCK:* This peptide has dominant contractile activity on gallbladder (and secretory activity on exocrine pancreas) and causes brief contraction of trachea.

4. *Leukotrienes C_4 and D_4:* The potent airway-constrictor activity of these peptide-leukotrines has been well documented and is reviewed elsewhere (Said, 1986).

5. *Neurokinins:* This family of chemically related neuropeptides is derived from a common precursor (Otsuka and Konishi, 1983; Nawa *et al.*, 1983). Originally named tachykinins (Erspamer and Anastasi, 1966), they induce rapid contractions of isolated intestine, compared with the slower contraction induced by bradykinin. Included in the family of neurokinins are neurokinin A, neurokinin B, and substance P, of which neurokinin A is the most potent airway constrictor. Neurokinins are predominantly localized in sensory neurons in the airway and elsewhere, often with more than one neurokinin contained within the same neuron (Lundberg *et al.*, 1984).

6. *Spasmogenic lung peptide:* Still incompletely identified but apparently distinct from all known peptides (Said and Mutt, 1977), spasmogenic lung peptide contracts trachea, small airways (peripheral lung), and other smooth muscle (Said and Mutt, 1977).

7. *CGRP:* Variously reported to exhibit potent constrictor activity (Palmer *et al.*, 1985) or no activity (Lundberg *et al.*, 1985) on airway smooth muscle, CGRP is relatively inactive on guinea pig trachea (H. D. Foda

and S. I. Said, unpublished observations). Like the neurokinins,and often coexisting with them in the same neurons, CGRP is present mainly in sensory neurons, sensory ganglia, and axons projecting to the trachea and lower airways (Lundberg and Saria, 1987).

Peptides that relax airway smooth muscle make up a shorter list:

1. *Vasoactive intestinal peptide (VIP):* First discovered in the intestine (Said and Mutt, 1970) and later rediscovered as a neuropeptide (Said and Rosenberg, 1976), VIP relaxes guinea pig (Wasserman *et al.,* 1982), rabbit, cat, and human airway smooth muscle (Saga and Said, 1984) and prevents or reduces bronchoconstriction induced by histamine, $PGF_{2\alpha}$ (Said *et al.,* 1982), serotonin (Diamond *et al.,* 1983), leukotriene D4 (Hamasaki *et al.,* 1983), or neurokinin. The most potent endogenous airway relaxant known, VIP, is 50 times as potent as isoproterenol (Barnes, 1986). VIP binds to specific receptors in lung membranes (Robberecht *et al.,* 1982). Its airway-relaxant activity is independent of prostaglandin release (Saga and Said, 1984) and is more pronounced on larger airways.
2. *Peptide histidine methionine (PHM) or peptide histidine isoleucine (PHI):* PHM in human tissues and PHI in subhuman mammals is cosynthesized (Itoh *et al.,* 1983), colocalized and coreleased with VIP. PHM/PHI shares many of the actions and receptors of VIP, including relaxation of tracheobronchial smooth muscle and binding to lung receptors (Robberecht *et al.,* 1982), but it is generally less potent than VIP.
3. *Atrial natriuretic peptide (ANP):* Localized mainly in cardiac atria and brain (Ballerman and Brenner, 1985), ANP is also present in the lung (Kubota *et al.,* 1986). It is a potent relaxant of airways and an even more potent relaxant of pulmonary vessels (Chou *et al.,* 1986).
4. *Vasopressin and oxytocin:* These peptides contract, respectively, systemic vascular and uterine smooth muscle, relax guinea pig trachea with moderate potency (H. D. Foda and S. I. Said, unpublished observations).

Other peptides occurring in the lung include the following:

1. *Neuropeptide Y (NPY):* Localized in nerve fibers and nerve terminals in airways and pulmonary vessels (Sheppard *et al.,* 1984), NPY appears to coexist with epinephrine or norepinephrine in adrenergic neurons (Lundberg *et al.,* 1982).
2. *Galanin:* Galanin-like immunoreactivity, demonstrated in nerve fibers in airways and pulmonary vessels, is also present in local airway ganglia (Cheung *et al.,* 1985).

F. Interactions between Neuropeptides and Classic Neurotransmitters

The presence of neuropeptides in central and peripheral neurons, and their frequent coexistence in the same neurons with other peptides and classic neurotransmitters provide a setting for functional interactions among these peptides and between the peptides and other neurotransmitters (Lundberg and Hökfelt, 1983). A growing number of such interactions have been demonstrated, as illustrated below for two of the principal families of neuropeptides with activity on airway smooth muscle.

1. VIP

Most of the interactions between VIP and other neurotransmitters have been reported in other tissues, but they are probably applicable to the regulation of airway smooth muscle.

1. VIP coexists and is co-released with PHI (or PHM), in many neurons, with ACh as well (Lundberg et al., 1984, 1979). VIP also coexists with substance P in some neurons within the lung (Dey et al., 1988).
2. VIP and ACh have mutually synergistic effects. By contrast, VIP enhances the affinity of ACh for muscarinic receptors and ACh-induced secretion of saliva (Lundberg et al., 1980). The blood flow-promoting effect of VIP in salivary glands is, in turn, potentiated by acetylcholine (Lundberg et al., 1980). VIP also stimulates the release of ACh from myenteric neurons (Yau et al., 1985) and induces depolarization and muscarinic excitation in sympathetic ganglionic synapses (Kawatani et al., 1985).
3. Both cholinergic agonists (Bitar et al., 1980; Wang et al., 1986) and pretreatment with atropine (Hedlund et al., 1983) stimulate VIP release. VIP release is also enhanced by dopaminergic agonists (Uvnäs-Moberg et al., 1982), γ-aminobutyric acid (GABA) antagonists (Wang et al., 1986), and opiate antagonists (Wang et al., 1986).
4. VIP acts synergistically with norepinephrine to stimulate cyclic adenosine monophosphate (cAMP) production in the cerebral cortex (Magistretti and Schorderet, 1984) and with ATP to relax gastric smooth muscle (Bitar and Makhlouf, 1982).

2. Neurokinins

Several important interactions are known to exist between neurokinins and the cholinergic system, among neurokinins, and with the functionally related peptide, CGRP. The contraction of rabbit tracheobronchial smooth muscle induced by substance P is accentuated by neostigmine, a cholinesterase inhibitor, and is substantially inhibited by atropine (Tanaka and Grunstein, 1984). The bronchoconstrictor action of substance P in that species is therefore partially mediated by the accelerated release of ACh at the airway neuromuscular junction (Tanaka and Grunstein, 1984). The bronchoconstrictor effects of neurokinins in the guinea pig and in humans, however, appear to be independent of muscarine receptor activation (Lundberg and Saria, 1987; H. D. Foda and S. I. Said, unpublished observations).

Selective release to the point of depletion of substance P from sensory nerve endings results from treatment with capsaicin, the irritant ingredient of hot peppers of the genus *Capsicum* (Jansco *et al.*, 1977; Lundberg *et al.*, 1985; Lundberg and Saria, 1983). This release is probably associated with the release of other neurokinins as well as of CGRP, all of which are contained within the same neurons (Lundberg *et al.*, 1985).

VI. NEUROPEPTIDE RECEPTORS ON SMOOTH MUSCLE

A. VIP and PHI/PHM

Specific receptors for VIP have been identified in membrane preparations of normal mammalian and human lungs and in human lung tumor cells (Taton *et al.*, 1981). As for VIP binding to receptors in all other cells and tissues (Frandsen *et al.*, 1978), its binding to airway sites is coupled to an adenylate cyclase. The secondary increase in cAMP levels is believed to mediate the airway relaxation and other biological effects of the peptide.

The VIP-induced rise in tissue cAMP content has also permitted the immunocytochemical localization of VIP receptors in airways: in submucosal serous and mucous glands of the ferret and in ciliated and basal cells of tracheal epithelium of dog (Lazarus *et al.*, 1986). VIP receptors have also been localized by autoradiography in bronchial smooth muscle and in alveolar cells (Leroux *et al.*, 1984; Carstairs and Barnes, 1986a,b; Said and Dey, 1988; Amiranoff and Rosselin, 1982).

The binding of VIP to smooth muscle receptors and the resultant smooth muscle relaxation have been successfully investigated in preparations of isolated gastric smooth muscle (Bitar and Makhlouf, 1982) and in vascular smooth muscle (Huang and Rorstad, 1983). Similar studies in tracheobronchial smooth muscle have not been reported. VIP receptors in the lung have recently been solubilized and partially characterized (Paul and Said, 1987; Provow and Veliçelebi, 1987). PHI and PHM also bind to bronchial receptors, resulting in smooth muscle relaxation, in much the same way as VIP does, but at lower potency.

B. Substance P and Other Neurokinins

Substance P-specific receptors have been identified in guinea pig and human airway smooth muscle (Carstairs and Barnes, 1986). The use of substance P antagonists has aided in defining the actions of this peptide, although the specificity of these antagonists is uncertain, owing to their local anesthetic effect (Regoli et al., 1987).

Subclasses of receptors for the three related neurokinins—substance P, neurokinin A, and neurokinin B—have been described in the airways as well as in the pulmonary artery and other sites based on the relative biological potencies of the peptides and their binding affinities (Karlsson et al., 1984).

C. Atrial Natriuretic Peptide

Pulmonary receptors for atrial natriuretic peptide have been purified and partially characterized (Kuno et al., 1986). As in other organs and tissues, the binding of atrial natriuretic peptide to these receptors is linked to stimulation of intracellular, particulate guanylate cyclase (Kuno et al., 1986). Specific binding of atrial natriuretic peptide has been demonstrated in membrane preparations of guinea pig bronchi and pulmonary artery (E. Kubota, L.-W. Liu, and S. I. Said, unpublished observations).

VII. MECHANISMS OF BRONCHIAL RELAXATION AND CONSTRICTION BY NEUROPEPTIDES

The mechanisms by which neuropeptides—and other bronchoactive agents—induce or facilitate relaxation or contraction of airway smooth muscle are incompletely understood but include one or more of the following:

1. *Binding to specific receptors:* Specific airway receptors have been identified for several groups of neuropeptides, including VIP/PHM, the neurokinins, and atrial natriuretic peptide. The binding of these peptides to their receptor sites determines their respective biological activities.
2. *Release of secondary mediators:* The contribution of secondary mediators is illustrated by two examples: (a) At least in some species, the bronchoconstrictor effect of the neurokinins is partially mediated by the release of ACh and is attenuated by muscarinic blockade (Tanaka and Grunstein, 1984). (b) The contraction of isolated guinea pig trachea induced by bradykinin is abolished by indomethacin and is thus attributable to the release of cyclo-oxygenase products (Said et al., 1980).
3. *Intracellular second messengers:* The binding of VIP to its receptors is linked to the stimulation of an adenylate cyclase and that of atrial

natriuretic peptide to a guanylate cyclase. These events led to the intracellular accumulation, respectively, of cAMP and cyclic guanyl monophosphate (cGMP). Despite their different intracellular second messengers, the two peptides share at least two major biological actions: bronchial relaxation and vascular relaxation. Other intracellular second messengers of importance in mediating neuropeptide actions include cytosolic calcium (Hassid, 1986) and inositol phosphates (Berridge and Irvine, 1984). The contributions of intracellular messengers to neuropeptide actions on airways remain inadequately understood.

4. *Modulation of adrenergic or cholinergic transmitters:* The modulation of certain muscarinic cholinergic activities by VIP and by substance P, and the potentiation of some adrenergic actions by VIP, demonstrate how neuropeptides can act as modulators of other transmitters.

VIII. IMPORTANCE OF AIRWAY EPITHELIUM IN RESPONSES TO BRONCHOACTIVE AGENTS

The vasoactivity of several peptides and of other compounds has been shown to depend on the presence of intact vascular endothelium (Furchgott, 1983). Similar observations on airways suggest that tracheobronchial epithelium play a correspondingly important role in airway responses to bronchoactive agents. In experiments on isolated strips of guinea pig trachea, removal of epithelium enhanced the contraction induced by ACh and attenuated the relaxation induced by VIP, as well as by norepinephrine and sodium nitroprusside (Chou and Said, 1987; Flavahan et al., 1985). The mechanisms by which airway epithelium can modify airway smooth muscle tone are unknown. In the same way that vascular endothelium generates one or more smooth muscle-relaxant factor(s) [endothelium-derived relaxant factor(s)], tracheobronchial epithelium may release one or more factors promoting airway relaxation [epithelium-derived relaxant factor(s)].

IX. NEUROPEPTIDES AS PHYSIOLOGICAL REGULATORS OF AIRWAY SMOOTH MUSCLE FUNCTION

Most of the data available on the possible physiological significance of neuropeptides in the regulation of airway function relate to VIP and to the neurokinins.

A. VIP

A physiological role for VIP in regulating airway smooth muscle function is suggested by (1) the presence of VIP-containing nerve fibers and nerve terminals in the smooth muscle layer of the tracheobronchial tree (Dey *et al.*, 1981), (2) its potent relaxant activity on airways *in vitro* and *in vivo,* and (3) its binding to specific receptors on smooth muscle and other airway structures.

The likely role for VIP is as a transmitter of the nonadrenergic, non-cholinergic (NANC) component of the autonomic nervous system of the lung (Coburn and Tomita, 1973; Richardson and Beland, 1976). Two functions that are mediated or influenced by this system are the relaxation of airways and pulmonary vessels and the secretion of bronchial water, chloride ion, and macromolecules (Said and Dey, 1988). VIP may also modulate the release of mast cell mediators of inflammation (Undem *et al.*, 1983).

With special reference to NANC relaxation of airway smooth muscle, the following observations are in favor of VIP as a likely transmitter:

1. VIP fulfills the criteria of a neurotransmitter (Giachetti *et al.*, 1977; Said *et al.*, 1980).
2. VIP is present in the vagus nerve (Lundberg *et al.*, 1978); one means of demonstrating NANC airway relaxation *in vivo* is by electrical vagal stimulation in the presence of cholinergic and adrenergic blockade (Irvin *et al.*, 1980).
3. VIP mimics the electrophysiological changes in airway smooth muscle elicited by NANC nerve stimulation (Cameron *et al.*, 1983; Ito and Takeda, 1982).
4. Prolonged incubation of airway smooth muscle with VIP reduces subsequent relaxation by VIP (tachyphylaxis) and also reduces the magnitude of NANC relaxation in cats (Ito and Takeda, 1982) and guinea pigs (Venugopalan *et al.*, 1984).
5. VIP was released on transmural stimulation of guinea pig tracheal segments, in the presence of adrenergic and muscarinic cholinergic receptor blockade, and the release was proportionate to the magnitude of relaxation of these segments (Matsuzaki *et al.*, 1980).
6. In the same experiments, both the VIP release and the associated tracheal relaxation were markedly inhibited by blockade of neurotransmission with tetrodotoxin (Matsuzaki *et al.*, 1980).
7. The relaxation was also greatly reduced following incubation of the tracheal segments with a specific VIP-antiserum (Matsuzaki, 1980).

A number of other findings have been interpreted against such a role for VIP. These findings and alternative interpretations are as follows:

1. Immunoreactive VIP is present in the canine tracheobronchial tree, but the NANC relaxant system is weak or absent in this species (Russell, 1980). This apparent discrepancy may be explained by the less intimate

anatomical relationship between VIP-containing nerves and airway smooth muscle in dog airways, compared with those of cats or humans (Dey and Said, 1985).

2. Guinea pig tracheal segments, already fully relaxed by application of VIP, relaxed further with transmural (electrical field) stimulation (Karlsson and Persson, 1983). This additional relaxation could be due to the release of other relaxant peptides, especially PHI (or PHM), which is contained within the same VIP-containing nerve terminals in the airways. Also, some relaxation induced by electrical field stimulation may, under certain conditions, result from direct (myogenic) relaxation (Coburn and Tomita, 1973).

3. The proteolytic enzyme α-chymotrypsin, which hydrolyzes VIP *in vitro* (Mutt and Said, 1974), was found not to inhibit the relaxation induced by electrical field stimulation (Diamond and Altiere, 1988). This lack of inhibition may have resulted from the failure of the enzyme to reach the sites of VIP release in sufficient concentrations to degrade the peptide.

4. Normal human airways show a major NANC relaxant response, but at least some human asthmatics improve inadequately with VIP as aerosol. Possible explanations for the latter results include (a) the inability of aerosolized VIP to reach its airway smooth muscle receptors in sufficient concentrations, due to impediment by mucus; (b) degradation or inactivation of the peptide by proteolytic enzymes in bronchial secretions or bronchial epithelium; and (c) altered responsiveness of asthmatic airways to VIP.

B. Substance P and Other Neurokinins

The localization of neurokinins in nerve fibers and nerve terminals in the airways (Lundberg *et al.*, 1984), the potent effects of these peptides on airway smooth muscle and on bronchial chloride and glycoprotein secretion (Al-Bazzaz *et al.*, 1985; Coles *et al.*, 1984), and their binding to specific sites in the airways, suggest a regulatory influence for this family of neuropeptides on airway function. This role, related to the mediation of neurogenic airway constriction (Karlsson and Persson, 1983), secretion, and edema, is further discussed below.

X. REGULATING THE REGULATORS: ENZYMATIC DEGRADATION OF AIRWAY PEPTIDES

The identification and characterization of peptidases in the lung are of considerable physiological significance, since these enzymes, by controlling peptide activity, may be key regulators of airway and other lung function. Relatively little is known, however, about the degradative mechanisms of many peptides in the lung, with the exception of angiotensin and bradykinin (Erdös, 1979).

The enzymatic inactivation of substance P and VIP has recently been examined. Indirect evidence, through the use of various protease inhibitors (Schwartz *et al.*, 1985), suggests that substance P is enzymatically inactivated in bronchial mucosa by a neutral endopeptidase enkephalinase: a selective inhibitor of this enzyme, phosphoramidon, enhanced the secretagogue activity of this peptide on bronchial mucus secretion (Borson *et al.*, 1987). Phosphoramidon also potentiated the tracheal-relaxant activity of VIP (Liu *et al.*, 1987b) and slowed its enzymatic degradation by lung extracts (Liu *et al.*, 1987a). Neutral metalloendopeptidase activity has been described in human lung tissue (Johnson *et al.*, 1985).

XI. NEUROPEPTIDES IN AIRWAY DISEASE

A. VIP in the Pathogenesis of Airway Disease

A number of experimental and clinical observations suggested the hypothesis that decreased biological activity of VIP (due to deficient biosynthesis, impaired binding to receptors, or neutralization by an inhibitor or antibody) may contribute to the pathogenesis of two major and common disorders of the airways and lungs: cystic fibrosis and bronchial asthma (Said, 1982; Matsuzaki *et al.*, 1980).

The possibility that cystic fibrosis may be causally related to a deficiency of VIP was prompted by the observations that (1) VIP-containing nerves richly supply all exocrine organs (Lundberg *et al.*, 1980), the principal organs affected in this disorder; (2) VIP influences all major exocrine function, stimulating water, intestinal and bronchial Cl^-, pancreatic HCO_3^-, and macromolecular secretion, and increasing blood flow (Said, 1986); and (3) VIP binds to specific receptors on exocrine glands, including salivary and sweat glands, bronchial glands, and exocrine pancreas (Amiranoff and Rosselin, 1982). Some of the manifestations of cystic fibrosis, notably the decreased Cl^- permeability in sweat gland ducts (Quinton and Bijman, 1983), seemed consistent with a deficiency of VIP in exocrine glands.

Support for this hypothesis came from immunohistochemical examination of VIP innervation of sweat glands in cystic fibrosis patients and in normal subjects. Whereas normal sweat gland acini and ducts are well supplied by VIP-immunoreactive nerves, VIP innervation of sweat glands in cystic fibrosis patients is significantly decreased around the acini and virtually absent around the ducts (Heinz-Erian *et al.*, 1985). Additional VIP-related abnormalities in cystic fibrosis are a relatively elevated plasma level of the peptide, as well as of VIP-binding antibodies (Said and Heinz-Erian, 1988). On the other hand, the VIP gene and the cystic fibrosis genes have recently been localized in different chromosomal sites (Gozes et al., 1987), a finding that renders improbable a primary defect in VIP synthesis in that disease.

A similar hypothesis, linking the pathogenesis of bronchial asthma to a postulated lack of VIP innervation of airway smooth muscle, is based on (1) the likelihood that VIP may be the transmitter of NANC relaxation of airways; (2) evidence that the NANC system is the dominant, if not exclusive, relaxant system in human airways (Richardson and Beland, 1976); and (3) the possibility that a lack of NANC relaxation could explain the airway hyperreactivity of bronchial asthma (Richardson, 1979; Matsuzaki et al., 1980).

B. Neurokinins P and Other Sensory Neuropeptides

The use of capsaicin has contributed to the understanding of the functional role of substance P and other neurokinins. The release of these peptides from sensory neurons (Jancso et al., 1977; Lundberg and Saria, 1983) probably accounts for the bronchoconstrictor and permeability-promoting effects of capsaicin and for the subsequent desensitization of airway mucosa to various mechanical and chemical stimuli (Lundberg and Saria, 1983).

On the basis of these and related observations, the sensory neuropeptides neurokinins are believed to participate in the pathogenesis of asthma (Barnes, 1986). Damage to airway epithelium (Laitinen, 1985) exposes afferent C-fiber nerve endings that are then stimulated by inflammatory mediators, such as bradykinin and prostaglandins, leading to reflex cholinergic bronchoconstriction and stimulation of local axon reflexes. The latter triggers the release of neurokinins (and CGRP) from sensory collaterals in the airway, resulting in further bronchoconstriction, mucus hypersecretion, and edema of the airway wall.

XII. CONCLUSION

The first bronchoactive (and vasoactive) peptide to be discovered in the lung was isolated and characterized in 1970, from an embryologically related organ, the small intestine. Since then, more than 20 additional peptides have been described in the tracheobronchial tree and their biological activities investigated. Most of these are neuropeptides serving as neurotransmitters or neuromodulators, and some are already known to have major influence on airway smooth muscle town and reactivity. More neuropeptides are known to exist in the brain and peripheral nervous system, and it is probably only a matter of time before these are identified in the lung. The coming few years should see a sharpened definition and increased appreciation of the role of neuropeptides in airway smooth muscle physiology and pathophysiology.

ACKNOWLEDGMENTS. The author would like to thank Marilyn Satkiewicz for preparation of the manuscript. This work was supported by grants HL-30450 and HL-35656 from the National Institutes of Health, and by research funds from the Veterans Administration.

XIII. REFERENCES

Al-Bazzaz, F. J., Kelsey, J. G., and Kaage, W. D., 1985, Substance P stimulation of chloride secretion by canine tracheal mucosa, Am. Rev. Respir. Dis. 131:86–89.

Amiranoff B., and Rosselin G., 1982, VIP receptors and control of cyclic AMP production, in: Vasoactive Intestinal Peptide (S. I. Said, ed.), pp. 307–322, Raven, New York.

Ballermann, B. J., and Brenner, B. M., 1985, Biologically active atrial peptides, J. Clin. Invest. 76: 2041–2048.

Barnes, P. J., 1986, Neural control of human airways in health and disease, Am. Rev. Respir. Dis. 134:1289–1314.

Becker, K. L., 1985, Peptide hormones and their possible functions in the normal and abnormal lung, Recent Results Cancer Res. 99:17–28.

Berridge, M. J., and Irvine, R. F., 1984, Inositol trisphosphate, a novel second messenger in cellular signal transduction, Nature (Lond.) 312:315–321.

Bitar, K. N., and Makhlouf, G. M., 1982, Relaxation of isolated gastric smooth muscle cells by vasoactive intestinal peptide, Science 216:531–533.

Bitar, K. N., Said, S. I., Weir, G. C., Saffouri, B., and Makhlouf, G. M., 1980, Neural release of vasoactive intestinal peptide from the gut, Gastroenterology 79:1288–1294.

Borson, D. B., Corrales, R., Varsano, S., Gold, M., Viro, N., Caughey, G., Ramachandran, J., and Nadel, J. A., 1987, Enkephalinase inhibitors potentiate substance P-induced secretion of $^{35}SO_4$-macromolecules from ferret trachea, Exp. Lung Res. 12:21–36.

Cameron, A. C., Johnson, C. F., Kirkpatrick, C. T., and Kirkpatrick, M. C. A., 1983, The quest for the inhibitory neurotransmitter in bovine tracheal smooth muscle, O. J. Exp. Physiol. 68:413–426.

Carstairs, J. R., and Barnes, P. J., 1986a, Autoradiographic mapping of substance P receptors in lung, Eur. J. Pharmacol. 127:295–296.

Carstairs, J. R., and Barnes, P. J., 1986b, Visualization of vasoactive intestinal peptide receptors in human and guinea pig lung, J. Pharmacol. Exp. Ther. 239:249–255.

Cheung, A., Polak, J. M., Bauer, F. E., Cadieux, A., Christofides, N. D., Springall, D. R., and Bloom, S. R., 1985, Distribution of galanin immunoreactivity in the respiratory tract of pig, guinea pig, rat, and dog, Thorax. 40:889–896.

Chou, J., Kubota, E., Sata, T., and Said, S. I., 1986, Comparative relaxant activities of atrial natriuretic peptides (ANPs) and vasoactive intestinal peptide (VIP) on smooth muscle structures in lung, Fed. Proc. 45:553.

Chou, J., and Said, S. I., 1987, Removal of epithelium attenuates airway relaxation and enhances constriction, Fed. Proc. 46:659.

Coburn, R. F., 1984, The anatomy of the ferret paratracheal parasympathetic nerve-ganglion plexus, *Exp. Lung Res.* **7**:1–9.

Coburn, R. F., and Tomita, T., 1973, Evidence for nonadrenergic inhibitory nerves in guinea pig trachealis muscle, *Am. J. Physiol.* **224**:1072–1080.

Coleridge, H. M., Coleridge, J. C. G., Dangel, A., Kidd, C., Luck, J. C., and Sleight, P., 1973, Impulses in slowly conducting vagal fibers from afferent endings in the veins, atria, and arteries of dogs and cats, *Circ. Res.* **33**:87–97.

Coles, S. J., Neill, K. H., and Reid, L. M., 1984, Potent stimulation of glycoprotein secretion in canine trachea by substance P, *J. Appl. Physiol.* **57**:1323–1327.

Cox, C. P., Lerner, M. R., Wells, J. J., and Said, S. I., 1983, Inhaled vasoactive intestinal peptide (VIP) prevents bronchoconstriction induced by inhaled histamine, *Am. Rev. Respir. Dis.* **127**: 249.

Dey, R. D., Hoffpauir, J., and Said, S. I., 1988, Co-localization of VIP- and SP-containing nerves in cat bronchi, *Neuroscience* **24**:275–281.

Dey, R. D., and Said, S. I., 1985, Lung peptides and the pulmonary circulation, in: *The Pulmonary Circulation and Acute Lung Injury* (S. I. Said, ed.), pp. 101–122, Futura, New York.

Dey, R. D., Shannon, W. A., Jr., and Said, S. I., 1981, Localization of VIP-immunoreactive nerves in airways and pulmonary vessels of dogs, cats, and human subjects, *Cell Tissue Res.* **220**:231–238.

Diamond, L., and Altiere, R. J., Nonadrenergic inhibitory nerves, 1988, in: *Neural Regulation of the Airways in Health and Disease* (M. A. Kaliner and P. Barnes, eds.), in: *Lung Biology in Health and Disease* (C. Lenfant, exec. ed.), pp. 343–394, Dekker, New York.

Diamond, L., Szarek, J. L., Gillespie, M. N., and Altiere, R. J., 1983, In vivo bronchodilator activity of vasoactive intestinal peptide in the cat, *Am. Rev. Respir. Dis.* **128**:827–832.

Dockray, G. J., 1979, Evolutionary relationships of the gut hormones, *Fed. Proc.* **38**:2295–2301.

Eipper, B. A., Mains, R. E., and Herbert, E., 1986, Peptides in the nervous system, *Trends Neurosci.* **9**:463–468.

Erdös, E. G., 1979, Kininases, in: *Bradykinin, Kallidin and Kallirein: Supplement. Handbook of Experimental Pharmacology,* Vol. 25 (Suppl.) (E. G. Erdös, ed.), pp. 427–487, Springer-Verlag, Berlin.

Erspamer, V., and Anastasi, A., 1966, Polypeptides active on plain muscle in the amphibian skin, in: *Hypotensive Peptides* (E. G. Erdös, N. Back, and F. Sicuteri, eds.), pp. 63–75, Springer-Verlag, New York.

Flavahan, N. A., Aarhus, L. L., Rimele, T. J., and Vanhoutte, P. M., 1985, Respiratory epithelium inhibits bronchial smooth muscle tone, *J. Appl. Physiol.* **58**:834–838.

Frandsen, E. K., Krishna, G. A., and Said, S. I., 1978, Vasoactive intestinal polypeptide promotes cyclic adenosine 3', 5'-monophosphate accumulation in guinea pig trachea, *Br. J. Pharmacol.* **62**:367–369.

Furchgott, R. F., 1983, Role of endothelium in responses of vascular smooth muscle, *Circ. Res.* **53**: 557–573.

Ghatei, M. A., Sheppard, M. N., O'Shaughnessy, D. J., Adrian, T. E., McGregor, G. P., Polak, J. M., and Bloom, S. R., 1982, Regulatory peptides in the mammalian respiratory tract, *Endocrinology* **111**:1248–1254.

Ghatei, M. A., Sheppard, M. N., Henzen-Logman, S., Blank, M. A., Polak, J. M., and Bloom, S. R., 1983, Bombesin and vasoactive intestinal polypeptide in the developing lung. Marked changes in acute respiratory distress syndrome, *J. Clin. Endocrinol. Metab.* **57**:1226–1232.

Giachetti, A., Said, S. I., Reynolds, R. C., and Koniges, F. C., 1977, Vasoactive intestinal polypeptide (VIP) in brain: Localization in, and release from, isolated nerve terminals, *Proc. Natl. Acad. Sci. USA* **74**:3424–3428.

Gozes, I., Avidor, R., Yahav, Y., Katznelson, D., Croce, C. M., and Huebner, K., 1987, The gene encoding vasoactive intestinal peptide is located on human chromosome 6p21→6qter, *Hum. Genet.* **75**:41–44.

Hamasaki, Y., Saga, T., Mojarad, M., and Said, S. I., 1983, VIP counteracts leukotriene D4 induced-contractions of guinea pig trachea, lung and pulmonary artery, *Trans. Assoc. Am. Physicians* **96**:406–411.

Hassid, A., 1986, Atriopeptin II decreases cytosolic free Ca in cultured vascular smooth muscle cells, *Am. J. Physiol.* **251**:C681–686.

Hedlund, B., Abens, J., and Bartfai, T., 1983, Vasoactive intestinal polypeptide and muscarinic receptors: Supersensitivity induced by long-term atropine treatment, *Science* **220**:519–521.

Heinz-Erian, P., Dey, R. D., and Said, S. I., 1985, Deficient vasoactive intestinal peptide innervation in sweat glands of cystic fibrosis patients, *Science* **229**:1407–1408.

Holtman, J. R., Jr., Buller, A. L., Hamosh, P., and Gillis, R. A., 1986, Central respiratory stimulation produced by thyrotropin-releasing hormone in the cat, *Peptides* **7**:207–212.

Huang, M., and Rorstad, O. P., 1983, Effects of vasoactive intestinal polypeptide, monoamines, prostaglandins, and 2-chloradenosine on adenylate cyclase in rat cerebral microvessels, *J. Neurochem.* **40**:719–726.

Irvin, C. G., Boileau, R., Tremblay, J., Martin, R. R., and Macklem, P. T., 1980, Bronchodilation: Noncholinergic, nonadrenergic mediation demonstrated in vivo in the cat, *Science* **207**:791–792.

Ito, Y., and Takeda, K., 1982, Non-adrenergic inhibitory nerves and putative transmitters in the smooth muscle of cat trachea, *J. Physiol. (Lond.)* **330**:497–511.

Itoh, N., Obata, K., Yanaihara, N., and Okamoto, H., 1983, Human prepro-vasoactive intestinal polypeptide contains a novel PHI-27-like peptide, PHM-27, *Nature (Lond.)* **304**:547–549.

Jancso, G., Kiraly, E., and Jancso-Gabor, A., 1977, Pharmacologically induced selective degeneration of chemosensitive primary sensory neurones, *Nature (Lond.)* **270**:741–743.

Johnson, A. R., Ashton, J., Schulz, W. W., and Erdös, E. G., 1985, Neutral metalloendopeptidase in human lung tissue and cultured cells, *Am. Rev. Respir. Dis.* **132**:564–568.

Karlsson, J. A., Finney, M. J. B., Persson, C. G. A., and Post, C., 1984, Substance P antagonists and the role of tachykinins in non-cholinergic bronchoconstriction, *Life Sci.* **35**:2681–2691.

Karlsson, J. A., and Persson, C. G. A., 1983, Evidence against vasoactive intestinal polypeptide (VIP) as a dilator in favour of substance P as a constrictor in airway neurogenic response, *Br. J. Pharmacol.* **79**:634–636.

Kawatani, M., Rutigliano, M., and De Groat, W. C., 1985, Depolarization and muscarinic excitation induced in a sympathetic ganglion by vasoactive intestinal polypeptide, *Science* **229**:879–881.

Kubota, E., Chou, J., Paul, S., Sata, T., and Said, S. I., 1986, Atrial natriuretic peptide: Presence in pulmonary artery and potent pulmonary vasodilator activity, *Clin. Res.* **34**:728 (abst.)

Kuno, T., Andresen, J. W., Kamisaki, Y., Waldman, S. A., Saheki, S., and Murad, F., 1986, Co-purification of atrial natriuretic factor receptor and particulate guanylate cyclase from rat lung, *Clin. Res.* **34**:638 (abst.)

Laitinen, L. A., Heino, M., Laitinen, A., Kava, T., and Haahtela, T., 1985, Damage of the airway epithelium and bronchial reactivity in patients with asthma, *Am. Rev. Respir. Dis.* **131**:599–606.

Lauweryns, J. M., Coklaere, M., and Theunynck, P., 1973, Serotonin-producing neuroepithelial bodies in rabbit respiratory mucosa, *Science* **180**:410–413.

Lazarus, S. C., Basbaum, C. B., Barnes, P. J., and Gold, W. M., 1986, Mapping of VIP receptors by use of an immunocytochemical probe for the intracellular mediator cyclic AMP, *Am. J. Physiol.* **251**:C115–119.

Leroux, P., Vaudry, H., Fournier, A., St.-Pierre, S., and Pelletier, G., 1984, Characterization and localization of vasoactive intestinal peptide receptors in the rat lung, *Endocrinology* **114**:1506–1512.

Liu, L. W., Sata, T., Kubota, E., Paul, S., Iwanaga, T., Foda, H., and Said, S. I., 1987a, VIP is enzymatically degraded in the trachea, probably by an enkephalinase, *Clin. Res.* **35**:647A.

Liu, L. W., Sata, T., Kubota, E., Paul, S., and Said, S. I., 1987b, Airway relaxant effect of vasoactive intestinal peptide (VIP): Selective potentiation by phosphoramidon, an enkephalinase inhibitor, *Am. Rev. Respir. Dis.* **135**:A86.

Lundberg, J. M., Änggård, A., Fahrenkrug, J., Hökfelt, T., and Mutt, V., 1980, Vasoactive intestinal polypeptide in cholinergic neurons of exocrine glands: Functional significance of co-existing transmitters for vasodilation and secretion, *Proc. Natl. Acad. Sci. USA* **77**:1651–1655.

Lundberg, J. M., Fahrenkrug, J., Hökfelt, T., Martling, C-R., Larsson, O., Tatemoto, K., and Änggård, A., 1984, Co-existence of peptide HI (PHI) and VIP in nerves regulating blood flow and bronchial smooth muscle tone in various mammals including man, *Peptides* **5**:593–606.

Lundberg, J. M., Franco-Cereceda, A., Hua, X., Hökfelt, T., Fischer, J. A., 1985, Co-existence of substance P and calcitonin gene-related peptide-like immunoreactivites in sensory nerves in relation to cardiovascular and bronchoconstrictor effects of capsaicin, *Eur. J. Pharmacol.* **108:** 315–319.

Lundberg, J., and Hökfelt, T., 1983, Co-existence of peptides and classical neurotransmitters, *Trends Neurosci.* **6**:325–333.

Lundberg, J. M., Hökfelt, T., Martling, C-R., Saria, A., and Cuello, C., 1984, Substance P-immunoreactive sensory nerves in the lower respiratory tract of various mammals including man, *Cell Tissue Res.* **235**:251–261.

Lundberg, J. M., Hökfelt, T., Nilsson, G., Terenius, L., Rehfeld, J., Elde, R., and Said, S. I., 1978, Peptide neurons in the vagus, splanchnic, and sciatic nerves, *Acta. Physiol. Scand.* **104:** 499–501.

Lundberg, J. M., and Saria, A., 1983, Capsaicin-induced desensitization of the airway mucosa to cigarette smoke, mechanical and chemical irritants, *Nature (Lond.)* **302**:251–253.

Lundberg, J. M., and Saria, A., 1987, Polypeptide-containing neurons in airway smooth muscle, *Annu. Rev. Physiol.* **49**:557–572.

Lundberg, J. M., Hökfelt, T., Schultzberg, M., Uvnas-Wallensten, K., Kohler, C., and Said, S. I., 1979, Occurrence of VIP-like immunoreactivity in certain cholinergic neurons of the cat: Evidence from combined immunohistochemistry and acetylocholi-nesterase staining, *Neuroscience* **4**:1539–1559.

Lundberg, J. M., Terenius, L., and Hökfelt, T., 1982, Neuropeptide Y (NPY)-like immunoreactivity in peripheral noradrenergic neurons and effects of NPY on sympathetic function, *Acta Physiol. Scand.* **116**:477–480.

Magistrett, P. J., and Schorderet, M., 1984, VIP and noradrenaline act synergistically to increase cyclic-AMP in cerebral cortex, *Nature (Lond.)* **308**:208–282.

Matsuzaki, Y., Hamasaki, Y., and Said, S. I., 1980, Vasoactive intestinal peptide: A possible transmitter of non-adrenergic relaxation of guinea pig airways, *Science* **210**:1252–1253.

McDonald, T. J., Jörnvall, H., Nilsson, G., Vagne, M., Ghatei, M., Bloom, S. R., and Mutt, V., 1979, Characterization of a gastrin releasing peptide from porcine non-antral gastric tissue, *Biochem. Biophys. Res. Commun.* **90**:227–233.

Mojarad, M., Grode, T. L., Cox, C. P., Kimmel, G., and Said, S. I., 1985, Differential responses of human asthmatics to inhaled vasoactive intestinal peptide (VIP), *Am. Rev. Respir. Dis.* **131**:281 (abst.)

Morice, A., 1983, Vasoactive intestinal peptide causes bronchodilation and pro-tects against bron-choconstriction in asthmatic subjects, *Lancet* **2**:1225–1227.

Mutt, V., 1983, New approaches to the identification and isolation of hormonal polypeptides, *Trends Neurosci.* **6**:357–360.

Mutt, V., and Said, S. I., 1974, Structure of the porcine vasoactive intestinal octacosapeptide, *Eur. J. Biochem.* **42**:581–589.

Nadel, J. A., and Barnes, P. J., 1984, Autonomic regulation of the airways, *Annu. Rev. Med.* **35:** 451–467.

Nawa, H., Hirose, T., Takashima, H., Inayama, S., and Nakanishi, S., 1983, Nucleotide sequences of cloned cDNAs for two types of bovine brain substance P precursor, *Nature (Lond.)* **306**:32–36.

Okamura, H., Murakami, S., Fukui, K., Uda, K., Kawamoto, K., Kawashima, S., Yanaihara, N., and Ibata, Y., 1986, Vasoactive intestinal peptide and peptide histidine isoleucine amide-like immunoreactivity colocalize with vasopressin-like immunoreactivity in the canine hypothalamo-neurohypophysial neuronal system, *Neurosci. Lett.* **69**:227–232.

Otsuka, M., Konishi, S., 1983, Substance P—The first peptide neurotransmitter?, *Trends Neurosci.* **6**:317–320.

Palmer, J. B., Cuss, F. M. C., and Barnes, P. J., 1986, VIP and PHM and their role in non-adrenergic inhibitory responses in isolated human airways, *J. Appl. Physiol.* **61**:1322–1328.

Palmer, J. B. D., Cuss, F. M. C., Mulderry, P. K., Ghatei, M. A., Bloom, S. R., and Barnes, P. J., 1985, Calcitonin gene related peptide is a potent constrictor of human airway smooth muscle, *Thorax* **40**:713.

Paul, S., and Said, S. I., 1987, Characterization of receptors for vasoactive intestinal peptide solubilized from the lung, *J. Biol. Chem.* **262**(1):158–162.

Provow, S., and Veliçelebi, G., 1987, Characterization and solubilization of vasoactive intestinal peptide receptors from rat lung membranes, *Endocrinology* **120**:2442–2452.

Quinton, P. M., and Bijman, J., 1983, Higher bioelectrical potentials due to decreased chloride absorption in the sweat glands of patients with cystic fibrosis, *N. Engl. J. Med.* **308**:1185–1189.

Regal, J. F., and Johnson, D. E., 1983, Indomethacin alters the effects of substance-P and VIP on isolated airway smooth muscle, *Peptides* **4**:581–584.

Regoli, D., Drapeau, G., Dion, S., and D'Orleans-Juste, P., 1987, Pharmacological receptors for substance P and neurokinins, *Life Sci.* **40**:109–117.

Richardson, J. B., 1979, Nerve supply to the lungs, *Am. Rev. Respir. Dis.* **119**:785–802.

Richardson, J., and Beland, J., 1976, Nonadrenergic inhibitory nervous system in human airways, *J. Appl. Physiol.* **41**:764–771.

Robberecht, P., Tatemoto, K., Chatelain, P., Waelbroeck, M., Delhay, M., Taton, G., De Neef, P., Camus, J-C., Heuse, D., and Christophe, J., 1982, Effects of PHI on vasoactive intestinal peptide receptors and adenylate cyclase activity in lung membranes. A comparison in man, rat, mouse and guinea pig, *Regul. Pept.* **4**:241–250.

Russell, J. A., 1980, Nonadrenergic inhibitory innervation canine airways, *J. Appl. Physiol. Respir. Environ. Exerc. Physiol.* **48**:16–22.

Ryan, U. S., 1985, Processing of angiotensin and other peptides by the lungs, in: *Circulation and Nonrespiratory Functions* (A. P. Fishman and A. B. Fisher, eds.), Vol. 1, in: *The Respiratory System,* Section 3 (A. P. Fishman, sect. ed.), in: *Handbook of Physiology* (S. R. Geiger, exec. ed.), pp. 351–364, American Physiological Society, Bethesda.

Saga, T., and Said, S. I., 1984, Vasoactive intestinal peptide relaxes isolated strips of human bronchus, pulmonary artery, and lung parenchyma, *Trans. Assoc. Am. Physicians* **97**:304–310.

Said, S. I., 1980, Peptides common to the nervous system and the gastrointestinal tract, in: *Frontiers in Neuroendocrinology,* Vol. 6 (L. Martini and W. F. Ganong, eds.), pp. 293–331, Raven, New York.

Said, S. I., 1967, Vasoactive substances in the lung, in: *Proceedings of the Tenth Aspen Emphysema Conference,* pp. 223–228, U.S. Public Health Service Publication 1787, Aspen, Colorado.

Said, S. I., 1982, Vasoactive peptides in the lung, with special reference to vasoactive intestinal peptide, *Exp. Lung Res.* **3**:343–348.

Said, S. I., 1984, Peptide hormones and neurotransmitters of the lung, in: *The Endocrine Lung in Health and Disease,* Vol. 15 (K. L. Becker and A. Gazdar, eds.), pp. 267–276, W. B. Saunders, Philadelphia.

Said, S. I., 1986, Regulation of pulmonary vascular tone by autonomic mediators, peptides, and leukotrienes, in: *Acute Lung Injury: Pathogenesis of Adult Respiratory Distress Syndrome* (H. Kazemi, A. L. Hyman, and P. J. Kadowitz, eds.), pp. 211–218, Littleton, Massachusetts.

Said, S. I., and Dey, R. D., VIP in the airways, 1988, in: *Neural Regulation of the Airways in Health and Disease* (M. A. Kaliner and P. Barnes, eds.), in: *Lung Biology in Health and Disease* (C. Lenfant, exec. ed.), pp. 395–416, Dekker, New York.

Said, S. I., Geumei, A., and Hara, N., 1982, Bronchodilator effect of VIP in vivo: Protection against bronchoconstriction induced by histamine or prostaglandin F2, in: *Vasoactive Intestinal Peptide* (S. I. Said, ed.), pp. 185–191, Raven, New York.

Said, S. I., Giachetti, A., and Nicosia, S., 1980, VIP: Possible functions as a neural peptide, in: *Neural Peptides and Neuronal Communication* (E. Costa and M. Trabucchi, eds.), pp. 75–82, Raven, New York.

Said, S. I., and Heinz-Erian, P., 1988, VIP and exocrine: possible role in cystic fibrosis, in: *Cellular and Molecular Basis of Cystic Fibrosis* (G. Mastella and P. M. Quinton, eds.), pp. 355–361, San Francisco Press, San Francisco.

Said, S. I., and Mutt, V., 1977, Relationship of spasmogenic and smooth muscle relaxant peptide from normal lung to other vasoactive compounds, *Nature (Lond.)* **265**:84–86.

Said, S. I., and Mutt, V., 1970, Polypeptide with broad biological activity: Isolation from small intestine, *Science* **169**:1217–1218.

Said, S. I., and Rosenberg, R. N., 1976, Vasoactive intestinal polypeptide: Abundant immunoreactivity in neural cell lines and normal nervous tissues, *Science* **192**:907–908.

Said, S. I., and Zfass, A. M., 1978, Gastrointestinal hormones, *Disease-A-Month* **24**:1–40.

Scharrer, B., 1978, Peptidergic neurons: Facts and trends, *Gen. Comp. Endocrinol.* **34**:50–62.

Schwartz, J-C., Costentin, J., and Lecomte, J-M., 1985, Pharmacology of enkephalinase inhibitors, *Trends Pharmacol. Sci.* **6**:472–476.

Shaw, J. O., Jr., Wetsel, R. A., and Kolb, W. P., 1985, Complement and the lung, in: *Acute Respiratory Failure* (W. M. Zapol and K. J. Falke, eds.), in: *Lung Biology in Health and Disease* (C. Lenfant, exec. ed.), pp. 407–433, Dekker, New York.

Sheppard, M. N., Polak, J. M., Allen, J. M., and Bloom, S. R., 1984, Neuropeptide tyrosine (NPY): A newly discovered peptide is present in the mammalian respiratory tract, *Thorax* **39**: 326–330.

Stimler, N. P., Bloor, C. M., and Hugli, T. E., 1983, Immunopharmacology of complement anaphylatoxins in the lung, in: *Immunopharmacology of the Lung*, Vol. 19 (H. H. Newball, ed.), in: *Lung Biology in Health and Disease* (C. Lenfant, exec. ed.), pp. 401–434, Dekker, New York.

Tanaka, D. T., and Grunstein, M. M., 1984, Mechanisms of substance P-induced contraction of rabbit airway smooth muscle, *J. Appl. Physiol.* **57**:1551–1557.

Taton, G., Delhaye, M., Camus, J-C., De Neef, P., Chatelain, P., Robberecht, P., and Christopher, J., 1981, Characterization of the VIP- and secretin-stimulated adenylate cyclase system from human lung, *Pflugers Arch.* **391**:178–182.

Till, G. O., and Ward, P. A., 1985, Complement-induced lung injury, in: *The Pulmonary Circulation and Acute Lung Injury* (S. I. Said, ed.), pp. 387–402, Futura, Mount Kisco, New York.

Uddman, R., Moghimzadeh, E., and Sundler, F., 1984, Occurrence and distribution of GRP-immunoreactive nerve fibers in the respiratory tract, *Arch. Otorhinolaryngol.* **239**:145–151.

Undem, B. J., Dick, E. C., and Buckner, C. K., 1983, Inhibition by vasoactive intestinal peptide of antigen-induced histamine release from guinea pig minced lung, *Eur. J. Pharmacol.* **88**:247–249.

Uvnäs-Moberg, K., Goiny, M., Posloncec, B., and Blomquist, L., 1982, Increased levels of VIP (vasoactive intestinal polypeptide)-like immunoreactivity in peripheral venous blood in dogs following injections of apomorphine and bromocriptine. Do dopaminergic agents induce gastric relaxation and hypotension by a release of endogenous VIP?, *Acta Physiol. Scand.* **115**:373–375.

Venugopalan, G. S., Said, S. I., and Drazen, J. M., 1984, Effect of vasoactive intestinal peptide on vagally mediated tracheal pouch relaxation, *Respir. Physiol.* **56**:205–216.

Wang, J., Yaksh, T., Harty, G., and Go, V., 1986, Neurotransmitter modulation of VIP release from cat cerebral cortex, *Am. J. Physiol.* **250:**R104–111.

Wasserman, M. A., Griffin, R. L., and Malo, P. E., 1982, Comparative in vitro tracheal-relaxant effects of porcine and hen VIP, in: *Vasoactive Intestinal Peptide* (S. I. Said, ed.), pp. 177–184, Raven, New York.

Widdicombe, J. G., 1982, Pulmonary and respiratory tract receptors, *J. Exp. Biol.* **100:**41–57.

Wood, J. D., 1984, Enteric neurophysiology, *Am. J. Physiol.* **247:**G585–G598.

Yau, W., Youther, M., and Verdun, P., 1985, A presynaptic site of action of substance P and vasoactive intestinal polypeptide on myenteric neurons, *Brain Res.* **330:**382–385.

Chapter 5

Cell-Surface Receptors in Airway Smooth Muscle

Peter J. Barnes

Department of Clinical Pharmacology
Cardiothoracic Institute
Brompton Hospital
London SW3 6HP, England

I. INTRODUCTION

Airway smooth muscle tone is influenced by many hormones, neurotransmitters, drugs, and mediators that produce their effects by binding to specific surface receptors on airway smooth muscle cells. Bronchoconstriction and bronchodilation may therefore be viewed in terms of receptor activation or blockade and the contractile state of airway smooth muscle is probably the resultant effect of interacting, excitatory, and inhibitory receptors.

It is important to recognize that airway caliber is not only the result of airway smooth muscle tone, but in asthma it is likely that airway narrowing may also be explained by edema of the bronchial wall (resulting from microvascular leakage) and to luminal plugging by viscous mucus secretions and extravasated plasma proteins, which may be produced by a mixture of mediators released from inflammatory cells, including mast cells, macrophages, and eosinophils. Activation of receptors on other target cells, such as submucosal glands, airway epithelium, postcapillary venules, mast cells, and other inflammatory cells, may therefore influence airway caliber as well.

II. INDIRECT REGULATION OF AIRWAY SMOOTH MUSCLE

There is a complex interaction between different cells in the airway and, whereas many stimuli may act directly on airway smooth muscle cells, others may affect smooth muscle tone indirectly, either through neural control mechanisms, through release of mediators from inflammatory cells, or through release of epithelial factors. Thus, bradykinin is a potent bronchoconstrictor when given by inhalation in humans but has little effect on human airway smooth muscle *in vitro* (Fuller *et al.*, 1986), suggesting an indirect action, in part due to activation of a cholinergic reflex, since the bronchoconstriction may be reduced by a cholinergic antagonist. Other mediators may have a bronchoconstrictor effect that, in the case of adenosine, is the result of mast cell-mediator release (Mann *et al.*, 1985), or in the case of platelet-activating factor (PAF), the result of platelet products (Morley *et al.*, 1985).

Epithelium-Derived Relaxant Factor

Recently there has been considerable interest in the possibility of a relaxant factor released from airway epithelial cells, which may be analogous to endothelial derived relaxant factor (Cuss and Barnes, 1987). The presence of airway epithelium in bovine airways *in vitro* appears to reduce the sensitivity and maximum contractile effect of spasmogens, such as histamine, acetylcholine (ACh), or serotonin, although not potassium, which depolarizes airway smooth muscle directly (Barnes *et al.*, 1985b). Similar results have been obtained in dog (Flavahan *et al.*, 1985) and guinea pig airways (Goldie *et al.*, 1986). This suggests that these spasmogens may release a factor from epithelium that directly relaxes airway smooth muscle. The nature of this putative factor is uncertain, but it does not appear to be influenced by either cyclo-oxygenase or lipoxygenase blockade (Barnes *et al.*, 1985b; Flavahan *et al.*, 1985). There is some evidence that this factor may be superfused, but its evanescent nature makes it difficult to study.

III. AIRWAY RECEPTORS AND DISEASE

Since surface receptors may determine tissue responsiveness, it is possible that alterations in receptors on airway smooth muscle might account for increased airway responsiveness seen in asthma, and, to a lesser extent, in chronic obstructive pulmonary disease. Many different factors are known to alter receptor expression and could change either receptor density, affinity, or coupling. Thus, inflammatory mediators formed in the airway wall may have effects on various receptors that could lead to increased responsiveness. Since in asthma the increased responsiveness is found with many different bronchoconstrictor stimuli, it is unlikely that there is an effect on a single type of receptor (e.g., muscarinic or histamine receptors). It is more probable that there is enhanced coupling of all receptors, perhaps by means of phosphoinositide hydrolysis or that there is a defect in inhibitory receptors (e.g., β-adrenoceptors).

IV. AUTONOMIC RECEPTORS

A. Autonomic Control of Airways

Autonomic innervation of the airways is complex (Nadel and Barnes, 1984; Barnes, 1986b). In addition to classic cholinergic pathways, which cause bronchoconstriction, and adrenergic mechanisms, which are usually bronchodilators, there is a more recently recognized component of autonomic control that is neither cholinergic nor adrenergic (Barnes, 1986a). Autonomic nerves influence airway tone by activating specific receptors on airway smooth muscle. In the case of cholinergic pathways, ACh released from postganglionic nerve endings stimulates muscarinic cholinergic receptors. Adrenergic mechanisms include sympathetic nerves, which release norepinephrine, and circulating epinephrine secreted from the adrenal medulla; these catecholamines activate α- or β-receptors. The neurotransmitters of the nonadrenergic noncholinergic (NANC) nervous system are not certain, but the most likely candidate for nonadrenergic inhibitory nerves is vasoactive intestinal peptide (VIP), whereas that of noncholinergic excitatory nerves is probably substance P or a related peptide. These neuropeptides interact with specific receptors on target cells.

The different components of the autonomic nervous system interact with each other in a complex way, both by affecting release of neurotransmitter (via prejunctional receptors), at ganglia in the airways, and by interaction at postjunctional receptors. Thus, airway tone may be determined by a complex interplay between different components of the autonomic nervous system.

B. β-Adrenoceptors

Both histochemical and functional studies indicate that there are few, if any, adrenergic nerve fibers directly supplying airway smooth muscle in human airways (Richardson, 1979), although in other species, such as cat and dog, adrenergic bronchodilator nerves have been described. This suggests that β-receptors in airway smooth muscle are under the control of circulating epinephrine (Barnes, 1986b).

Direct receptor binding studies indicate that β-receptors are present in high density in lung of many species, including humans (Rugg et al., 1978; Barnes et al., 1980b). Autoradiographic studies have shown that β-receptors are found on many different cell types within lung, including airway smooth muscle from trachea down to terminal bronchioles (Barnes et al., 1982; Carstairs et al., 1984). This is consistent with functional studies, indicating that β-agonists are potent relaxants of bronchi, bronchioles, and peripheral lung strips (Goldie et al., 1982; Zaagsma et al., 1983). Whereas a direct relaxant effect of β-agonists on airway smooth muscle is undoubtedly their major mode of action as bronchodilators, they may also lead to bronchodilation indirectly, either by inhibiting release of bronchoconstrictor mediators from airway mast cells (Peters *et al.*, 1982), by inhibiting ACh release from cholinergic nerves (Vermeire and Vanhoutte, 1979), or by stimulating release of an epithelial relaxant factor (Goldie *et al.*, 1986). Indeed, a very high density of β-receptors is seen in airway epithelium, which far exceeds the density of receptors in smooth muscle itself (Barnes *et al.*, 1982).

C. β-Receptor Subtypes

Although β-receptors of airway smooth muscle were originally classified as β_2-receptors, later studies showed that, in several species, relaxation of tracheal smooth muscle was intermediate between a β_1- and β_2-mediated response. This suggests the presence of β_1- in addition to β_2-receptors. Using direct receptor-binding techniques and selective β-antagonists, the coexistence of β_1- and β_2-receptors was confirmed in animal and human lung (Rugg *et al.*, 1978; Engel, 1981). In dog tracheal smooth muscle, while β_2-receptors predominate, 20% of receptors are of the β_2-subtype (Barnes *et al.*, 1983b). Functional studies of the same tissue *in vitro* show that relaxation to exogenous β-agonists is mediated by β_2-receptors, but relaxation to sympathetic nerve stimulation is mediated by β_1-receptors. These findings are consistent with the hypothesis that β_1-receptors are regulated by sympathetic nerves (neuronal β-receptors), whereas β_2-receptors are regulated by circulating epinephrine (hormonal β-receptors).

Further support for this idea is provided by studies of airway β-receptor function in other species. Thus, in guinea pig trachea, which has a sparse sympathetic innervation, mixed β_1- and β_2-responses are found, and binding studies show that approximately 15% of β-receptors are of the β_1-subtype (Carswell and Nahorski, 1983b). In cat trachea, which has a dense sympathetic nerve supply, relaxation to β-agonists is mediated predominantly by β_1-receptors, whereas in lung strips, which contain bronchioles devoid of sympathetic nerves, responses are mediated by β_2-receptors (Lulich *et al.*, 1976). In human airway smooth muscle, with its absence of significant sympathetic innervation, no β_1-receptor-mediated effects would be expected. This has been confirmed in functional studies *in vitro*, in which relaxation of central and peripheral airways is mediated by β_2-receptors (Goldie *et al.*, 1982; Zaagsma *et al.*, 1983). Similarly, *in vivo* prenalterol, a β_1-selective agonist, has no bronchodilator effect in asthmatic subjects, despite significant cardiac effects (Lofdahl and Svedmyr, 1982). Recent autoradiographic studies of human lung have confirmed that the β-receptors of human airway smooth muscle from bronchi to terminal bronchioles are entirely of the β_2-subtype (Carstairs *et al.*, 1985).

D. β-Receptor Dysfunction in Asthma

The suggestion that there may be a defect in β-receptor function in asthma (Szentivanyi, 1968) provided a great impetus to research and the question is still unresolved. While the defects in peripheral β-receptor function described in asthmatic subjects can largely be ascribed to the effects of prior adrenergic therapy (Barnes, 1986b), it is still not certain whether airway smooth muscle β-receptor function is impaired, largely because of the difficulties in obtaining asthmatic airways to study *in vitro*. Asthmatic subjects are apparently less responsive to inhaled β-agonists than normal subjects (Barnes and Pride, 1983), but this could be explained by reduced aerosol penetration or by functional antagonism (a larger dose of β-agonist is reburied to reverse a greater initial degree of bronchoconstriction), rather than a defect in airway β-receptors. Recent observations in asthmatic airways *in vitro*, however, suggest that there is a reduced relaxant response to isoproterenol with no evidence of an increase in responsiveness to spasmogens (Paterson *et al.*, 1982; Cerrina *et al.*, 1986).

A reduction in lung β-receptor density is found in a guinea pig model of asthma (Barnes *et al.*, 1980a), although whether this is found in airway smooth muscle is unknown. In bovine tracheal smooth muscle cholinergic stimulation reduces β-receptor density and uncouples β-receptors (Grandordy *et al.*, 1987c). This is probably mediated by activation of protein kinase C, since phorbol esters, which also activate this enzyme, have a similar effect. While this reduction in β-receptor density did not have any effect on function, it is possible that more prolonged stimulation, and a combination of several spasmogens that stimulate PI turnover may eventually result in a functional defect in β-receptors. In guinea pigs exposure to the inflammatory mediator, PAF, increases bronchial responsiveness and reduces the bronchial responsiveness to isoproterenol *in vivo*, although tracheal smooth muscle response normally to isoproterenol *in vitro*, and the density or affinity of β-receptors is not altered (Barnes *et al.*, 1987b).

E. α-Adrenoceptors

α-Receptors that mediate contraction of airway smooth muscle have been demonstrated in many species, including humans (Kneussl and Richardson, 1978), although it may only be possible to demonstrate their presence under certain conditions. Human peripheral lung strips contract with α-agonists, although it is likely that contractile elements other than airway smooth muscle are responsible (Black *et al.*, 1981). Autoradiographic studies confirm a very low density of α_1-receptors in smooth muscle of large airways but have shown a surprisingly high density in small airways (Barnes *et al.*, 1983a).

F. α-Receptor Subtypes

The classic α-receptor that mediates contractile effects is known as an α_1-receptor, selectively blocked by prazosin, whereas prejunctional α-receptors that mediate negative feedback of norepinephrine release are selectively blocked by yohimbine are termed α_2-receptors. More recently, α_2-receptors have also been found postjunctionally. It was suggested that α_1-receptors were regulated physiologically by norepinephrine and α_2-receptors by epinephrine, but there is now little evidence to support this idea. Few studies have been performed on α-receptor subtypes in airways; in dog tracheal smooth muscle the contractile response to both sympathetic nerve stimulation and to exogenous α-agonists is mediated almost entirely by α_2-receptors, most α-receptors measured by radioligand binding in the same tissue are of this subtype (Barnes et al., 1983d).

G. α-Receptors in Asthma

There is some evidence that α-adrenergic responses may be increased in asthma and may therefore contribute to bronchial hyperresponsiveness. The α-agonist methoxamine causes bronchoconstriction in asthmatic but not in normal subjects (Snashall et al., 1978), even in the absence of β-blockade (Black et al., 1982). This finding suggests that α-adrenergic responses may be increased in the airways of asthmatics.

No α-adrenergic response can be demonstrated in normal canine or human airway smooth muscle in vitro, even after β-blockade, but in diseased human airways or after pretreatment of normal canine airways with histamine or serotonin a marked α-adrenergic contractile response is seen (Kneussl and Richardson, 1978; Barnes et al., 1983c), suggesting activation of α-adrenergic responses by mediators or disease. This is not an artifact of the in vitro situation, as similar activation of α-adrenergic responses can also be demonstrated in dogs in vivo (Brown et al., 1983). This finding suggests that inflammatory mediators may "turn on" α-adrenergic responses in asthma. The mechanism for this activation does not involve any change in either density or affinity of α-receptors in airway smooth muscle and is likely to be a postreceptor mechanism, possibly involving voltage-dependent calcium channels (Barnes et al., 1983c). A guinea pig model of asthma showed an increase in lung α_1-receptors (Barnes et al., 1980a), but no increased α-adrenergic responsiveness has been found in airway smooth muscle (Turner et al., 1983).

If exaggerated α-adrenergic responsiveness of airway smooth muscle were an important factor in bronchial hyperresponsiveness, α-blockers would be beneficial in asthma. α-Antagonists, such as phentolamine and thymoxamine, have been shown to inhibit bronchoconstriction induced by histamine, allergen, and exercise, but such drugs lack specificity and their protective effects may be explained by their pharmacological actions, such as antihistamine activity (Barnes, 1986b). The specific α_1-blocker prazosin given by inhalation has no bronchodilator effect in asthmatics, who readily bronchodilate with a β-agonist, suggesting that α-hyperresponsiveness does not contribute to resting bronchomotor tone in asthma (Barnes et al., 1981a). Similarly, prazosin has no effect on histamine-induced bronchoconstriction but has a weak protective effect against exercise-induced bronchospasm (Barnes et al., 1981b), which may be explained by an effect on bronchial blood flow rather than on airway smooth muscle. The effects of specific α_2-blockers on airway function in humans have not yet been reported.

H. Cholinergic Receptors

Acetylcholine released from cholinergic nerves causes contraction of airway smooth muscle by activation of muscarinic receptors, which are blocked by atropine (Barnes, 1987c). There is a close association between receptor occupation and stimulation of PI hydrolysis, but the contraction response curve is well to the left (Grandordy *et al.,* 1986b). Thus, maximum contraction is obtained when only about 20% of receptors are occupied, indicating the existence of "spare" receptors. This is confirmed by the use of phenoxybenzamine, which irreversibly alkylates muscarinic receptors; with progressive loss of receptors, despite a rightward shift in the concentration response curve, the maximum response is only reduced when receptor density falls below 20% (Fig. 1).

Direct receptor-binding studies have demonstrated a high density of muscarine receptors in the smooth muscle of large airways (Murlas *et al.,* 1982; Cheng and Townley, 1982); this has been confirmed autoradiographically (Barnes *et al.,* 1983a). There is a gradient of receptor density toward the lumen, with the highest density at the site of cholinergic innervation away from the lumen (Basbaum *et al.,*

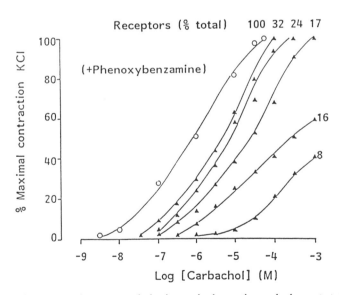

FIGURE 1. Spare muscarine receptors in bovine tracheal smooth muscle demonstrated by progressive reduction in receptors by phenoxybenzamine. The maximal contractile response is reduced only when receptor density falls below 17%.

1984). The density of muscarinic receptors decreases as airways become smaller, so that terminal bronchioles are almost devoid of muscarinic receptors (Barnes *et al.*, 1983a). This is consistent with physiological studies in dogs using tantalum bronchography, showing that vagal stimulation has a marked effect on large airways but little effect on bronchioles (Nadel *et al.*, 1971). In humans anticholinergic drugs have more effect on large than on small airways, as measured by helium–oxygen flow–volume curves, whereas β-agonists relax all airways (Ingram *et al.*, 1977).

I. Muscarinic Receptor Subtypes

Recent evidence from binding and functional studies with selective antagonists suggests that muscarinic receptors may be subclassified into at least two types, although there is still considerable confusion about terminology (Hammer and Giachetti, 1982). In gut, pirenzepine selectively blocks muscarinic receptors on ganglia, which are termed M_1 receptors, but not those on smooth muscle, which are designated M_2 receptors. As expected, in airway smooth muscle there is no evidence for M_1 receptors, since pirenzepine has low affinity in both inhibition of receptor binding and in blocking PI hydrolysis (Grandordy *et al.*, 1986b). *In vivo* it has been possible to demonstrate a different type of muscarinic receptor in airway nerves, since in cat and guinea pig the noncompetitive antagonist gallamine selectively inhibits a muscarinic receptor on cholinergic nerve terminals, enhancing ACh release and having little inhibitory effect on ACh in airway smooth muscle (Fryer and MacLagan, 1984; Blaber *et al.*, 1985). This suggests that ACh inhibits its release from cholinergic nerves by means of an autoreceptor that differs from the receptor on airway smooth muscle. Similar observations were recently made in human airway smooth muscle (Minette and Barnes, 1988). Since the gallamine-sensitive receptor in heart is classified as an M_2 receptor, it follows that the receptor on airway smooth muscle must be of a different subtype (designated M_3 receptor).

V. NEUROPEPTIDE RECEPTORS

Many different neuropeptides have now been localized to airway nerves in several species, including humans (Polak and Bloom, 1986; Barnes, 1987b). There is increasing evidence that these peptides play a role as a neurotransmitter or co-transmitter, and may be the neurotransmitters of nonadrenergic noncholinergic (NANC) nerves. These neuropeptides have effects on airway smooth muscle mediated by specific surface receptors. The functional significance of airway neuropeptides is still uncertain and will remain so until specific receptor antagonists become available. Both inhibitory and excitatory NANC nerves have been described; evidence is in favor of VIP as a neurotransmitter of the inhibitory nerves and suggests substance P and related tachykinins as the neurotransmitters of excitatory nerves.

A. VIP Receptors

Vasoactive intestinal peptide is a potent relaxant of animal and human airways *in vitro* and is 50–100 times more potent than isoproterenol as a bronchodilator (Palmer *et al.*, 1986a). VIP is a potent bronchodilator in cats when given intravenously, but in human subjects the cardiovascular effects of the peptide prevent the infusion of a dose high enough to bronchodilate (Palmer *et al.*, 1986b). Inhaled VIP also has no bronchodilator effect in humans, probably because it is unable to reach receptors on smooth muscle (Barnes and Dixon, 1984). VIP is the most favored candidate as neurotransmitter of nonadrenergic inhibitory nerves in airways, since in many respects it mimics NANC inhibitory nerve effects. VIP produces its effects by activation of specific receptors on airway smooth muscle, and it is unaffected by β-blockers, indomethacin, or tetrodotoxin. Like β-agonists, VIP stimulates adenylate, cyclase in target cells and therefore increases cyclic adenosine monophosphate (cAMP) in lung tissue (Frandsen *et al.*, 1978). Using an immunocytochemical method, it has been possible to demonstrate increases in the cAMP content of airway smooth muscle cells in several species (Lazarus *et al.*, 1986). VIP receptors have been identified by [^{125}I]-VIP binding in lung homogenates (Robberecht *et al.*, 1981); the distribution of these receptors in lung was recently studied by autoradiographic methods (Carstairs and Barnes, 1986a). VIP receptors are found in several cell types, including smooth muscle of large airways, but not of small airways. This finding is consistent with the lack of effect of VIP in relaxing small airways, although isoproterenol relaxes both large and small airways to an equal extent (Palmer *et al.*, 1986a) (Fig. 2). The lack of VIP receptors in small airway smooth muscle is also in keeping with the paucity of VIP-immunoreactive nerves and with the lack of NANC inhibitory nerves in small airways.

Peptide histidine isoleucine (PHI), which exists in humans, with a terminal methionine (PHM), is closely related in structure to VIP and is coded by the same gene. Its effects are similar to those of VIP, and it is equipotent in relaxing airway smooth muscle (Fig. 2). It may activate the same receptors, although evidence now suggests that it may have different receptors, since it has a different potency on vascular smooth muscle.

B. Tachykinin Receptors

Substance P (SP) is localized to sensory nerves in the airways and may be released as part of an axon reflex. SP constricts airway smooth muscle of animals and humans *in vitro* (Lundberg *et al.*, 1983) by activating specific receptors. *In vivo* SP infusion causes bronchoconstriction in animals that may be partially

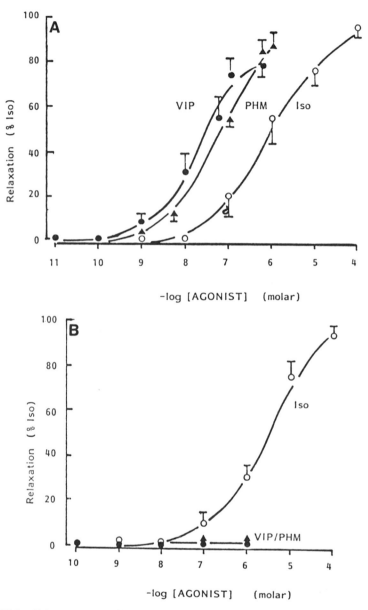

FIGURE 2. Relaxant response of human airways to vasoactive intestinal peptide (VIP) and peptide histidine methionine (PHM) compared with isoprotenerol (Iso). There is a marked difference in response to the bronchi (A) and bronchioles (B). (From Palmer *et al.*, 1986a.)

blocked with atropine, suggesting that SP release of ACh may contribute to its bronchoconstrictor effect (Tanaka and Grunstein, 1984). Autoradiographic studies using labeled SP have demonstrated high densities of SP receptors on smooth muscle of guinea pig and human airways from large airways down to terminal bronchioles (Carstairs and Barnes, 1986b).

Recently related peptides (tachykinins) have been isolated from the mammalian nervous system. Neurokinins A and B appear to activate distinct receptors termed NK_2 and NK_3 receptors, respectively, whereas SP activates NK_1 receptors. In airway smooth muscle of several species, including humans, the order of potency is NKA > NKB > SP (NKA being about 100-fold more potent than SP), indicating an NK_2 receptor (Karlsson et al., 1984; Palmer et al., 1986c) (Fig. 3). This finding suggests that NKA is the endogenous bronchoconstrictor tachykinin. This has recently been demonstrated in human subjects in vivo, since infused NKA causes bronchoconstriction with little cardiovascular effect, whereas SP has profound cardiovascular actions but no bronchoconstrictor effect (Fuller et al., 1987; Clarke et al., 1987). Tachykinins appear to cause airway smooth muscle contraction by stimulating PI hydrolysis and, as expected, NKA is more potent than NKB or SP (Grandordy et al., 1988).

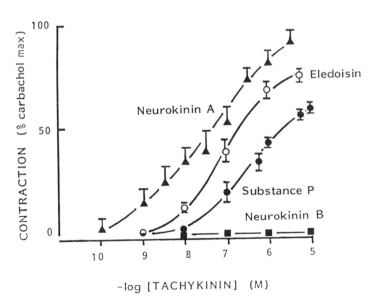

FIGURE 3. Response of human bronchi to tachykinins. The order of potency neurokinin A > substance P is characteristic of an NK_2 receptor.

C. Other Neuropeptide Receptors

Calcitonin gene-related peptide is also localized to sensory nerves in airways and has a potent bronchoconstrictor effect on human bronchi *in vitro* (Barnes *et al.*, 1987a). This effect is presumed to be by means of specific receptors, as it is not antagonized by any specific blockers.

Galanin, which is localized to VIP-immunoreactive nerves in airways, has no effect on airway smooth muscle of guinea pig (Ekblad *et al.*, 1985) or humans (D. Stretton and P. J. Barnes, unpublished observations). Neuropeptide Y is a co-transmitter of norepinephrine but has no effect on airway smooth muscle (D. Stretton and P. J. Barnes, unpublished observations), although it is very potent as a constrictor of pulmonary vessels.

VI. MEDIATOR RECEPTORS

Many inflammatory mediators have effects on airway smooth muscle; their effects are produced by the activation of specific receptors on airway smooth muscle cells (Barnes, 1987a). The effects of several mediators are produced indirectly on airway smooth muscle, either by activating bronchoconstrictor nerves or by releasing bronchoconstrictor mediators from other inflammatory cells.

A. Histamine Receptors

Histamine produces its effects by activation of H_1 and H_2 receptors. Inhaled histamine causes bronchoconstriction *in vivo* and contraction of large and small human airways *in vitro* by activating H_1 receptors, which are antagonized by the classic antihistamines, such as chlorpheniramine (White and Eiser, 1983; Persson and Ekman, 1976). H_1 receptors have been identified in guinea pig and human lung homogenates (Carswell and Nahorski, 1983a; Grandordy and Barnes, 1987; Grandordy *et al.*, 1987a); autoradiographic studies indicate that they are localized to smooth muscle of all airways. H_1 receptors have been characterized in bovine tracheal smooth muscle by [^{3}H]pyrilamine binding. There is a close relationship between H_1-receptor occupancy and stimulation of PI turnover by histamine, indicating that H_1 receptors may lead to contraction by stimulating PI hydrolysis and release of intracellular Ca^{2+}, as in other tissues (Grandordy *et al.*, 1987a). There is also a close association between the contractile response and H_1-receptor occupancy in this tissue, indicating that there are no spare receptors (Fig. 4).

The role of H_2 receptors in airways is less certain. In some species H_2 receptors mediate bronchodilation (Chand and De Roth, 1979). Human peripheral lung strips may also relax with histamine (Vincenc *et al.*, 1984), although this is likely to reflect the presence of vascular smooth muscle in these

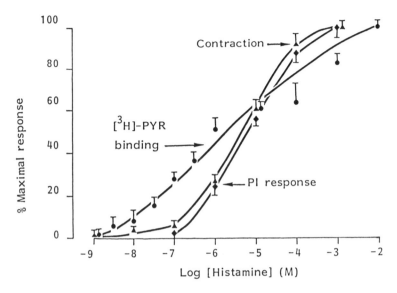

FIGURE 4.   Histamine H_1 receptors on bovine tracheal smooth muscle. There is a concordance of contractile response to histamine stimulation of phosphoinositide (PI) hydrolysis and occupation of lower-affinity H_1 receptors measured by [^{3}H]pyrilamine binding.

preparations. *In vivo* H_2-receptor antagonists have no effect on the bronchoconstrictor effect of histamine, suggesting that H_2 receptors do not play a role in regulating human airway smooth muscle tone (Braude *et al.*, 1984). H_2 receptors have been identified in guinea pig lung homogenates by receptor binding using [^{3}H]tiotidine, but their localization is uncertain (Foreman *et al.*, 1985).

B. Prostanoid Receptors

Prostaglandin receptors have not been well characterized, as few specific antagonists are available (Gardiner and Collier, 1980). Prostacyclin receptors have been identified in guinea pig lung by measuring activation of adenylate cyclase (MacDermot and Barnes, 1980) and by direct receptor binding (MacDermot *et al.*, 1981). These receptors are probably localized to pulmonary vessels, rather than to airways, since prostacyclin has little effect on airway smooth muscle (Hardy *et al.*, 1985). Prostaglandins PGD_2 and $PGF_{2\alpha}$ are potent constrictors of human airways *in vitro* and *in vivo*, presumably by activation of specific receptors in airway smooth muscle. Thromboxane A_2 (TXA_2) is also a potent bronchoconstrictor that activates specific receptors. Specific thromboxane receptor antagonists have recently been developed.

C. Leukotriene Receptors

There has been considerable interest in the role of leukotriene B_4 (LTB_4) and the sulfidopeptide leukotrienes C_4, D_4, and E_4 (which comprise slow-reacting substance of anaphylaxis) in the pathogenesis of asthma. Leukotrienes produce their effects by activating specific, and probably distinct, receptors that have recently been identified using ^3H-labeled leukotrienes. LTB_4 exhibits potent chemotactic activity, particularly for neutrophils, and has little direct effect on airway smooth muscle.

Functional studies with sulfidopeptide leukotrienes have indicated that there may be discrete receptors for LTC_4 and LTD_4 (Snyder and Krell, 1984) and that the compound FPL 55712 selectively antagonizes the action of LTD_4. Since LTC_4 is rapidly converted to LTD_4 by the enzyme γ-glutamyl transpeptidase, it is necessary to inhibit this enzyme with serine borate complex to demonstrate the two receptors. Radioligand binding studies have confirmed that there are two distinct binding sites in lung homogenates that correspond with the functional LTC_4 and LTD_4 receptors (Kuehl et al., 1984). Studies of isomers of LTC_4 have demonstrated that differences in binding correspond closely with differences in contractile potency, which supports the concept that the lung-binding site has the properties of a specific receptor. Further evidence in favor of different receptors for LTC_4 and D_4 is the differential sensitivity of binding to cations and GTP (Bruns et al., 1983). LTE_4 appears to bind to the D_4 receptor but, although it is less potent than LTD_4, it is more slowly metabolized and so may have a more significant functional effect and contribute to the prolonged duration of bronchoconstriction of leukotrienes.

The biochemical mechanisms by which LT receptors lead to contraction of airway smooth muscle have recently been investigated. Both LTC_4 and LTD_4 stimulate PI hydrolysis in guinea pig trachea, but LTC_4 is significantly more potent and has a greater stimulatory effect (Grandordy et al., 1986a). Functionally, the two LTs have a similar potency, however, and the greater effect of LTC_4 may be related to the much higher number of specific binding sites. Both LTs also inhibit adenylate cyclase but, again, LTC_4 is more potent than LTD_4 (Barnes and Grandordy, 1986).

Autoradiographic studies have mapped the distribution of LTC_4 and D_4 receptors in guinea pig lung (Barnes et al., 1985a). There appears to be a differential distribution, with LTC_4 receptors being more widely distributed and present in higher density, particularly in airway smooth muscle, than LTD_4 receptors.

D. Platelet Activating Factor Receptors

Platelet activating factor is a phospholipid that is a potent bronchoconstrictor in several animal species, including humans (Morley et al., 1985; Cuss et al.,

1986). In guinea pigs, bronchoconstriction is dependent on platelet activation, since it may be prevented by depletion of platelets or with drugs that inhibit the release reaction of platelets (Lefort *et al.*, 1984). PAF produces its effects by activation of specific receptors identified by direct binding assays on human platelets, neutrophils (O'Flaherty *et al.*, 1986), and lung membranes (Huang *et al.*, 1985). PAF receptors are presumably not present on human airway smooth muscle, since PAF has no direct constrictor effect, but may cause constriction in the presence of platelets (Schellenberg *et al.*, 1983).

E. Adenosine Receptors

At least two cell-surface receptors for adenosine have been recognized with the development of a series of adenosine analogues (Londos *et al.*, 1980). A_1 receptors are usually excitatory and associated with a fall in intracellular cAMP, whereas A_2 receptors are usually inhibitory and associated with a rise in cAMP. Inhaled adenosine causes bronchoconstriction in asthmatic subjects (Cushley *et al.*, 1983). Adenosine relaxes guinea pig airways by an A_2 receptor and has little effect on isolated human airways *in vitro* (Finney *et al.*, 1985), suggesting that the bronchoconstrictor effect may be indirect, possibly by potentiating inflammatory mediator release. Theophylline is a specific antagonist of both A_1 and A_2 receptors. However, there is evidence against this as its major mechanism of bronchodilation, since an analogue of theophylline, enprofylline, is a more potent bronchodilator, without significant adenosine antagonism (Persson, 1982).

F. Bradykinin Receptors

Bradykinin is a potent bronchoconstrictor when given by inhalation to asthmatic subjects, being significantly more potent than methacholine (Simonsson *et al.*, 1973; Fuller *et al.*, 1986), but it has little effect on human airways *in vitro*, suggesting that its bronchoconstrictor action is indirect and that human airway smooth muscle does not have a significant population of bradykinin receptors. Bradykinin releases prostaglandins in several species, suggesting that its effects might be mediated via bronchoconstrictor prostaglandins, but aspirin has no inhibitory effect (Fuller *et al.*, 1986). To some extent, the bronchoconstrictor effect is diminished by anticholinergic drugs, implicating a vagal cholinergic reflex. There is evidence that bradykinin may activate C-fiber afferent nerve endings in bronchi (Kaufman *et al.*, 1980), so that the effect of bradykinin may be attributable to a cholinergic reflex as well as to release of bronchoconstrictor sensory neuropeptides by means of a local (axon) reflex. Two receptor subtypes (designated B_1 and B_2) have been recognized using peptide analogues of bradykinin; there is some evidence that experimental inflammation may enhance the responsiveness of B_1-receptors (Regoli and Barabé, 1980).

VII. CONCLUSIONS

Many different receptors have now been characterized in airway smooth muscle by functional studies and by direct receptor-binding methods. Recently the mechanisms by which receptors are coupled to contraction of airway smooth muscle have been elucidated. The development of techniques to measure electrophysiological changes after receptor activation and the study of isolated airway smooth muscle cells should further increase our understanding of airway smooth muscle receptors. The demonstration that several agonists may have different effects on different sizes of airways is important, since the receptor population of smooth muscle in peripheral airways may be quite different from that in proximal airways. As most studies have been performed in larger airways, this may give misleading information about the responsiveness of peripheral airways, which may be involved in airway disease. The question of whether receptor populations may be changed in airway disease remains unanswered. Stimulation of one receptor may change the expression of another. Thus, stimulation of muscarinic receptors may reduce the number and coupling of β-adrenoceptors, possibly via the enzyme protein kinase C, which is stimulated by PI breakdown. This implies that other inflammatory mediators, which also lead to receptor-mediated PI hydrolysis, may produce marked changes in surface receptor populations, resulting in altered responsiveness in asthma. Further studies of the molecular pharmacology of airway receptors should lead to greater understanding of the pathogenesis and treatment of airway disease.

ACKNOWLEDGMENTS. The author is grateful to Madeleine Wray for typing the manuscript. This work was supported by grants from the Medical Research Council and Asthma Research Council.

VIII. REFERENCES

Barnes, P. J., 1986a, Non-adrenergic non-cholinergic neural control of human airways, *Arch Intern. Pharmacodyn.* **280**:208–228.

Barnes, P. J., 1986b, Neural control of human airways in health and disease. State of the art, *Am. Rev. Respir. Dis.* **134**: 1289–1314.

Barnes, P. J., 1987a, Inflammatory mediator receptors and asthma, *Am. Rev. Respir. Dis.* **135**:S26–S31.

Barnes, P. J., 1987b, Neuropeptides in the lung: Localization, function and pathophysiological implications, *J. Allergy Clin. Immunol.* **79**:285–295.

Barnes, P. J., 1987c, Muscarinic receptors in lung, *Postgrad. Med. J.* **63**(Suppl.):13–19.

Barnes, P. J., and Dixon, C. M. S., 1984, The effect of inhaled vasoactive intestinal peptide on bronchial hyperreactivity in man, *Am. Rev. Respir. Dis.* **130**:162–166.

Barnes, P. J., and Grandordy, B., 1986, Leukotriene C4 elicits inositol phosphate formation and inhibits adenylate cyclase in guinea-pig lung, *Br. J. Pharmacol.* **89**:783P.

Barnes, P. J., Pride, N. B., 1983, Dose–response curves to inhaled beta-adrenoceptor agonists in normal and asthmatic subjects, *Br. J. Clin. Pharmacol.* **15**:677–682.

Barnes, P. J., Dollery, C. T., and MacDermot, J., 1980a, Increased pulmonary alpha-adrenergic and reduced beta-adrenergic receptors in experimental asthma, *Nature (Lond.)* **285:**569–571.

Barnes, P. J., Karliner, J. S., and Dollery, C. T., 1980b, Human lung adrenoceptors studied by radioligand binding, *Clin. Sci.* **58:**457–461.

Barnes, P. J., Ind, P. W., and Dollery, C. T., 1981a, Inhaled prazosin in asthma, *Thorax* **36:**378–381.

Barnes, P. J., Wilson, N. M., and Vickers, H., 1981b, Prazosin, an alpha 1-adrenoceptor antagonist partially inhibits exercise-induced asthma, *J. Allergy. Clin. Immunol.* **68:**411–419.

Barnes, P. J., Basbaum, C. B., Nadel, J. A., and Roberts, J. M., 1982, Localization of beta-adrenoceptors in mammalian lung by light microscopic autoradiography, *Nature (Lond.)* **299:** 444–447.

Barnes, P. J., Basbaum, C. B., and Nadel, J. A., 1983a, Autoradiographic localization of autonomic receptors in airway smooth muscle: Marked differences between large and small airways, *Am. Rev. Respir. Dis.* **127:**758–762.

Barnes, P. J., Nadel, J. A., Skoogh, B.-E., and Roberts, J. M., 1983b, Characterization of beta-adrenoceptor subtypes in canine airway smooth muscle by radioligand binding and physiologic responses, *J. Pharmacol. Exp. Ther.* **225:**456–461.

Barnes, P. J., Skoogh, B.-E., Brown, J. K., and Nadel, J. A., 1983c, Activation of alpha-adrenergic responses in tracheal smooth muscle: A post-receptor mechanism, *J. Appl. Physiol.* **54:**1469–1476.

Barnes, P. J., Skoogh, B.-E., Nadel, J. A., and Roberts, J. M., 1983d, Postsynaptic alpha$_2$-adrenoceptors predominate over alpha$_1$-adrenoceptors in canine tracheal smooth muscle and mediate neuronal and humoral alpha-adrenergic contraction, *Mol. Pharmacol.* **23:**570–575.

Barnes, P. J., Carstairs, J. R., Norman, P., and Abram, T. S., 1985a, Autoradiographic localization of leukotriene receptors in guinea pig lung and trachea, *Am. Rev. Respir. Dis.* **131:**29A.

Barnes, P. J., Cuss, F. M. C., and Palmer, J. B. D., 1985b, The effect of airway epithelium on smooth muscle contractility in bovine trachea, *Br. J. Pharmacol.* **86:**685–691.

Barnes, P. J., Bloom, S. R., Cadieux, A., Cuss, F. M. C., Ghatei, M. A., Mulderry, P. K., Palmer, J. B. D., Polak, J. M., and Springall, D. R., 1987a, Calcitonin gene-related peptide is localised to human airway nerves and potently constricts human airway smooth muscle, *Br. J. Pharmacol.* **91:**95–101.

Barnes, P. J., Grandordy, B. M., Page, C. P., Rhoden, K. J., and Robertson, D. N., 1987b, The effect of platelet activating factor on pulmonary beta-adrenoceptors, *Br. J. Pharmacol.* **90:**709–715.

Basbaum, C. B., Grillo, M. A., and Widdicombe, J. H., 1984, Muscarinic receptors: Evidence for a non uniform distribution in tracheal smooth muscle and exocrine glands, *J. Neurosci.* **4:**508–520.

Blaber, L. C., Fryer, A. D., and Maclagan, J., 1985, Neuronal muscarinic receptors attenuate vagally-induced contraction of feline bronchial smooth muscle, *Br. J. Pharmacol.* **86:**723–728.

Black, J., Turner, A., and Shaw, J., 1981, Alpha-adrenoceptors in human peripheral lung, *Eur. J. Pharmacol.* **72:**83–86.

Black, J. L., Salome, C. M., Yan, K., and Shaw, J., 1982, Comparison between airways response to an alpha-adrenoceptor agonist and histamine in asthmatic and non-asthmatic subjects, *Br. J. Clin. Pharmacol.* **14:**464–465.

Braude, S., Royston, D., Coe, C., and Barnes, P. J., 1984, Histamine increases lung permeability by an H$_2$-receptor mechanism, *Lancet* **2:**372–374.

Brown, J. K., Shields, R., Jones, C., and Gold, W. M., 1983, Augmentation of alpha-adrenergic responsiveness in trachealis muscle of living dogs, *J. Appl. Physiol.* **54:**1558–1566.

Bruns, R. T., Thomsen, W. J., and Pugsley, T. A., 1983, Binding of leukotrienes C4 and D4 to membranes from guinea pig lung: Regulation by ions and guanine nucleotides, *Life Sci.* **33:**645–653.

Carstairs, J. R., and Barnes, P. J., 1986a, Visualization of vasoactive intestinal peptide receptors in human and guinea pig lung, *J. Pharmacol. Exp. Ther.* **239:**249–255.

Carstairs, J. R., and Barnes, P. J., 1986b, Autoradiographic mapping of substance P receptors in lung, *Eur. J. Pharmacol.* **127**:295–296.

Carstairs, J. R., Nimmo, A. J., and Barnes, P. J., 1984, Autoradiographic localisation of beta-adrenoceptors in human lung, *Eur. J. Pharmacol.* **103**:189–190.

Carstairs, J. R., Nimmo, A. J., and Barnes, P. J., 1985, Autoradiographic visualization of beta-adrenoceptor subtypes in human lung, *Am. Rev. Respir. Dis.* **132**:541–547.

Carswell, H., and Nahorski, S. R., 1983a, Distribution and characteristics of histamine H1-receptors in guinea-pig airways identified by [^{3}H]mepyramine, *Eur. J. Pharmacol.* **81**:301–307.

Carswell, H., and Nahorski, S. R., 1983b, Beta-adrenoceptor heterogeneity in guinea-pig airways: Comparison of functional and receptor labelling studies, *Br. J. Pharmacol.* **79**:965–971.

Cerrina, J., Ladurie, M. L., Labat, C., Raffestin, B., Bayol, A., and Brink, C., 1986, Comparison of human bronchial muscle responses to histamine in vivo with histamine and isoprotenerol agonists in vitro, *Am. Rev. Respir. Dis.* **134**:57–61.

Chand, N., and De Roth, L., 1979, Dual histamine receptor mechanism in guinea pig lung, *Pharmacology* **19**:185–190.

Cheng, J. B., and Townley, R. G., 1982, Comparison of muscarinic and beta-adrenergic receptors between bovine peripheral lung and tracheal smooth muscle: A striking difference, *Life Sci.* **30**:2079–2086.

Clarke, B., Evans, T. W., Dixon, C. M. S., Conradson, T.-B., and Barnes, P. J., 1987, Comparison of the cardiovascular and respiratory effects of substance P and neurokinin A in man, *Clin Sci.* **72**:41.

Cushley, M. J., Tattersfield, A. E., and Holgate, S. T., 1983, Inhaled adenosine and guanosine on airway resistance in normal and asthmatic subjects, *Br. J. Clin. Pharmacol.* **15**:161–165.

Cuss, F. M., and Barnes, P. J., 1987, Epithelial mediators, *Am. Rev. Respir. Dis.* **136**:S32–S35.

Cuss, F. M., Dixon, C. M. S., and Barnes, P. J., 1986, Effects of inhaled platelet activating factor on pulmonary function and bronchial responsiveness in man, *Lancet* **2**:189–192.

Ekblad, E., Hakanson, R., Sundler, F., and Wahlestedt, C., 1985, Galanin: Neuromodulatory and direct contractile effects on smooth muscle preparations, *Br. J. Pharmacol.* **86**:241–246.

Engel, G., 1981, Subclasses of beta-adrenoceptor—A quantitative estimation of beta$_1$- and beta$_2$-adrenoceptors in guinea pig and human lung, *Postgrad. Med. J.* **57**:77–83.

Finney, M. J. B., Karlsson, J.-A., and Persson, C. G. A., 1985, Effects of bronchoconstrictors and bronchodilators on a novel human small airway preparation, *Br. J. Pharmacol.* **85**:29–36.

Flavahan, N. A., Aarhus, L. L., Rimele, T. J., and Vanhoutte, P. M., 1985, Respiratory epithelium inhibits bronchial smooth muscle tone, *J. Appl. Physiol.* **58**:834–838.

Foreman, J. C., Norris, D. B., Rising, T. J., and Webber, S. E., 1985, The binding of [^{3}H]-tiotidine to homogenates of guinea-pig lung parenchyma, *Br. J. Pharmacol.* **86**:475–482.

Frandsen, E. K., Krishna, G. A., and Said, S. I., 1978, Vasoactive intestinal polypeptide promotes cyclic adenosine 3', 5'-mono-phosphate accumulation in guinea pig trachea, *Br. J. Pharmacol.* **62**:367–369.

Fryer, A. D., and MacLagan, J., 1984, Muscarinic inhibitory receptors in pulmonary parasympathetic nerves in the guinea pig, *Br. J. Pharmacol.* **83**:973–978.

Fuller, R. W., Dixon, C. M. S., Cuss, F. M. C., and Barnes, P. J., 1986, Bradykinin-induced bronchoconstriction in man: Mode of action, *Am. Rev. Respir. Dis.* **135**:176–180.

Fuller, R. W., Maxwell, D. L., Dixon, C. M. S., McGregor, G. P., Barnes, V. F., Bloom, S. R., and Barnes, P. J., 1987, The effects of substance P on cardiovascular and respiratory function in human subjects, *J. Appl. Physiol.* **62**:1473–1479.

Gardiner, P. J., and Collier, H. O. J., 1980, Specific receptors for prostaglandins in airways, *Prostaglandins* **19**:819–841.

Goldie, R. G., Paterson, J. W., and Wale, J. L., 1982, Pharmacological responses of human and porcine lung parenchyma, bronchus and pulmonary artery, *Br. J. Pharmacol.* **76**:515–521.

Goldie, R. G., Papadmitriou, J. M., Paterson, J. W., Rigby, P. J., Self, H. M., and Spina, D.,

1986, Influence of epithelium on responsiveness of guinea-pig isolated trachea to contractile and relaxant agonists, *Br. J. Pharmacol.* **87:**5–14.

Grandordy, B., and Barnes, P. J., 1987, Phosphoinositide turnover in airway smooth muscle, *Am. Rev. Respir. Dis.* **136:**S17–S20.

Grandordy, B. M., Cuss, F. M., Meldrum, L., Sturton, R. G., and Barnes, P. J., 1986a, Leukotriene C4 and D4 induce contraction and formation of inositol phosphates in airways and lung parenchyma, *Am. Rev. Respir. Dis.* **133:**A113.

Grandordy, B. M., Cuss, F. M., Sampson, A. S., Palmer, J. B., and Barnes, P. J., 1986b, Phosphatidylinositol response to cholinergic agonists in airway smooth muscle: Relationship to contraction and muscaranic receptor occupancy, *J. Pharmacol. Exp. Ther.* **238:**273–279.

Grandordy, B. M., Rhoden, K., and Barnes, P. J., 1987a, Histamine H1-receptors in human lung: Correlation of receptor binding and function, *Am. Rev. Respir. Dis.* **135:** A274.

Grandordy, B., Rhoden, K., and Barnes, P. J., 1987b, Effects of protein kinase C activation on adrenoceptors in airway smooth muscle, *Am. Rev. Respir. Dis.* **135:**A272 .

Grandordy, B. M., Rhoden, K. J., Frossard, N., and Barnes, P. J., 1988, Tachykinin induced phosphoinositide breakdown in airway smooth muscle and epithelium: relationship to contraction, *Mol. Pharmacol.* **33:**515–519.

Hammer, P., and Giachetti, A., 1982, Muscarinic receptor subtypes: M1 and M2 biochemical and functional characterization, *Life Sci.* **31:**2991–2998.

Hardy, C., Robinson, C. Lewis, R. A., Tattersfield, A. E., and Holgate, S. T., 1985, Airway and cardiovascular responses to inhaled prostaglandin in normal and asthmatic subjects, *Am. Rev. Respir. Dis.* **131:**18–21.

Huang, S.-B., Lam, M.-H., and Shen, T. Y., 1985, Specific binding sites for platelet activating factor in human lung tissues, *Biochem. Biophys. Res. Commun.* **128:**972–979.

Ingram, R. H., Wellman, J. J., McFadden, E. R., and Mead, J., 1977, Relative contribution of large and small airways to flow limitation in normal subjects before and after atropine and isoproterenol, *J. Clin. Invest.* **59:**696–703.

Karlsson, J.-A., Finney, M. J. B., Persson, C. G. A., and Post, C., 1984, Substance P antagonist and the role of tachykinins in non-cholinergic bronchoconstriction, *Life Sci.* **35:** 2681–2691.

Kaufman, M. P., Coleridge, H. M., Coleridge, J. C. G., and Baker, D. G., 1980, Bradykinin stimulates afferent vagal C-fibers in intrapulmonary airways of dogs, *J. Appl. Physiol.* **48:**511–517.

Kneussl, M. P., and Richardson, J. B., 1978, Alpha-adrenergic receptors in human and canine tracheal and bronchial smooth muscle, *J. Appl. Physiol.* **45:**307–311.

Kuehl, F. A., De Haven, R. N., and Pong, S. S., 1984, Lung tissue receptors for sulfidopeptide leukotrienes, *J. Allergy Clin. Immunol.* **74:**378–381.

Lazarus, S. C., Basbaum, C. B., Barnes, P. J., and Gold, W. M., 1986, cAMP immunocytochemistry provides evidence for functional VIP receptors in trachea, *Am. J. Physiol.* **251:**C115–119.

Lefort, J., Rotilio, D., and Vargaftig, B. B., 1984, The platelet-independent release of thromboxane A2 by PAF-acether for guinea-pig lungs involves mechanisms distinct from those for leukotriene C4 and bradykinin, *Br. J. Pharmacol.* **82:**525–531.

Lofdahl, C.-G., and Svedmyr, N., 1982, Effects of prenalterol in asthmatic patients, *Eur. J. Clin. Pharmacol.* **23:**297–303.

Londos, C., Cooper, D. M. F., and Wolff, J., 1980, Subclass of external adenosine receptors, *Proc. Natl. Acad. Sci. USA* **77:**2551–2554.

Lulich, V. M., Mitchell, H. W., and Sparrow, M. P., 1976, The cat lung strip as an in vitro preparation of peripheral airways: A comparison of beta-adrenoceptor agonists, autocoids and anaphylactic challenge of the lung strip and trachea, *Br. J. Pharmacol.* **58:**71–79.

Lundberg, J. M., Martling, C.-R., and Saria, A., 1983, Substance P and capsaicin-induced contraction of human bronchi, *Acta Physiol. Scand.* **119:**49–53.

MacDermot, J., and Barnes, P. J., 1980, Activation of guinea pig pulmonary adenylate cyclase by prostacyclin, *Eur. J. Pharmacol.* **67**:419–425.

MacDermot, J., Barnes, P. J., Wadell, K., Dollery, C. T., and Blair, I. A., 1981, Prostacylin binding to guinea pig pulmonary receptors, *Eur. J. Pharmacol.* **68**:127–130.

Mann, J. S., Cushley, M. J., and Holgate, S. T., 1985, Adenosine-induced bronchoconstriction in asthma. Role of parasympathetic stimulation and adrenergic inhibition, *Am. Rev. Respir. Dis.* **132**:1–6.

Minette, P., and Barnes, P. J., 1988, Prejunctional muscarinic receptors on cholinergic nerves to human and guinea pig airways, *J. Appl. Physiol.* **64**:2532–2537.

Morley, J., Page, C. P., and Sanjar, S., 1985, Pulmonary responses to platelet-activating factor, *Prog. Respir. Res.* **19**:117–123.

Murlas, C., Nadel, J. A., and Roberts, J. M., 1982, The muscaranic receptors of airway smooth muscle: Their characterization in vitro, *J. Appl. Physiol.* **52**:1084–1091.

Nadel, J. A., and Barnes, P. J., 1984, Autonomic regulation of the airways, *Annu. Rev. Med.* **35**:451–467.

Nadel, J. A., Cabezas, G. A., and Austin, J. H. M., 1971, In vivo roentgenographic examination of parasympathetic innervation of small airways: Use of powdered tantulum and a fine focal spot x-ray, *Invest. Radiol.* **6**:9–17.

O'Flaherty, J. T., Surles, J. R., Redman, J., Jacobson, D., Piantadosi, C., and Wykle, R. L., 1986, Binding and metabolism of platelet-activating factor by human neutrophils, *J. Clin. Invest.* **78**:381–388.

Palmer, J. B., Cuss, F. M. C., and Barnes, P. J., 1986a, VIP and PHM and their role in non-adrenergic inhibitory responses in isolated human airways, *J. Appl. Physiol.* **61**:1322–1328.

Palmer, J. B. D., Cuss, F. M. C., Warren, J. B., and Barnes, P. J., 1986b, The effect of infused vasoactive intestinal peptide on airway function in normal subjects, *Thorax* **41**:663–666.

Palmer, J. B., Cuss, F. M. C., and Barnes, P. J., 1986c, Sensory neuropeptides and human airway function, *Am. Rev. Respir. Dis.* **133**:A239.

Paterson, J. W., Lulich, K. M., and Goldie, R. G., 1982, The role of beta-adrenoceptors in bronchial hyperreactivity, in: *Bronchial Hyperreactivity* (J. Morley, ed.), pp. 19–38, Academic, London.

Persson, C. G. A., 1982, Universal adenosine receptor antagonism is neither necessary nor desirable with xanthine anti-asthmatics, *Med. Hypotheses* **8**:515–526.

Persson, C. G. A., and Ekman, M., 1976, Contractile effects of histamine in large and small airways, *Agents Actions* **6**:389–393.

Peters, S. P., Schulman, E. S., Schleimer, R. P., Macglashan, D. W., Newball, H. H., and Lichtenstein, L. M., 1982, Dispersed human lung mast cells. Pharmacologic aspects and comparison with human lung tissue fragments, *Am. Rev. Respir. Dis.* **126**:1034–1039.

Polak, J. M., and Bloom, S. R., 1986, Regulatory peptides of the gastrointestinal and respiratory tracts, *Arch. Intern. Pharmacodyn.* **280**:16–49.

Regoli, D., and Barabe, J., 1980, Pharmacology of bradykinin and related kinins, *Pharmacol. Rev.* **32**:1–46.

Richardson, J. B., 1979, Nerve supply to the lungs, *Am. Rev. Respir. Dis.* **119**:785–802.

Robberecht, P., Chatelain, P., De Neef, P., Camus, J.-C., Waelbroeck, M., and Christophe, J., 1981, Presence of vasoactive intestinal peptide receptors coupled to adenylate cyclase in rat lung membranes, *Biochim. Biophys. Acta* **678**:76–82.

Rugg, E. L., Barnett, D. B., and Nahorski, S. R., 1978, Coexistence of beta 1 and beta 2 adrenoceptors in mammalian lung: Evidence from direct binding studies, *Mol. Pharmacol.* **14**:996–1005.

Schellenberg, R. R., Walker, B., and Snyder, F., 1983, Platelet-dependent contraction of human bronchus by platelet activating factor, *J. Allergy Clin. Immunol.* **71**:145 (abstr.).

Simonsson, B. G., Skoogh, B. E., Bergh, N. P., Anderson, R., and Svedmyr, N., 1973, In vivo and in vitro effect of bradykinin on bronchial motor tone in normal subjects and in patients with airway obstruction, *Respiration* **30**:378–388.

Snashall, R., Boother, F. A., and Sterling, G. M., 1978, The effect of alpha-adrenergic stimulation on the airways of normal and asthmatic man, *Clin. Sci.* **54**:283–289.

Snyder, D. W., and Krell, R. D., 1984, Pharmacological evidence for a distinct leukotriene C4 receptor in guinea-pig trachea, *J. Pharmacol. Exp. Ther.* **231**:616–622.

Szentivanyi, A., 1968, The beta adrenergic theory of the atopic abnormality in bronchial asthma, *J. Allergy* **42**:203–232.

Tanaka, D. T., and Grunstein, M. M., 1984, Mechanisms of substance P-induced contraction of rabbit airway smooth muscle, *J. Appl. Physiol.* **57**:1551–1557.

Turner, A. J., Seale, J. P., and Shaw, J., 1983, Antigen sensitization does not alter response of guinea pig lung strips to noradrenaline, *Eur. J. Pharmacol.* **87**:141–144.

Vermeire, P. A., and Vanhoutte, P. M., 1979, Inhibitory effects of catecholamines in isolated canine bronchial smooth muscle, *J. Appl. Physiol.* **46**:787–791.

Vincenc, K., Black, J., and Shaw, J., 1984, Relaxation and contraction responses to histamine in the human lung parenchymal strip, *Eur. J. Pharmacol.* **84**:201–210.

White, J., and Eiser, N. M., 1983, The role of histamine and its receptors in the pathogenisis of asthma, *Br. J. Dis. Chest.* **77**:215–226.

Zaagsma, J., van der Heijden, P. J. C. M., van der Schaar, M. W. G., and Blank, C. M. C., 1983, Comparison of functional beta-adrenoceptor heterogeneity in central and peripheral airway smooth muscle of guinea pig and man, *J. Recept. Res.* **3**:89–106.

Chapter 6

Cellular Control Mechanisms in Airway Smooth Muscle

Primal de Lanerolle

Department of Physiology and Biophysics
College of Medicine
University of Illinois at Chicago
Chicago, Illinois 60680

I. INTRODUCTION

The regulation of smooth muscle contraction is really an exercise in understanding the regulation of a number of protein phosphorylation/dephosphorylation reactions. By this I mean the enzyme-catalyzed transfer (by a class of enzymes known as protein kinases) of the terminal phosphate of ATP to a serine or threonine* residue on a protein with the formation of a covalent phosphoester linkage and, equally important, the enzyme-catalyzed dephosphorylation (by enzymes known as phosphoprotein phosphatases) of the phosphorylated proteins. This type of reversible phosphorylation is known to be important in regulating cellular processes (reviewed by Krebs and Beavo, 1979). As we shall see, smooth muscles contain Ca^{2+}/calmodulin-dependent, cyclic nucleotide-dependent, and Ca^{2+}/phospholipid-dependent protein kinases and at least three different phosphatases. In many ways, the contractile properties of smooth muscles can be visualized as being determined by the balance between the activities of these kinases and phosphatases. Consequently, much of the research emphasis on smooth muscle regulation has been directed at understanding the properties of kinases and phosphatases purified from smooth muscles. These studies have proved important for two reasons. First, they have resulted in a greater under-

---

*Tyrosine residues are also known to be phosphorylated by an important class of protein kinases associated with the regulation of cell growth. Little is known about tyrosine kinases in smooth muscle, and these are not discussed. However, the study of tyrosine kinases is an important and exciting area of research, and the reader is directed to Hunter and Sefton (1985), for a discussion of this subject.

standing of the regulation of smooth muscle contraction. Second, they serve as an important model for studying the role of protein phosphorylation reactions in other biological systems.

II. BIOCHEMISTRY OF SMOOTH MUSCLE CONTRACTION

The two most important contractile proteins are actin and myosin. These proteins are virtually ubiquitous, having been found in every mammalian cell studied to date. Their importance in the contractile process relates to the fact that myosin is an enzyme that hydrolyzes ATP. But, in the presence of 100–150 mM KCl and $MgCl_2$ in excess of 1 mM, purified striated muscle myosin hydrolyzes ATP at a low rate (Carlson and Wilkie, 1974). The removal of Mg^2 stimulates the rate of ATP hydrolysis greatly. Thus, Mg^{2+} at physiological concentrations inhibits ATP hydrolysis by myosin. By contrast, purified actin greatly stimulates the rate of ATP hydrolysis by skeletal or cardiac myosin in the presence of Mg^{2+}. Consequently, the physiologically relevant ATPase of any myosin is the actin-activated Mg^{2+}-ATPase activity of that myosin (abbreviated actomyosin ATPase).

The special function of Ca^{2+} is to regulate the contractile process. Ringer (1883) demonstrated during the nineteenth century that amphibian hearts would not contract when perfused with a buffer that did not contain Ca^{2+}. Nevertheless, a mechanistic understanding of the role of Ca^{2+} in regulating striated muscles became clear only with the discovery of troponin by Setsuro Ebashi and colleagues (Ebashi and Endo, 1968). Troponin is a complex of three proteins, one of which binds Ca^{2+} (Ebashi, 1980). Troponin, through its effect on the location of tropomyosin in the thin filament, inhibits the interaction of actin and myosin in the absence of Ca^{2+}. When intracellular Ca^{2+} rises in response to a contractile stimulus, Ca^{2+} binds to troponin and initiates a change in the structure of the thin filament that permits the interaction of actin and myosin and the hydrolysis of ATP (Squire, 1981). The energy released following the hydrolysis of ATP is used in a way that we do not fully understand (Aderstein and Eisenberg, 1980; Eisenberg and Hill 1985), to move thick and thin filaments past each other. Muscle contraction, then, is the result of filament movement as described by the sliding filament model (Huxley and Niedergerke, 1954). Relaxation results from the reversal of the entire process following the dissociation of Ca^{2+} from troponin.

Initially, experiments designed to understand the contractile process in smooth muscles were modeled after studies on striated muscles. This approach proved of limited value because smooth muscles differ from striated muscles in two important respects. First, unlike striated muscle myosin, the mixing of actin and ATP with smooth muscle myosin does not result in ATP hydrolysis. Second, there is no troponinlike complex in smooth muscles. The importance of these points is that they raise two fundamentally important questions regarding smooth muscle contractility: (1) What is the mechanism for actin-activation of smooth muscle myosin? (2) How is Ca^{2+} regulation mediated in smooth muscle? The answer to both questions resides in the phosphorylation of the 20,000-dalton subunit (LC_{20}) of smooth muscle myosin. Phosphorylation of smooth muscle myosin LC_{20} was not unexpected because Perry and colleagues had demonstrated the phosphorylation of the comparable light chain of skeletal muscle myosin (Perrie and Perry, 1970), and Adelstein had demonstrated the phosphorylation of platelet LC_{20} (Adelstein and Conti, 1975). In addition, Adelstein had found that phosphorylation stimulated the actomyosin ATPase activity of platelet myosin. Consequently, Sobieszek's demonstration that phosphorylation of LC_{20} stimulated the actin-activated Mg^{2+}-ATPase activity of smooth muscle myosin (Sobieszek, 1977) was met with great interest. This observation gained even greater significance when Hartshorne and co-workers demonstrated a few years later that myosin light-chain kinase, the enzyme that phosphorylates myosin, requires Ca^{2+}-calmodulin for activity (Adelstein and Klee, 1981; Dabrowska et al., 1977, 1978).

Many investigators have reproduced and expanded on these observations in the intervening years (Chacko et al., 1977; Chacko and Rosenfeld, 1980; Nag and Siedel, 1983; Persechini and Hartshorne, 1981; Rees and Fredericksen, 1981; Sellers et al., 1983). The data in Table I are from one such study (Sellers et al., 1981). In this experiment, myosin and myosin phosphatase 1 were purified from turkey gizzard smooth muscle and used in a reconstitution assay. The addition of actin does not appreciably increase the Mg^{2+}-ATPase activity of

TABLE I
Correlation of the Actin-Activated Mg^{2+}-ATPase Activity with Phosphorylation of Smooth Muscle Myosin[a]

Smooth muscle myosin[b]	Phosphate incorporation (moles PO_4/mole myosin)[c]	Mg^{2+} ATPase	
		Plus actin[d]	Minus actin
		(nmoles/min/mg)	
Unphosphorylated	0	4	<2
Phosphorylated	1.9	51	<2
Dephosphorylated	0.1	5	—
Rephosphorylated	2.0	46	—

[a]Modified from Sellers et al. (1981).
[b]Myosin purified from turkey gizzard smooth muscle was phosphorylated, dephosphorylated, and rephosphorylated using myosin light-chain kinase (Adelstein and Klee, 1981) and smooth muscle phosphatase 1 (Pato and Adelstein, 1980) purified from turkey gizzard smooth muscle.
[c]The extent of myosin phosphorylation was determined by urea–polyacrylamide gel electrophoresis as described by Perrie and Perry (1970).
[d]Note that the actin-activated Mg^{2+}-ATPase activity of phosphorylated and rephosphorylated myosin is much greater than that of the unphosphorylated or dephosphorylated myosin.

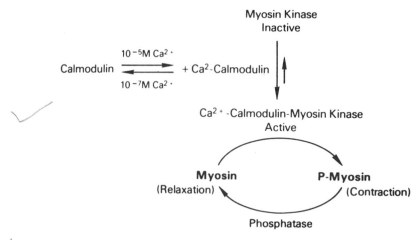

FIGURE 1. Regulation of smooth muscle contraction by calcium. Calcium, at a concentration of 10^{-5} M, binds to calmodulin, which then binds to the inactive form of myosin light-chain kinase. The active kinase then catalyzes the phosphorylation of myosin, which results in a form of the myosin that can interact with actin, hydrolyze ATP, and cause smooth muscle contraction. A decrease in the Ca^{2+} concentration to 10^{-7} M leads to a dissociation of calmodulin from myosin light-chain kinase, which results in the inactive form of the enzyme. Under these conditions, myosin phosphatase dephosphorylates myosin, causing smooth muscle relaxation. (Adapted from Adelstein and Eisenberg, 1980).

unphosphorylated myosin. However, the incorporation of 2 moles PO_4/moles myosin* leads to an 11-fold increase in the rate of ATP hydrolysis. This experiment is particularly important because it demonstrates the reversibility of both phosphorylation and actin-activated Mg^{2+}-ATPase activity. Incubation of the phosphorylated myosin with phosphatase I (Pato and Adelstein, 1980) almost completely dephosphorylates the myosin with an accompanying decrease in the actomyosin ATPase activity; rephosphorylation of this myosin again stimulates the actomyosin ATPase activity.

These types of biochemical experiments have led to the suggestion that myosin phosphorylation is the Ca^{2+}-dependent mechanism for regulating smooth muscle contraction. The idea, as depicted in Fig. 1, is that an increase in intracellular Ca^{2+} leads to the binding of Ca^{2+} to calmodulin. The Ca^{2+}/calmodulin complex then binds to myosin light-chain kinase and activates this

---

*The stoichiometry of phosphorylation is somewhat confusing. It is possible to incorporate 1 mole PO_4/mole LC_{20} when this reaction is catalyzed by purified myosin light-chain kinase. Because there are 2 moles LC_{20}/mole myosin, the stoichiometry of phosphorylation is 2 mole PO_4/mole myosin. As a general rule, I prefer to use moles PO_4/mole LC_{20}, as this is the experimentally determined value. It is also important to keep in mind that smooth muscle myosin can be phosphorylated by other protein kinases and that this can affect the stoichiometry of phosphorylation.

enzyme. The activated enzyme then phosphorylates LC_{20}, which transforms the myosin into a form that can interact with actin and hydrolyze ATP. ATP hydrolysis results in the movement of filaments past each other as well as muscle shortening. By contrast, relaxation is thought to result from a decrease in intracellular Ca^{2+}. The decay in Ca^{2+}/calmodulin complexes leads sequentially to the inactivation of myosin light-chain kinase, dephosphorylation of the myosin due to myosin phosphatase activity, termination of ATP hydrolysis, and relaxation.

What, then, is the physiological significance of this mechanism? The answer to this question can only come from studies on contracting smooth muscles. Many such experiments, including those on tracheal smooth muscles, have been performed, and these are described in the next section.

III. PHYSIOLOGICAL EXPERIMENTS ON MYOSIN PHOSPHORYLATION

The basic approach in studying myosin phosphorylation in contracting muscles has been to correlate myosin phosphorylation and dephosphorylation with contraction and relaxation of smooth muscles. Clearly, if myosin phosphorylation is the mechanism for Ca^{2+} regulation, myosin phosphorylation should precede force production when a smooth muscle is stimulated to contract. The demonstration that phosphorylation lags behind the contractile response would disprove the hypothesis. The rate-limiting step in performing these types of correlative experiments has been the availability of accurate methods to quantitate the myosin phosphate content. The two most important aspects of the myosin phosphorylation assay must be (1) the assurance that myosin, and not a contaminating or extraneous protein, is being assayed; and (2) that there is no change in the myosin phosphate content during the purification and/or while performing the assay.

The earliest report on the temporal relationship between force and phosphorylation is a study on canine trachealis smooth muscle (de Lanerolle, 1979; de Lanerolle and Stull, 1980). In this study, the trachealis muscularis was separated from the adventitia and the mucosa, mounted in water baths, contracted with the muscarinic agonist methacholine, and quick-frozen at various times. The critical element in this study was the use of antibodies to tracheal smooth muscle myosin to immunoprecipitate the myosin from tissue extracts under conditions that inhibited both phosphorylation and dephosphorylation of the myosin (de Lanerolle, 1979). The phosphate content of the immunoprecipitated myosin was quantitated using isoelectric focusing polyacrylamide gel electrophoresis. As shown in Fig. 2, myosin is rapidly phosphorylated upon addition of 10^{-4} M methacholine; this concentration of methacholine results in the generation of maximal force. The myosin phosphate content peaks at 0.65 mole PO_4/moles

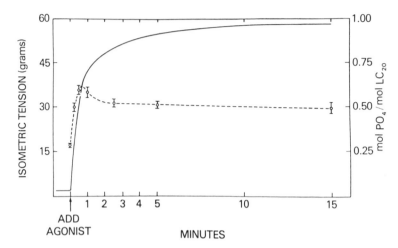

FIGURE 2. Myosin phosphorylation and tension generated by tracheal smooth muscles contracted with 10^{-4} M methacholine. Tracheal muscles from six consecutive tracheal rings were isolated and mounted in six myobaths. The data at 15 min were obtained from separate experiments. (——) Tension generated by a representative muscle contracted for 15 min; (---) phosphate content of muscles frozen at the times indicated as described by de Lanerolle and Stull (1980). Error bars indicate SEM for phosphate determinations on at least five different pieces of tissue. (From de Lanerolle *et al.*, 1982.)

LC_{20} and then declines to 0.5 mole PO_4/moles LC_{20}. This level of phosphorylation is maintained for the duration of the experiment. Note that the peak in myosin phosphorylation precedes the generation of maximal force.*

   Other experiments were performed to study the role of external Ca^{2+} on both force and phosphorylation in canine tracheal muscles by contracting the muscles in a Ca^{2+}-free buffer (de Lanerolle and Stull, 1980). The addition of 10^{-4} M methacholine to the Ca^{2+}-free buffer leads to a reversible contraction that appears to be due to the presence of sufficient intracellular Ca^{2+} to support a single contraction. However, the muscles relax with time as the intracellular

*It must be pointed out that the data presented in Fig. 2 are different from those obtained from studies on bovine tracheal muscles (Silver and Stull, 1982) and rabbit tracheas (Gerthoffer and Murphy, 1983) stimulated with carbachol. However, the bovine tracheal muscles produce about 30% of the force generated by canine tracheal muscles (1.1×10^3 versus 3.4×10^3 g/cm², respectively). It is possible that the bovine tracheal muscles were damaged during dissection, which would account for the observed decline in phosphorylation and the low level of force production. The study on rabbit tracheas (Gerthoffer and Murphy, 1983a) was performed with a very inhomogeneous preparation, in which the smooth muscle occupied only $19 \pm 2.5\%$ of the cross-sectional area of the tissue. Moreover, the mucosa, a rich source of myosin-containing secretory cells, was not removed. More recently, Gerthoffer (1986) studied myosin phosphorylation in canine tracheal muscles in which the adventitia and mucosa were removed and reported data virtually identical to those depicted in Figs. 2 and 3.

Ca^{2+} concentration flows out of the cell. Interestingly, the myosin phosphate content decreases below the resting level of 0.1 mole PO_4/moles LC_{20} in these muscles. Upon readdition of Ca^{2+}, the myosin is slowly phosphorylated, and the muscles contract slowly until they generate maximal tension. This is an important experiment because it demonstrates that (1) myosin phosphorylation in an intact smooth muscle is a Ca^{2+}-dependent process, and (2) there might be a relationship between the rate of phosphorylation and the rate of force production.

It was also of interest to determine the relationship between steady-state levels of force and phosphorylation when tracheal smooth muscles were contracted with varying concentrations of agonist (Fig. 3). The methacholine concentration dependencies of force and phosphorylation are virtually identical. It is worth noting that Gerthoffer (1986) recently published results very similar to those shown in Fig. 3. These data support the prediction, based on data from biochemical experiments, of a direct relationship between the level of myosin phosphorylation and the force generated by a smooth muscle.

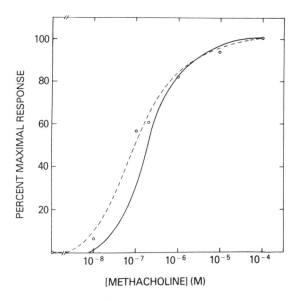

[METHACHOLINE] (M)

FIGURE 3. Dose–response relationship for myosin phosphorylation and isometric tension in tracheal smooth muscle stimulated with methacholine. Myosin phosphorylation was determined on tracheal smooth muscle contracted for 15 min using the agonist concentrations shown. Phosphorylation and tension data were expressed as the percentage maximal increase using the equation X-R/M − R, where X = myosin phosphate content or tension at 15 min of muscles stimulated with various methacholine concentrations; R = the resting phosphate content (29%) or resting tension; and M = the myosin phosphate content (49%) or tension at 15 min of muscles stimulated with 10^{-4} M methacholine. (——) Rise in tension after stimulation at each agonist concentration; (---) increase in myosin phosphorylation at each agonist concentration. Each point represents the mean from phosphate assays on at least five different pieces of tissue. (From de Lanerolle et al., 1982.)

On reflection, these data are very consistent with the idea that myosin phosphorylation regulates smooth muscle contraction. However, the relationship between steady-state myosin phosphorylation and force is a complicated and controversial one requiring further discussion. The temporal relationship between myosin phosphorylation and force has been studied in tracheal muscles (de Lanerolle, 1979; de Lanerolle and Stull, 1980; de Lanerolle et al., 1982; Gerthoffer, 1986; Kamm and Stull, 1985; Silver and Stull, 1982), vascular smooth muscles (Aksoy et al., 1982, 1983; Barron et al., 1980; Dillon et al., 1981; Driska et al., 1981; Gerthoffer and Murphy, 1983b; Ledvora et al., 1983), uterine smooth muscles (Csabina et al., 1986; Haeberle et al., 1985b; Janis et al., 1981; Nishikori et al., 1983), and intestinal smooth muscles (Butler et al., 1983). There is unanimity of opinion regarding the fact that myosin phosphorylation is essential for initiating the contractile process. That is, most investigators agree that myosin phosphorylation is a switch needed to turn on the contractile process. Whether myosin phosphorylation is involved in determining the force maintained by smooth muscles is the subject of intense debate. The basis of this debate is the observation, originally made by Richard Murphy and co-workers that force can be maintained despite dephosphorylation of the myosin to almost the resting level. On the basis of this observation, they suggested that a special actomyosin cross-bridge, known as a *latch bridge,* was responsible for force maintenance in smooth muscles (Dillon et al., 1981).

The original observation that led to the latch-bridge hypothesis is shown in Fig. 4. Murphy and colleagues found, using an isoelectric focusing-sodium dodecyl sulfate (SDS) two-dimensional polyacrylamide gel electrophoresis (PAGE) technique (Driska et al., 1981), that the myosin phosphate content reached a peak value and then declined almost to the resting level when hog carotid artery muscles were stimulated with KCl (compare Figs. 2 and 4). This observation was important for two reasons. First, it seriously questioned the role of myosin phosphorylation in determining the force generated by smooth muscles. Second, the data raised the possibility of an intrinsic change in the nature of the actomyosin cross-bridge with time. The latter statement is based on the observation, shown in Fig. 4, that myosin phosphorylation correlates more closely with the unloaded velocity of shortening (Vus) than with isometric force.

Vus is an important measurement because it represents an indirect estimation of the rate at which actin and myosin hydrolyze ATP (i.e., the cross-bridge cycling rate). Although it is important not to overinterpret the significance of this measurement (see Arner and Hellstrand, 1985, for a more complete discussion of the significance of Vus), it does provide insights into the nature of the cross-bridge. The data in Fig. 4 suggest that myosin phosphorylation is essential for initiating contraction but that the cross-bridge undergoes a change during contraction and hat a new type of cross-bridge, which Murphy labeled a latch bridge (Dillon et al., 1981), is responsible for maintaining force. Murphy defined latch-bridges as being force maintenance in the presence of dephosphorylated myosin

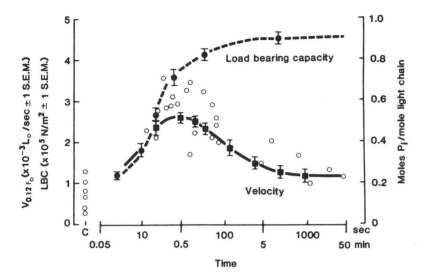

FIGURE 4. Dissociation of force from phosphorylation and shortening velocity in K^+-stimulated carotid media. Load-bearing capacity (●), which is a representation of force; shortening velocity (■) and fractional phosphorylation of tissue LC_{20} (○) are plotted as a function of time on a log scale after K^+ stimulation of carotid medial preparations. Mechanical data points are averages for a series of five preparations (± 1 SEM), with extensions of the curves based on time-course studies in other tissues. Each phosphorylation determination represents a single tissue, frozen at the indicated times ($N = 30$); C designates unstimulated tissues. (From Dillon et al., 1981.)

(Chatterjee and Murphy, 1983). He also suggested that latch-bridges hydrolyzed ATP at a relatively slow rate (i.e., compared with phosphorylated cross-bridges) and that latch-bridges are more sensitive to Ca^{2+} than the myosin phosphorylation mechanism (Chatterjee and Murphy, 1983). This last point raises the possibility of a second, Ca^{2+}-dependent regulatory mechanism in smooth muscles.

As might be expected, the suggestion that there may be multiple Ca^{2+}-dependent regulatory mechanisms in smooth muscles was greeted with great excitement. Many experiments have been performed to test the validity of this hypothesis. Some of the most intriguing are those performed on skinned fibers. Skinned fibers are muscle preparations in which the plasma membrane has been made permeable to both small molecules and proteins. Muscle fibers can be permeabilized by mechanical means (Cassidy et al., 1981; Hoar et al., 1979) or by treatment with various detergents (Endo et al., 1977; Gordon, 1978; Paul et al., 1983; Ruegg and Paul, 1982; Saida and Nonomura, 1978; Sparrow et al., 1984). These fibers are contracted by increasing Ca^{2+} ($pCa^{2+} = 5$) in the bathing medium and relaxed by decreasing Ca^{2+} ($pCa > 8$). Ideally, the fibers should be able to contract and relax repeatedly without loss of active tension, generate as much force as a comparable intact muscle, and contract and relax in a time frame similar to that of the intact muscle. The major advantage of this preparation is that it is possible to manipulate the internal environment of the cell and thereby manipulate the contractile properties of the muscle fiber.

Walsh *et al.* (1982a) used skinned fibers prepared from chicken gizzard muscle to perform an interesting experiment. They purified myosin light-chain kinase from turkey gizzard smooth muscle and, by enzymatic digestion, prepared a form of the enzyme that does not require Ca^{2+} or calmodulin for activity (Walsh *et al.*, 1982b). Addition of the Ca^{2+}-independent myosin light-chain kinase to skinned fibers in the presence of relaxing solution (i.e., in the absence of Ca^{2+}), resulted in myosin phosphorylation and contraction of the fibers. This experiment provides direct support for the importance of myosin phosphorylation in force production by smooth muscles.

If the experiments by Walsh *et al.* (1982a) demonstrated the importance of myosin phosphorylation in force production, experiments by Haeberle *et al.* (1985a) have demonstrated the importance of myosin phosphorylation in force maintenance. These investigators purified a phosphoprotein phosphatase that is capable of dephosphorylating the myosin light chain. Incubation of contracted skinned uterine smooth muscles with the catalytic subunit of this phosphatase leads to the relaxation of these fibers (Fig. 5). Note that the relaxation occurs despite the continued presence of Ca^{2+} at a concentration that maintains full contraction in control (untreated) muscles. The fact that the myosin is dephosphorylated and that the fibers relax following the addition of the phosphatase suggests tight coupling between dephosphorylation and relaxation. Moreover, if latch-bridges are a real phenomenon, the fibers should have remained contracted as the myosin was dephosphorylated due to the generation of latch-bridges. Since the fibers relaxed, this experiment seriously questions the latch hypothesis and the presence of multiple Ca^{2+}-dependent mechanisms in this muscle.

What, then, can be said about the relationship between myosin phosphorylation and the contractile properties of smooth muscles. Most investigators agree that phosphorylation is essential for force production. Moreover, the relationship between the steady-state levels of force and phosphorylation in skinned fibers is quite clear. In addition to the experiments described above, studies by Bialojan *et al.* (1984), Chatterjee and Murphy (1983), Hoar *et al.* (1979), and Ruegg and Pfitzer (1985) demonstrated that the two processes are tightly coupled. The relationship between these two parameters is less obvious in intact smooth muscles. There is an impressive body of data demonstrating that steady-state phosphorylation correlates with maintained force (de Lanerolle and Stull, 1980; de Lanerolle *et al.*, 1982; Haeberle *et al.*, 1985; Gerthoffer, 1986). Even Murphy found that force and phosphorylation correlate when hog carotid artery muscles are contracted with histamine (Aksoy *et al.*, 1983). Nevertheless, there is a body of data indicating a dissociation between force and phosphorylation (Dillon *et al.*, 1981; Aksoy *et al.*, 1981, 1983; Gerthoffer and Murphy, 1983a; Silver and Stull, 1982; Kamm and Stull, 1985). Consequently, it is not possible

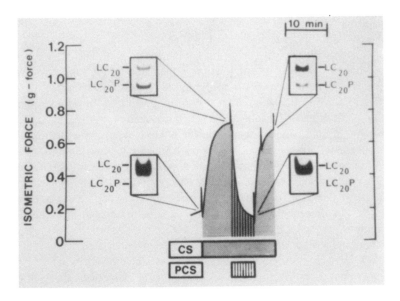

FIGURE 5. Effect of purified skeletal muscle phosphatase catalytic subunit (PCS) on isometric force maintenance by skinned smooth muscle. Muscles were contracted in a contracting solution (CS) containing saturating calcium and 10 μM calmodulin. The presence of CS is shown by the stippled horizontal bar and was present throughout the incubation with PCS. (Insets) Autoradiograms of myosin light-chain radioimmunoblots from representative muscle samples. The unphosphorylated light-chain band is designated LC_{20}; the phosphorylated light-chain band is designated LC_{20} P. Each autoradiogram within a particular panel corresponds to one of several muscles, all treated identically up to the point at which the muscles were frozen for phosphorylation analysis. PCS-induced relaxation (final concentration of PCS: 0.28 μM) and subsequent recontraction following washout of PCS are demonstrated. The fact that the muscle relaxes following dephosphorylation, despite the continued presence of Ca^{2+}, suggests that myosin phosphorylation is the sole mechanism for force maintenance in these muscles. (From Haeberle *et al.*, 1985.)

to define a single consistent relationship between the steady-state levels of myosin phosphorylation and force generated by smooth muscles. Whether the specific relationships reported to date represent true differences in the regulation of different smooth muscles, agonist-specific differences in smooth muscle contraction and/or merely vagaries in the tissue preparations or in the phosphate assays remains to be seen.

IV. OTHER POSSIBLE REGULATORY MECHANISMS

Another way to visualize the regulation of the contractile process is to think in terms of the site of action of Ca^{2+} in muscles. Using this line of thinking, regulatory mechanisms can be divided into thick-filament and thin-filament mechanisms. The myosin phosphorylation mechanism in smooth muscle is classified as a thick-filament mechanism in that contraction is regulated by direct modification of myosin, the major thick filament protein. This is in contrast to striated muscle, where Ca^{2+} regulation is mediated through the troponin–tropomyosin system and changes in the structure of the thin filament.

Thin-filament regulation had also been postulated in smooth muscles. Ebashi (1980) suggested that a protein called leiotonin may play a role in regulating smooth muscle contractility. However, there is a paucity of data regarding this mechanism, and its precise role is unclear. Marston and Smith (1985) also suggested the presence of a thin-filament regulatory mechanism in smooth muscle. This suggestion was based on the observation that partially purified thin filaments retained Ca^{2+} regulatory properties. This observation can be explained in part by the presence of a protein known as caldesmon in the thin-filament preparations.

Caldesmon is a protein with apparent molecular weights of 120,000–150,000 (Bretscher and Lynch, 1985). Caldesmon derives its name from its ability to bind to calmodulin (Sobue et al., 1981). Moreover, caldesmon also binds to F-actin in a Ca^{2+}-dependent manner and inhibits the actomyosin ATPase activity of phosphorylated myosin (Sobue et al., 1982). This inhibition is Ca^{2+} dependent, since the addition of Ca^{2+}/calmodulin can relieve the inhibition by caldesmon. In addition, phosphorylation of caldesmon by a Ca^{2+}/calmodulin-dependent protein kinase converts caldesmon into a form that does not inhibit the actomyosin ATPase (Ngai and Walsh, 1984).

Consequently, caldesmon is a prime candidate as an alternative regulatory protein in smooth muscles. However, a number of points need to be clarified before we can be sure of its role (Bretscher, 1986). First, the molar ratio of caldesmon to actin is surprisingly low. Even with a highly asymmetric protein such as caldesmon, it is hard to visualize how a relatively few caldesmon molecules can regulate a relatively large number of actin molecules. Second, the ability of Ca^{2+}/calmodulin to inhibit caldesmon binding to F-actin is dependent on a large molar excess of Ca^{2+}/calmodulin compared with caldesmon (Sobue et al., 1982). This ratio is greater than that found in chicken gizzards. Finally, caldesmon is only capable of inhibiting the actomyosin ATPase of phosphorylated myosin. Thus, myosin phosphorylation is required to initiate contraction and caldesmon, at best, can only act to modulate the force generated by smooth muscle. In any event, additional experiments are needed to determine the precise role of caldesmon and other possible thin-filament proteins in regulating smooth muscle contraction.

V. CYCLIC NUCLEOTIDE EFFECTS

Cyclic adenosine monophosphate (cAMP) plays a crucial role in regulating the contractile properties of airway smooth muscle. Studies by various investigators have demonstrated that an increase in cAMP levels is associated with relaxation of canine tracheal smooth muscles (Gold, 1980; Katsuki and Murad, 1979; Rinard et al., 1979, 1983). This, and most other biological responses to cAMP, are thought to be mediated through the activation of cAMP-dependent protein kinase (Krebs and Beavo, 1979). The demonstration of an increase in cAMP-dependent protein kinase activity when canine tracheal smooth muscles are exposed to isoproterenol has supported this notion (Torphy et al., 1982). Other investigators have suggested that the airways constriction associated with asthma is related to a derangement in the mechanism(s) by which an increase in cAMP leads to relaxation of airway muscle (Gold, 1980; Rinard et al., 1979; Szentivanyi, 1968).

There is, however, some controversy regarding the role of cAMP in mediating relaxation in smooth muscles. This controversy, reviewed by Gold (1980), arose because of the absence of a consistent relationship between changes in tension and cAMP levels. For instance, cAMP has been reported to increase in bovine tracheal smooth muscle contracted with cholinergic agents (Katsuki and Murad, 1977) and during relaxation of airway smooth muscles (de Lanerolle et al., 1984; Gold, 1980; Katsuki and Murad, 1977; Rinard et al., 1979; Torphy et al., 1982). Perhaps the most telling experiments on the role of cAMP are those performed on skinned smooth muscles contracted with high Ca^{2+}. The addition of the catalytic subunit of cAMP-dependent protein kinase to these muscles (Kerrick and Hoar, 1981), including guinea pig tracheal smooth muscle (Sparrow et al., 1984), relaxed them. These experiments demonstrate, very directly, that cAMP initiates a series of reactions that culminate in smooth muscle relaxation. The precise reactions involved in this process, however, have not been elucidated.

There are currently two hypotheses regarding cAMP-mediated relaxation of smooth muscles. One hypothesis is based on the observation that phosphorylation by cAMP-dependent protein kinase decreases the activity of myosin light-chain kinase (Adelstein et al., 1978; Conti and Adelstein, 1981; Nishikawa et al., 1984). The experimental basis of this suggestion is shown in Fig. 6. Myosin light-chain kinase can be phosphorylated by cAMP-dependent protein kinase at two sites. Phosphorylation of both sites decreases the affinity of myosin light-chain kinase for the activating Ca^{2+}–calmodulin complex. Thus, at any given concentration of Ca^{2+}–calmodulin complexes, there will be fewer activated myosin light-chain kinase molecules, less myosin phosphorylation, and, hence, less force. Note two fundamental aspects of this mechanism: (1) it assumes that force is coupled to myosin phosphorylation; (2) it is based on the idea that any steady-state level of myosin phosphorylation represents an equilibrium between myosin light-chain kinase and phosphatase activities. Factors that affect this equilibrium cause a change in the myosin phosphate content and a comparable change in tension. Therefore, a decrease in myosin light-chain kinase activity due to phosphorylation can lead to myosin dephosphorylation and relaxation.

The other hypothesis asserts that a decrease in cytoplasmic calcium is the

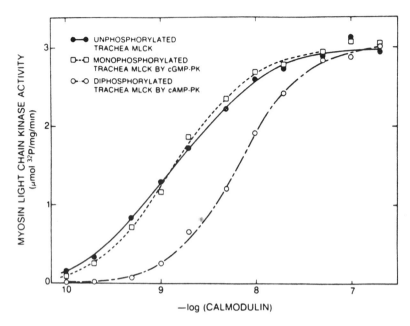

FIGURE 6. Effect of phosphorylation on the affinity of bovine tracheal smooth muscle myosin light-chain kinase (MLCK) for calmodulin. Tracheal MLCK (0.5 μM) was phosphorylated with 4 μg/ml of the catalytic subunit of cAMP-dependent protein kinase or with 5 μg/ml of cGMP-dependent protein kinase in the presence of EGTA. The activities of unphosphorylated myosin light-chain kinase (●) and myosin light-chain kinase monophosphorylated with cGMP-dependent protein kinase (□) or diphosphorylated with the catalytic subunit of cAMP-dependent protein kinase (○) were immediately assayed in the presence of various concentrations of calmodulin, at a Ca^{2+} concentration of 0.2 mM, as described by Adelstein and Klee (1981). Note that diphosphorylation changes the affinity for Ca^{2+}–calmodulin and that relatively more Ca^{2+}–calmodulin is required for half-maximal activation of the diphosphorylated myosin light-chain kinase. (From Nishikawa *et al.*, 1984.)

underlying mechanism of cAMP-mediated relaxation. This reduction in cytoplasmic calcium is postulated to occur through a reduction in calcium influx (Meisheri and Van Breeman, 1982) and/or stimulation of calcium efflux (Bulbring and der Hertog, 1980; Scheid and Fay, 1984; Scheid *et al.*, 1979; Van Breeman, 1977) and/or intracellular sequestration of calcium (Itoh *et al.*, 1982; Meuller and Van Breeman, 1979). This decrease in calcium can have one of two effects. It can lead to relaxation by a mechanism that involves the inactivation of (1) myosin light-chain kinase and myosin dephosphorylation, or (2) latch-bridges by an undefined mechanism. In fact, only the calcium depletion mechanism can cause relaxation in the presence of latch bridges.

Myosin light-chain kinase phosphorylation may be an important part of the mechanism of cAMP-mediated relaxation in those muscles, such as tracheal smooth muscle, in which phosphorylated myosin cross-bridges maintain tension. In fact, studies on intact canine tracheal smooth muscles (de Lanerolle *et al.*, 1984) have demonstrated (1) an increase in ^{32}P-labeling of myosin light-chain kinase, compared with control muscles, when canine tracheal smooth muscle is treated with forskolin, an agent that increases cAMP levels; (2) an increase in cAMP levels and a stoichiometric increase in myosin light-chain kinase phosphorylation in muscles treated with forskolin; (3) an increase in cAMP levels, myosin light-chain kinase phosphorylation, myosin dephosphorylation, and relaxation when methacholine-contracted muscles were relaxed by adding forskolin; and (4) myosin dephosphorylation and relaxation in the absence of an increase in cAMP levels and myosin light-chain kinase phosphorylation when methacholine-contracted muscles were relaxed by adding atropine. These data, summarized in Table II, suggest the presence of two pathways for relaxing canine tracheal muscle: one involving an increase in cAMP levels and one that does not. Moreover, they provide strong support for the idea that myosin light-chain kinase phosphorylation is an important part of cAMP-mediated relaxation of canine tracheal smooth muscle.

Although the data in Table II demonstrate that the myosin light-chain kinase phosphorylation can be part of the mechanism of cAMP-mediated relaxation, they do not demonstrate that myosin light-chain kinase phosphorylation is an

TABLE II
cAMP Content, Myosin Light-Chain Kinase (MLCK) Phosphorylation, Myosin Dephosphorylation, and Relaxation of Tracheal Smooth Muscle[a–f]

Treatment	cAMP (pmoles/mg protein)	MLCK PO$_4$ (mole/mole)	Myosin PO$_4$ (mole/mole LC$_{20}$)	Percent relaxation
None (control)	4.4 ± 0.2	1.1 ± 0.1	0.27 ± 0.01	—
Forskolin (10 min)	71.0 ± 12.4[b]	1.7 ± 0.1[b]	0.23 ± 0.01[c]	—
Methacholine (15 min)	3.2 ± 0.5	1.1 ± 0.14	0.45 ± 0.01[b]	—
Methacholine (15 min) and atropine (10 min)	5.1 ± 1.5	1.3 ± 0.1	0.30 ± 0.02	100
Methacholine (15 min) and forskolin (10 min)	38.4 ± 3.1[b]	1.9 ± 0.15[c]	0.34 ± 0.01[c]	73 ± 4.6

[a]From de Lanerolle *et al.* (1984).
[b]$p < 0.001$.
[c]$p < 0.01$.
[d]Tracheal smooth muscle strips were prepared and treated with 4×10^{-5} M forskolin, 10^{-6} M methacholine, or 2×10^{-7} M atropine, individually or sequentially, for various times. Control muscles were not treated with pharmacological agents.
[e]Details of the assays for cAMP content, MLCK phosphorylation, and myosin phosphorylation are given in the original publication.
[f]The data demonstrate two important points: (1) MLCK phosphorylation is specific to an increase in cAMP levels, i.e., there is no incrase in MLCK–PO$_4$ in muscles treated with methacholine or methacholine plus atropine; and (2) an increase in MLCK–PO$_4$ is accompanied by myosin dephosphorylation and relaxation. Both relationships must exist if MLCK phosphorylation is part of the mechanism of cAMP-mediated relaxation of smooth muscles.

obligatory event in cAMP-mediated relaxation. Many more studies are required to establish the precise relationship between myosin light-chain kinase phosphorylation and relaxation. It is particularly important to determine whether myosin light-chain kinase activity is altered in muscles following an increase in cAMP levels. This is a tricky experiment to perform because it involves inhibiting all other protein kinases without affecting myosin light-chain kinase activity. Not unexpectedly, these types of experiments have provided conflicting results (Miller *et al.*, 1983; Nishikori *et al.*, 1983). It is also important to emphasize that these two hypotheses for cAMP-mediated relaxation are not mutually exclusive. It is quite possible, if not likely, that these two mechanisms operate at the same time, that they are synergistic, and/or that one or the other of these mechanisms is relatively more important, depending on the level of stimulation prior to relaxation (as suggested by Meisheri and Ruegg, 1983). In fact, biochemical data on myosin light-chain kinase phosphorylation (Conti and Adelstein, 1981; Nishikawa *et al.*, 1984) predict a complementary relationship between these two mechanisms.

The mode of action of cyclic guanosine monophosphate (cGMP) in smooth muscles is less clear. Initially, cAMP and cGMP were postulated to mediate opposite functions—the Ying–Yang hypothesis proposed by Goldberg *et al.* (1975). That is, cGMP was thought to mediate contraction, while cAMP was postulated to mediate relaxation in smooth muscles. This idea, however, is not consistent with data available in the literature. For instance, nitroso compounds are known to increase cellular cGMP levels and to relax both canine (Katsuki and Murad, 1977) and bovine (Torphy *et al.*, 1985) tracheal smooth muscles. In addition, 8-bromo cGMP, which is a membrane permeable analogue of cGMP, also causes relaxation of bovine tracheal muscles (Torphy *et al.*, 1985). Other studies on skinned fibers have demonstrated that the addition of cGMP (Pfitzer *et al.*, 1984) or cGMP-dependent protein kinase (Pfitzer *et al.*, 1986) leads to the relaxation of contracted skinned mammalian smooth muscles.

The actions of cAMP are thought to be mediated through the activation of its dependent protein kinase and the phosphorylation of cellular proteins. This is very similar to the activation of cAMP-dependent protein kinase by an increase in cellular cAMP (Krebs and Beavo, 1979). Consequently, it is quite reasonable to ask whether cAMP- and cGMP-mediated relaxation of smooth muscles share common mechanisms or substrate proteins. cGMP-mediated relaxation does not appear to involve the phosphorylation of myosin light-chain kinase. Unlike cAMP-dependent protein kinase, cGMP-dependent protein kinase phosphorylates only one site on myosin light-chain kinase (Nishikawa *et al.*, 1984) (Fig. 6). This site is also phosphorylated by cAMP-dependent protein kinase. However, phosphorylation of this site alone by either cyclic nucleotide-dependent protein kinase does not affect the affinity of the enzyme for Ca^{2+}–calmodulin or the activity of the enzyme (Fig. 6). By contrast, recent studies have suggested that sodium nitroprusside decreases myoplasmic Ca^{2+} concentrations in K^+

contracted vascular smooth muscle (Morgan and Morgan, 1984). Thus, there appear to be correlations among an increase in cGMP, a decrease in cell Ca^{2+}, and relaxation of smooth muscles.

Finally, another interesting mechanism of action of cGMP has been described through studies on the vasculature. Furchgott (1983) and others (Ganz et al., 1986; Rapoport and Murad, 1983; Singer and Peach, 1982; Van de Vorde and Leusen, 1983) have suggested that the vasodilatory effects of a number of compounds is related to the release of endothelium-derived relaxing factor(s). The mechanism of action of this factor on vascular smooth muscle is far from clear but appears to involve the formation of cGMP and the phosphorylation of several proteins (Diamond and Chu, 1983; Ganz et al., 1986; Rapoport and Murad, 1983). By extension, it is of interest to determine whether tracheal epithilium produces an analogous agent that affects cGMP metabolism and the contractile properties of tracheal smooth muscle. This is a particularly interesting topic, since recent reports have suggested that removing the airway epithelium affects the responsiveness of airway smooth muscle to contractile agents (Flavahan et al., 1985; Hay et al., 1986). As suggested by Torphy and Gerthoffer (1986), this observation demonstrates a possible connection between the tracheal epithelium and tracheal smooth muscle as well as the possible existence of an epithelial factor that regulates airway muscle cGMP content and contractility in a manner analogous to endothelium-dependent relaxing factor.

VI. MYOSIN PHOSPHORYLATION BY OTHER PROTEIN KINASES

LC_{20} is a substrate for a number of protein kinases. LC_{20} purified from smooth muscles contains the consensus sequence, previously identified by Krebs and his associates (Krebs and Beavo, 1979), required for phosphorylation by cAMP-dependent protein kinase. Based on the presence of this sequence, Emily Noiman demonstrated some years ago that the catalytic subunit of cAMP-dependent protein kinase does, indeed, phosphorylate isolated LC_{20} at the same site phosphorylated by myosin light-chain kinase (Noiman, 1980). This is probably not a physiologically significant reaction because the rate at which LC_{20} is phosphorylated when LC_{20} is part of the myosin molecule is far too low to have a role in affecting contraction (Walsh et al., 1981).

A far more interesting reaction is the phosphorylation of smooth muscle myosin by protein kinase C. Protein kinase C is a calcium phospholipid-dependent kinase that is thought to be involved in a variety of physiological processes (see Nishizuka, 1986, for a more complete review of the properties and functions of protein kinase C). Nishikawa et al. (1983) demonstrated that protein kinase C incorporates 2 moles PO_4/mole myosin with all the PO_4 bound to LC_{20}. Moreover, these workers demonstrated by peptide mapping that protein kinase C phosphorylates a different site on LC_{20} than does myosin light-chain kinase. The most important aspect of this reaction is shown in Fig. 7, which demonstrates that the sequential phosphorylation of smooth muscle myosin by myosin light-chain kinase and protein kinase C leads to the incorporation of 4 moles PO_4/mole myosin. Moreover, phosphorylation by protein kinase C of myosin previously phosphorylated with myosin light-chain kinase decreases the actomyosin ATPase

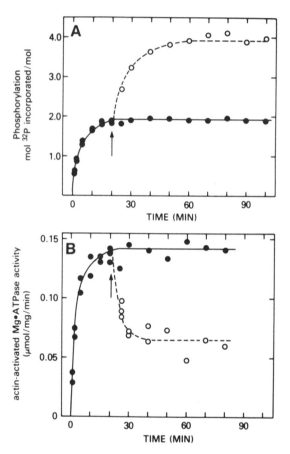

FIGURE 7. Sequential phosphorylation of smooth muscle heavy meromyosin by myosin light-chain kinase (MLCK) and protein kinase C (A), and the effect of phosphorylation on the actin-activated Mg^{2+} ATPase activity (B). 2 mg/ml of HMM was phosphorylated initially with 0.3 μg/ml of MLCK kinase (●——●) as described by Adelstein and Klee (1981). At 20 min (indicated by arrow), 20 μg/ml of protein kinase C and 50 μg/ml of phosphatidylserine and 0.8 μg/ml of diolein were added to part of the reaction mixture (○———○). Aliquots (90 μl) of the reaction mixtures withdrawn at the indicated times, quenched by adding a solution (10 μl) containing 50 mM Tris HCl (pH 7.5), 10 mM EGTA, and 12 mM EDTA and the protein-bound ^{32}P and the actin-activated Mg^{2+}-ATPase activities quantitated. The data demonstrate that phosphorylation by protein kinase attenuates the effect of MLCK phosphorylation of ATPase activity. (From Nishikawa *et al.*, 1983.)

activity. Thus, multiple site phosphorylation by different protein kinases can differentially regulate the activity of smooth muscle myosin.

The physiological significance of this reaction in smooth muscles is unclear, but it does appear to play an important role in platelets. Platelet myosin is similar to smooth muscle myosin in that platelet myosin LC_{20} must also be phosphorylated by platelet myosin light-chain kinase for stimulation of the actomyosin ATPase activity (Adelstein and Conti, 1975). Moreover, a variety of platelet functions such as degranulation and clot retraction are thought to be dependent on actin and myosin. Therefore, Hidaka and colleagues have used this system to study the effect of protein kinase C. Since protein kinase C activity is known to be stimulated by phorbol esters (Nishizuka, 1986), platelets were pretreated with varying concentrations of a protein kinase C inhibitor, cells were stimulated with a standard concentration of phorbol ester, and the effect of these combinations of drugs on serotonin secretion was studied. Inagaki et al. (1984) found that secretion increased as the inhibitor concentration was increased. Moreover, these workers found a reduction in phosphorylation of the LC_{20} site phosphorylated by protein kinase C, without any effect on the site phosphorylated by myosin light-chain kinase, in the presence of the inhibitor. Since there is an inverse relationship between actomyosin ATPase activity and protein kinase C-catalyzed phosphorylation of myosin, this is precisely the relationship predicted from the biochemical experiments. Comparable experiments have not been performed on smooth muscles, and the effect of myosin phosphorylation by protein kinase C on the contractile properties of smooth muscles is unknown.

VII. MYOSIN PHOSPHATASES

Phosphoprotein phosphatases are a class of enzymes that hydrolyze phosphoester bonds to dephosphorylate proteins. These enzymes are of obvious importance in regulating smooth muscle contraction, since phosphorylated myosin must be dephosphorylated before another contractile cycle can be initiated. At least three different phosphatases have been purified from smooth muscles (Pato and Adelstein, 1983a,b; Werth et al., 1982). In addition, Haeberle et al. (1985) (see Fig. 5) and Bialojan et al. (1985) demonstrated that incubation of contracted skinned fibers with purified phosphatases leads to myosin dephosphorylation and relaxation. The principal question regarding phosphatases is therefore the nature of their regulation in vivo.

Two phosphatases are of considerable interest in this respect. One of these is an enzyme called protein phosphatase 1. The catalytic subunit of this enzyme has an apparent molecular weight of 37,000 (Tung et al., 1984), and its activity is regulated by another protein, called inhibitor 1 (Huang and Glinsmann, 1976). Phosphorylation of a specific threonine residue on inhibitor 1 by cAMP-dependent protein kinase leads to the inhibition of protein phosphatase 1 (Aitken et al., 1982). Cohen (1985) pointed out that there are three lines of evidence suggesting that phosphorylation of inhibitor 1 is physiologically significant: (1) cAMP-dependent protein kinase phosphorylates inhibitor 1, *in vitro*, at a rate comparable to its other physiological substrates; (2) inhibitor 1 is present at about 3.5-fold M excess than protein phosphatase; and (3) epinephrine increases the phosphorylation of inhibitor 1 in skeletal muscle, and this effect can be antagonized by insulin.

The second phosphatase is an enzyme known as calcineurin (Klee and Krinks, 1978). Calcineurin is composed of a catalytic subunit of 61,000 M_r and a regulatory subunit of 19,000 M_r present in equimolar ratio. The regulatory subunit is a Ca^{2+}-binding protein that has a high degree of sequence homology with calmodulin (Aitken *et al.*, 1984). Calcineurin binds tightly to calmodulin in the presence of Ca^{2+} with an accompanying increase in the catalytic rate of the enzyme (Stewart *et al.*, 1983). Thus, calcineurin represents the first Ca^{2+}-regulated phosphatase that has been described. Nevertheless, a detailed understanding of how this, or any other phosphatase, is regulated *in vivo* is not available.

VIII. SUMMARY

The central theme of this chapter is the idea that the contractile properties of a smooth muscle are determined by a balance between the activities of various protein kinases and phosphoprotein phosphatases. This is shown graphically in Fig. 8. Changes in the contractile properties of a smooth muscle are the result of a complicated series of biochemical reactions that are initiated by the activation of cellular receptor(s). The signals detected by the receptors are communicated intracellularly through changes in the levels of second messengers and are mediated through the activation of protein kinases and phosphatases. The process culminates with the hydrolysis of ATP by actin and myosin. Those types of stimuli that lead to an increase in ATP hydrolysis lead to contraction (i.e., an increase in force); those stimuli that decrease ATP hydrolysis lead to relaxation.

Consequently, one of the most important questions regarding smooth muscles is whether phosphorylated myosin cross-bridges or latch-bridges are responsible for steady-state force maintenance. As reviewed, there are data supporting both types of mechanisms. An important point to keep in mind is that smooth muscles have heterogeneous responses to pharmacological stimuli. Therefore, there is no *a priori* reason to believe that identical mechanisms regulate all smooth muscles. For instance, the most convincing experiments on latch-bridges have been performed on vascular smooth muscles. This raises the interesting possibility that vascular smooth muscles contain a unique calcium regulatory mechanism. Many more experiments are needed to test this hypothesis adequately. But, if it proves correct, agents that inhibit latch-bridges in vascular smooth muscles would be site-directed antihypertensives devoid of the many side effects associated with currently available drugs.

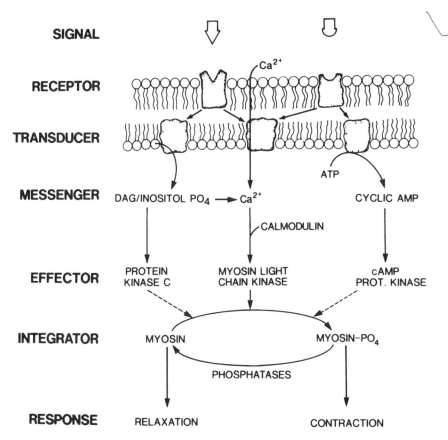

FIGURE 8. Schematic presentation of the molecular events involved in regulating smooth muscle contraction. Stimulation of receptors can lead to simultaneous modulation of transducer proteins involved in generating cellular messengers (DAG stands for diacylglycerol). Note that both inositol phosphates and cAMP can affect cellular Ca^{2+} levels. These messengers then activate the effector molecules, i.e., their dependent protein kinases. The protein kinases either directly (e.g., myosin light-chain kinase and protein kinase or indirectly (cAMP-dependent protein kinase through phosphorylation of myosin light-chain kinase) affect the myosin phosphate content. Myosin is labeled integrator because the level of myosin phosphorylation, based on what transpires between the signal and the effector, determines the rate of ATP hydrolysis and, apparently, the contractile state of the muscle.

Another important aspect of this issue is that smooth muscles must have mechanisms to integrate antagonistic stimuli. Assume, for a moment, that the cell depicted in Fig. 8 is a tracheal smooth muscle cell that is being stimulated with acetylcholine and epinephrine at the same time. The contractile response of this cell has to be a balance between the level of stimulation by these agonists. The importance of the scheme shown in Fig. 8 is that it permits integration at three levels: (1) at the level of the transducer through the interaction of GTP-binding proteins (see Gilman, 1984, for a description of these proteins and their actions), (2) at the level of the messengers (because inositol phosphates and cAMP can affect intracellular Ca^{2+} levels), and (3) at the level of the effector (e.g., by changing myosin light-chain kinase activity by changing the $Ca^{2+}-$ calmodulin concentration or by phosphorylating it with cAMP-dependent protein kinase). Ultimately, all these affect the level of myosin phosphorylation and the rate at which actin and myosin hydrolyze ATP. Thus, the level of myosin phosphorylation may be the ultimate determinant of an integrated response to receptor stimuli.

Other areas that require further investigation are the mechanisms involved in cAMP-mediated relaxation and the interaction between epithelial and smooth muscle cells in the trachea. It is also important to understand the regulation of phosphoprotein phosphatases in relationship to their effects on the contractile responses of smooth muscles. An important aspect of these studies will relate to how cAMP, potential epithelial–smooth muscle cell interactions and phosphatases affect the kinetics and stoichiometry of myosin phosphorylation. In many respects, the major importance of the myosin phosphorylation mechanism is that it gives us a basis on which to propose hypotheses and to design experiments to study the mechanisms that regulate smooth muscle contraction. Ultimately, regardless of the validity of the hypotheses we pose, we will learn a great deal about the regulation of smooth muscles by testing these hypotheses, which will be reflected in improved therapies for the multitude of diseases that afflict smooth muscles.

IX. REFERENCES

Adelstein, R. S., and Conti, M. A., 1975, Phosphorylation of platelet myosin increases actin-activated myosin ATPase activity, *Nature (Lond.)* **256:**597–598.

Adelstein, R. S., Conti, M. A., Hathaway, D. R., and Klee, C. B., 1978, Phosphorylation of smooth muscle myosin light chain kinase by the catalytic subunit of adenosine 3':5'-monophosphate-dependent protein kinase, *J. Biol. Chem.* **253:**8347–8350.

Adelstein, R. S., and Eisenberg, E., 1980, Regulation and kinetics of the actin–myosin–ATP interaction, *Annu. Rev. Biochem.* **49:**921–956.

Adelstein, R. S., and Klee, C. B., 1981, Purification and characterization of smooth muscle myosin light chain kinase, *J. Biol. Chem.* **256:**7501–7509.

Aitken, A., Bilham, T., and Cohen, P., 1982, Complete primary structure of protein phosphatase inhibitor-1 from rabbit skeletal muscle, *Eur. J. Biochem.* **126:**235–246.

Aitken, A., Klee, C. B., and Cohen, P., 1984, The structure of the B subunit of calcineurin, *Eur. J. Biochem.* **139:**663–671.

Aksoy, M. O., Murphy, R. A., and Kamm, K. E., 1982, Role of Ca^{2+} and myosin light chain phosphorylation in regulation of smooth muscle, *Am. J. Physiol.* **242:**C109–C116.

Aksoy, M. O., Mras, S., Kamm, K. E., and Murphy, R. A., 1983, Ca^{2+}, cAMP, and changes in myosin phosphorylation during contraction of smooth muscle, *Am. J. Physiol.* **245:**C255–C270.

Arner, A., and Hellstrand, P., 1985, Effects of calcium and substrate on force–velocity relation and energy turnover in skinned smooth muscle of the guinea pig, *J. Physiol. (Lond.)* **360:**347–365.

Barron, J. T., Barany, M., Barany, K., and Storti, R. V., 1980, Reversible phosphorylation and dephosphorylation of the 20,000 dalton light chain of myosin during the contraction–relaxation–contraction cycle of arterial smooth muscle, *J. Biol. Chem.* **255**:6238–6244.

Bialojan, C., Merkel, L., Ruegg, J. C., Gifford, D., and D. Salvo, J., 1985, Prolonged relaxation of detergent-skinned smooth muscle involves decreased endogenous phosphatase activity, *Proc. Soc. Exp. Biol. Med.* **178**:648–652.

Bretscher, A., 1986, Thin filament regulatory proteins of smooth and non-muscle cells, *Nature (Lond.)* **321**:726–727.

Bretscher, A., and Lynch, W. J., 1985, Identification and localization of immunoreactive forms of caldesmon in smooth and nonmuscle cells: A comparison with the distributions of tropomyosin and of α-actinin. *J. Cell. Biol.* **100**:1656–1663.

Bulbring, E., and den Hertog, A., 1980, The action of isoprenaline on the smooth muscle of the guinea-pig taenia coli, *J. Physiol. (Lond.)* **304**:277–296.

Butler, T. M., Siegman, M. J., and Mooers, S. U., 1983, Chemical energy usage during shortening and work production in mammalian smooth muscle, *Am. J. Physiol.* **244**:C234–C242.

Carlson, F. D., and Wilkie, D. R., 1974, *Muscle Physiology*, Prentice-Hall, Englewood Cliffs, New Jersey.

Cassidy, P. S., Kerrick, W. G. L., Hoar, P. E., and Malencik, D. A., 1981, Exogenous calmodulin increases Ca^{2+} sensitivity of isometric tension, activation and myosin phosphorylation in skinned smooth muscle, *Pflugers Arch.* **392**:115–120.

Chacko, S., and Rosenfeld, A., 1982, Regulation of actin-activated ATP hydrolysis by arterial myosin, *Proc. Natl. Acad. Sci. USA* **79**:292–296.

Chacko, S., Conti, M. A., and Adelstein, R. S., 1977, Effect of phosphorylation of smooth muscle myosin on actin activation and Ca^{2+} regulation, *Proc. Natl. Acad. Sci. USA* **74**:129–133.

Chatterjee, M., and Murphy, R. A., 1983, Calcium-dependent stress maintenance without myosin phosphorylation in skinned smooth muscle, *Science* **221**:464–466.

Cohen, P., 1985, The coordinated control of metabolic pathways by broad-specificity protein kinases and phosphatases, *Current Topics Cell. Regul.* **27**:23–37.

Conti, M. A., and Adelstein, R. S., 1981, The relationship between calmodulin binding and phosphorylation of smooth muscle myosin kinase by the catalytic subunit of 3':5' cAMP-dependent protein kinase, *J. Biol. Chem.* **256**:3178–3181.

Csabina, S., Barany, M., and Barany, K., 1986, Stretch-induced myosin light chain phosphorylation in rat uterus, *Arch. Biochem. Biophys.* **249**:374–381.

Dabrowska, R., Aromatorio, D., Sherry, J. M. F., and Hartshorne, D. J., 1977, Composition of the myosin light chain kinase from chicken gizzard, *Biochem. Biophys. Res. Commun.* **78**:1263–1272.

Dabrowska, R., Sherry, J. M. F., Aromatorio, D. K., and Hartshorne, D. J., 1978, Modulator protein as a component of myosin light chain kinase from chicken gizzard, *Biochemistry* **17**:253–258.

de Lanerolle, P., 1979, *Myosin Phosphorylation and the Contractile State of Tracheal Smooth Muscle*, Ph.D. thesis, University of California, San Diego.

de Lanerolle, P., and Stull, J. T., 1980, Myosin phosphorylation during contraction and relaxation of tracheal smooth muscle, *J. Biol. Chem.* **255**:9993–10000.

de Lanerolle, P., Condit, J. R., Jr., Tanenbaum, M., and Adelstein, R. S., 1982, Myosin phosphorylation, agonist concentration and contraction of tracheal smooth muscle, *Nature (Lond.)* **298**:871–872.

de Lanerolle, P., Nishikawa, M., Yost, D. A., and Adelstein, R. S., 1984, Increased phosphorylation of myosin light chain kinase after an increase in cyclic AMP in intact smooth muscle, *Science* **223**:1415–1417.

Diamond, J., and Chu, E. B., 1983, Possible role for cyclic GMP in endothelium-dependent relaxation of rabbit aorta by acetylcholine. Comparison with nitroglycerin, *Res. Commun. Chem. Pathol. Pharmacol.* **41**:369–381.

Dillon, P. F., Aksoy, M. O., Driska, S. P., and Murphy, R. A., 1981, Myosin phosphorylation and the cross-bridge cycle in arterial smooth muscle, *Science* **211**:495–497.

Driska, S. P., Aksoy, M. O., and Murphy, R. A., 1981, Myosin light chain phosphorylation associated with contraction in arterial smooth muscle, *Am. J. Physiol.* **240**:C222–C233.

Ebashi, S., 1980, The Croonian Lecture, 1979: Regulation of muscle contraction, *Proc. R. Soc. Biol. B.***207**:259–286.

Ebashi, S., and Endo, M., 1968, Calcium ion and muscle contraction, *Prog. Biophys. Mol. Biol.* **18**: 123–183.

Eisenberg, E., and Hill, T. L., 1985, Muscle contraction and free energy transduction in biological systems, *Science* **227**:999–1006.

Endo, M., Kitazawa, T., Yagi, S., Iino, M., and Kakuta, Y., 1977, Some properties of chemically skinned smooth muscle fibers, in: *Excitation–Contraction Coupling in Smooth Muscle* (R. Casteels, T. Godfraind, J. C. Ruegg, eds.), pp. 199–209, Elsevier/North-Holland, Amsterdam.

Flavahan, N. A., Aarhus, L. L., Rimele, T. J., and Vanhoutte, P. M., 1985, Respiratory epithelium inhibits bronchial smooth muscle tone, *J. Appl. Physiol.* **58**:834–838.

Furchgott, R., 1983, Role of endothelium in responses of vascular smooth muscle, *Circ. Res.* **53**: 557–573.

Ganz, P., Davies, P. F., Leopold, J. A., Gimbrone, M. A., Jr., and Alexander, R. W., 1986, Short and long-term interactions of endothelium and vascular smooth muscle in coculture: Effects on cyclic GMP production, *Proc. Natl. Acad. Sci. USA* **83**:3552–3556.

Gerthoffer, W. T., 1986, Calcium dependence of myosin phosphorylation and airway smooth muscle contraction and relaxation, *Am. J. Physiol.* **250**:C597–C604.

Gerthoffer, W. T., and Murphy, R. S., 1983a, Myosin phosphorylation and regulation of the cross-bridge cycle in tracheal smooth muscle, *Am. J. Physiol.* **244**:C182–C187.

Gerthoffer, W. T., and Murphy, R. A., 1983b, Ca^{2+}, myosin phosphorylation, and relaxation of arterial smooth muscle, *Am. J. Physiol.* **245**:C271–C277.

Gilman, A. G., 1984, Proteins and dual control of adenylate cyclase, *Cell* **36**:577–579.

Gold, W. M., 1980, The role of cyclic nucleotides in airway smooth muscle, in: *Physiology and Pharmacology of the Airways* (J. A. Nadel, ed.), pp. 123–190, Dekker, New York.

Goldberg, N. D., Haddox, M. K., Nicol, S. E., Glass, D. B., Sanford, C. H., Kuehl, F. A., Jr., and Estensen, R., 1975, Biological regulation through opposing influences of cyclic GMP and cyclic AMP: The Yin–Yang hypothesis, *Adv. Cyclic Nucleotide Res.* **5**:307–330.

Gordon, A. R., 1978, Contraction of detergent-treated smooth muscle, *Proc. Natl. Acad. Sci. USA* **75**:3527–3530.

Haeberle, J. R., Hathaway, D. R., and DiPauli-Roach, A. A., 1985a, Dephosphorylation of myosin by the catalytic subunit of a type-2 phosphatase produces relaxation of a chemically skinned uterine smooth muscle, *J. Biol. Chem.* **260**:9965–9968.

Haeberle, J. R., Hott, J. W., and Hathaway, D. R., 1985b, Regulation of isometric force and isotonic shortening velocity by phosphorylation of the 20,000 dalton myosin light chain of rat utering smooth muscle, *Pflugers Arch.* **403**:215–219.

Hay, D. W. P., Robinson, V. A., Fleming, W. W., and Fedan, J. S., 1985, Role of the epithelium in contractile responses of guinea pig isolated trachea, *Fed. Proc.* **44**:506A.

Hoar, P. E., Kerrick, W. G. L., and Cassidy, P. S., 1979, Chicken gizzard: Relation between calcium activated phosphorylation and contraction, *Science* **204**:503–506.

Huang, F. L., and Glinsman, W. H., 1976, Separation and characterization of two phosphorylase phosphatase inhibitors from rabbit skeletal muscle, *Eur. J. Biochem.* **70**:419–426.

Hunter, T., and Cooper, J. A., 1985, Protein-tyrosine kinases, *Annu. Rev. Biochem.* **54**:897–930.

Huxley, A. F., and Niedergerke, R., 1954, Structural change in muscle during contraction, *Nature (Lond.)* **173**:971–973.

Inagaki, M., Kawamoto, S., and Hidaka, H., 1984, Serotonin secretion from human platelets may be modified by a Ca^{2+}-activated, phospholipid-dependent myosin phosphorylation, *J. Biol. Chem.* **259**:14321–14323.

Itoh, T., Izumi, H., and Kuriyama, H., 1982, Mechanisms of relaxation induced by activation of β-adrenoreceptors in smooth muscle cells of the guinea-pig mesenteric artery, *J. Physiol. (Lond.)* **326**:475–493.

Janis, R. A., Barany, K., Barany, M., and Sarmiento, J. G., 1981, Association between myosin light chain phosphorylation and contraction of rat uterine smooth muscle, *Mol. Physiol.* **1**:3–11.

Kamm, K. E., and Stull, J. T., 1985, Myosin phosphorylation, force, and maximal shortening velocity in neurally stimulated tracheal smooth muscle, *Am. J. Physiol.* **249**:C238–C247.

Katsuki, S., and Murad, F., 1977, Regulation of adenosine cyclic 3',5'-monophosphate and guanosine cyclic 3',5' monophosphate levels and contractility in bovine tracheal smooth muscle, *Mol. Pharmacol.* **13**:330–341.

Kerrick, W. G. L., and Hoar, P. E., 1981, Inhibition of smooth muscle tension by cyclic AMP-dependent protein kinase, *Nature (Lond.)* **292**:253–255.

Kerrick, W. G. L., Hoar, P. E., and Cassidy, P. S., 1980, Calcium-activated tension: The role of myosin light chain phosphorylation, *Fed. Proc.* **39**:1558–1563.

Klee, C. B., and Krinks, M. H., 1978, Purification of cyclic 3',5'-nucleotide phosphodiesterase inhibitory protein by affinity chromatography on activator protein coupled to sepharose, *Biochemistry* **17**:120–126.

Krebs, E. G., and Beavo, J. A., 1979, Phosphorylation–dephosphorylation of enzymes, *Annu. Rev. Biochem.* **48**:923–959.

Ledvora, R. F., Barany, K., Van der Meulen, D. L., Barron, J. T., and Barany, M., 1983, Stretch-induced phosphorylation of the 20,000-dalton light chain of myosin in arterial smooth muscle, *J. Biol. Chem.* **258**:14080–14083.

Marston, S. B., and Smith, C. W. J., 1985, The thin filaments of smooth muscles, *J. Muscle Res. Cell Motil.* **6**:669–708.

Meisheri, K. D., and Ruegg, J. C., 1983, Dependence of cyclic-AMP-induced relaxation on Ca^{2+} and calmodulin in skinned smooth muscle of guinea pig taenia coli, *Pflugers Arch.* **399**:315–320.

Meisheri, K. D., and van Breemen, C., 1982, Effects of β-adrenergic stimulation on calcium movements in rabbit aortic smooth muscle: Relationship with cyclic AMP, *J. Physiol. (Lond.)* **331**:429–441.

Miller, J. R., Silver, P. J., and Stull, J. T., 1983, The role of myosin light chain kinase phosphorylation in β-adrenergic relaxation of tracheal smooth muscle, *Mol. Pharmacol.* **24**:235–242.

Morgan, J. P., and Morgan, K. G., 1984, Alteration of cytoplasmic ionized calcium levels in smooth muscle by vasodilators in the ferret, *J. Physiol. (Lond.)* **357**:539–551.

Mueller, E., and van Breeman, C., 1979, Role of intracellular Ca^{2+} sequestration in β-adrenergic relaxation of a smooth muscle, *Nature (Lond.)* **281**:682–683.

Nag, S., and Seidel, J. C., 1983, Dependence on Ca^{2+} and tropomyosin of the actin-activated ATPase activity of phosphorylated gizzard myosin in the presence of low concentrations of Mg^{2+}, *J. Biol. Chem.* **258**:6444–6449.

Ngai, P. K., and Walsh, M. P., 1984, Inhibition of smooth muscle actin-activated myosin Mg^{2+}-ATPase activity by caldesmon, *J. Biol. Chem.* **259**:13656–13659.

Nishikawa, M., Hidaka, H., and Adelstein, R. S., 1983, Phosphorylation of smooth muscle heavy meromyosin by calcium-activated phospholipid-dependent protein kinase, *J. Biol. Chem.* **258**:14069–14072.

Nishikawa, M., de Lanerolle, P., Lincoln, T. M., and Adelstein, R. S., 1984, Phosphorylation of mammalian myosin light chain kinases by the catalytic subunit of cyclic AMP-dependent protein kinase and by cyclic GMP-dependent protein kinase, *J. Biol. Chem.* **259**:8429–8436.

Nishikori, K., Weisbrodt, N. W., Sherwood, O. D., and Sanborn, B. M., 1983, Effects of relaxin on rat uterine myosin light chain kinase activity and myosin light chain phosphorylation, *J. Biol. Chem.* **258**:2468–2474.

Nishizuka, Y., 1986, Studies and perspectives of protein kinase C, *Science* **233**:305–312.

Noiman, E. S., 1980, Phosphorylation of smooth muscle myosin light chains by cAMP-dependent protein kinase, *J. Biol. Chem.* **255**:11067–11070.

Pato, M. D., and Adelstein, R. S., 1980, Dephosphorylation of the 20,000 dalton light chain of myosin by two different phosphatases from smooth muscle, *J. Biol. Chem.* **255**:6535–6538.

Pato, M. D., and Adelstein, R. S., 1983a, Characterization of a Mg^{2+}-dependent phosphatase from turkey gizzard smooth muscle, *J. Biol. Chem.* **258**:7055–7058.

Pato, M. D., and Adelstein, R. S., 1983b, Purification and characterization of a multisubunit phosphatase from turkey gizzard smooth muscle: The effect of calmodulin binding to myosin light chain kinase on dephosphorylation, *J. Biol. Chem.* **258**:7047–7054.

Paul, R. J., Doerman, G., Zeugner, C., and Ruegg, J. C., 1983, The dependence of unloaded shortening velocity on Ca^{++}, calmodulin, and duration of contraction in "chemically skinned" smooth muscle, *Circ. Res.* **53**:342–351.

Perrie, W. T., and Perry, S. V., 1970, An electrophoretic study of the low-molecular-weight components of myosin, *Biochem. J.* **119**:31–38.

Persechini, A., and Hartshorne, D. J., 1981, Phosphorylation of smooth muscle myosin: Evidence for cooperativity between the myosin heads, *Science* **213**:1383–1385.

Pfitzer, G., Hofmann, F., DiSalvo, J., and Ruegg, J. C., 1984, cGMP and cAMP inhibit tension development in skinned coronary arteries, *Pflugers Arch.* **401**:277–280.

Pfitzer, G., Merkel, L., Ruegg, J. C., and Hoffman, F., 1986, Cyclic GMP-dependent protein kinase relaxes skinned fibers from guinea pig taenia coli but not from chicken gizzard, *Pflugers Arch.* **407**:87–91.

Rapoport, R. M., and Murad, F., 1983, Agonist-induced endothelium-dependent relaxation in rat thoracic aorta may be mediated through cGMP, *Circ. Res.* **52**:352–357.

Rees, D. D., and Fredericksen, D. W., 1981, Calcium regulation of porcine aortic myosin, *J. Biol. Chem.* **256**:357–364.

Rinard, G. A., Jensen, A., and Puckett, M., 1983, Hydrocortisone and isoproterenol effects on trachealis, cAMP and relaxation, *J. Appl. Physiol.* **55**:1609–1613.

Rinard, G. A., Rubinfeld, A. R., Brunton, L. L., and Mayer, S. E., 1979, Depressed cyclic AMP levels in airway smooth muscle from asthmatic dogs, *Proc. Natl. Acad. Sci. USA* **76**:1472–1476.

Ringer, S., 1883, A further contribution regarding the influence of the different constituents of the blood on the contraction of the heart, *J. Physiol. (Lond.)* **4**:29–42.

Ruegg, J. C., and Paul, R. J., 1982, Vascular smooth muscle. Calmodulin and cyclic AMP-dependent protein kinase alter calcium sensitivity in porcine carotid skinned fibers, *Circ. Res.* **50**:394–399.

Ruegg, J. C., and Pfitzer, G., 1985, Modulation of calcium sensitivity in guinea pig taenia coli: skinned fiber studies, *Experientia* **41**:997–1001.

Saida, K., and Nonomura, Y., 1978, Characteristics of Ca^{2+}- and Mg^{2+}-induced tension development in chemically skinned smooth muscle fibers, *J. Gen. Physiol.* **72**: 1–14.

Scheid, C. R., and Fay, F. S., 1984, β-Adrenergic effects on transmembrane ^{45}Ca fluxes in isolated smooth muscle cells, *Am. J. Physiol.* **246**:C431–C438.

Scheid, C. R., Honeyman, T. W., and Fay, F. S., 1979, Mechanism of β-adrenergic relaxation of smooth muscle, *Nature (Lond.)* **277**:32–36.

Sellers, J. R., Pato, M. D., and Adelstein, R. S., 1981, Reversible phosphorylation of smooth muscle myosin, heavy meromyosin, and platelet myosin, *J. Biol. Chem.* **256**:13137–13142.

Sellers, J. R., Chock, P. B., and Adelstein, R. S., 1983, The apparently negatively cooperative phosphorylation of smooth muscle myosin at low ionic strength is related to its filamentous state, *J. Biol. Chem.* **258**:14181–14188.

Silver, P. J., and Stull, J. T., 1982, Regulation of myosin light chain and phosphorylase phosphorylation in tracheal smooth muscle, *J. Biol. Chem.* **257**:6145–6150.

Singer, H. A., and Peach, M. J., 1982, Calcium- and endothelial-mediated vascular smooth muscle relaxation in rabbit aorta, *Hypertension* **2**(Suppl.):19–25.

Sobieszek, A., 1977, Vertebrate smooth muscle myosin. Enzymatic and structural properties, in: *The Biochemistry of Smooth Muscle* (N. L. Stephens, ed.), pp. 413–443, University Park Press, Baltimore.

Sobue, K., Muramoto, Y., Fujita, M., and Kakiuchi, S., 1981, Purification of a calmodulin-binding protein from chicken gizzard that interacts with F-actin, *Proc. Natl. Acad. Sci. USA* **78**:5652–5655.

Sobue, K., Morimoto, K., Inui, M., Kanda, K., and Kakiuchi, S., 1982, Control of actin–myosin interaction of gizzard smooth muscle by calmodulin and caldesmon-linked flip-flop mechanism, *Biomed. Res.* **3**:188–196.

Sparrow, M. P., Pfitzer, G., Gagelmann, M., and Ruegg, J. C., 1984, Effect of calmodulin, Ca^{2+}, and cAMP protein kinase on skinned tracheal smooth muscle, *Am. J. Physiol.* **246**:C308–C314.

Squire, J., 1981, Muscle regulation: A decade of the steric blocking model, *Nature (Lond.)* **291**:614–615.

Stewart, A. A., Ingebritsen, T. S., and Cohen, P., 1983, The protein phosphatase involved in cellular regulation: purification and properties of a Ca^{2+}/Calmodulin-dependent protein phosphatase (2B) from rabbit skeletal muscle, *Eur. J. Biochem.* **132**:289–295.

Szentivanyi, A., 1968, The β-adrenergic theory of the atopic abnormality in bronchial asthma, *J. Allergy* **42**:203–232.

Torphy, T. J., and Gerthoffer, W. T., 1986, Biochemical mechanisms of airway smooth muscle contraction and relaxation, in: *Current Topics in Pulmonary Pharmacology and Toxicology*, Vol. 1 (M. A. Hollinger, ed.), pp. 23–56, Elsevier, New York.

Torphy, T. J., Freese, W. B., Rinard, G. A., Brunton, L. L., and Mayer, S. E., 1982, Cyclic nucleotide-dependent protein kinases in airway smooth muscle, *J. Biol. Chem.* **257**:11609–11616.

Torphy, T. J., Zheng, L., Peterson, S. M., Fiscus, R. R., Rinard, G. A., and Mayer, S. E., 1985, Inhibitory effect of methacholine on drug-induced relaxation, cyclic AMP accumulation, and cyclic AMP-dependent protein kinase activation in canine tracheal smooth muscle, *J. Pharm. Exp. Ther.* **233**:409–417.

van Breeman, C., 1977, Calcium requirement for activation of intact aortic smooth muscle, *J. Physiol. (Lond.)* **272**:317–329.

Van de Vorde, J., and Leusen, I., 1983, The role of endothelium in the vasodilator response of rat thoracic aorta to histamine, *Eur. J. Pharmacol.* **87**:113–120.

Walsh, M. P., Bridenbaugh, R., Hartshorne, D. J., and Kerrick, W. G. L., 1982a, Phosphorylation-dependent activated tension in skinned gizzard muscle fibers in the absence of Ca^{2+}, *J. Biol. Chem.* **257**:5987–5990.

Walsh, M. P., Dabrowska, R., Hinkins, S., and Hartshorne, D. J., 1982b, Calcium-independent myosin light chain kinase from smooth muscle. Preparation by limited chymotryptic digestion of the calcium ion dependent enzyme, purification and characterization, *Biochemistry* **21**:1919–1925.

Walsh, M. P., Persechini, A., Hinkins, S., and Hartshorne, D. J., 1981, Is smooth muscle myosin a substrate for cAMP dependent protein kinase?, *FEBS Lett.* **126**:107–110.

Werth, D. K., Haeberle, J. R., and Hathaway, D. R., 1982, Purification of a myosin phosphatase from bovine aortic smooth muscle, *J. Biol. Chem.* **257**:7306–7309.

Chapter 7

Transduction and Signaling in Airway Smooth Muscle

Carl B. Baron

Department of Physiology
University of Pennsylvania School of Medicine
Philadelphia, Pennsylvania 19104-6085

I. INTRODUCTION

Advances in knowledge of the metabolic pathways of transduction systems, over the past 5 years, have given us a fairly detailed picture of the nature of the chemical reactions associated with physiological functions. Most of these reactions have been studied in systems other than airway smooth muscle, e.g., blowfly salivary glands, liver, platelets, neutrophils, secretory cells, and vascular and iris smooth muscle. A number of scientific meetings on this subject have taken place over recent years covering more general areas of signal transduction (Poste and Crooke, 1985; Everd and Whelen, 1986; Poste and Crooke, 1986; Strand, 1986) and one that focused on airway smooth muscle (Nadel *et al.*, 1985). In addition, there have been a number of reviews and a book that deal with aspects of smooth muscle biochemistry and signal transduction (Stephens, 1977; Takenawa, 1982; Cauvin and Van Breemen, 1984; Barnes and Cuss, 1986; Russell, 1986).

Historically, the early studies implicating agonist-stimulated inositol phospholipid turnover in pancreas and brain showed that $[^{32}P]$-Pi was incorporated into phosphatidylinositol (PI) to a very much greater extent than other tissue phospholipids (Hokin and Hokin, 1953; Hokin and Hokin, 1955). This was followed by similar studies in a variety of smooth muscles: vas deferens (Canessa de Scanessa de Scarnati and Lapetina, 1974), rabbit iris (Abdel-Latiff, 1974), and guinea pig ileum (Jafferji and Michell, 1976). In addition, $[^{32}P]$-Pi was found to be incorporated into the rabbit iris smooth muscle polyphosphoinositides, phosphatidylinositol 4,5-bisphosphate (PI-4,5-P_2) (Fig. 1), and phosphatidylinositol 4-phosphate (PI-4-P) (Abdel-Latiff *et al.*, 1977) and into toad

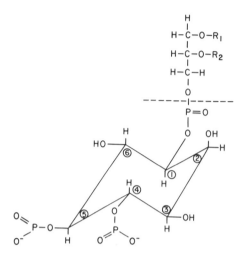

FIGURE 1. Structure of phosphatidylinositol-4,5-bisphosphate (PI-4,5-P_2). R_1 is usually stearyl ($C_{18:0}$), and R_2 is usually arachidonyl ($C_{20:4}$). The dashed line shows where the molecule is cleaved by phospholipase C to yield I-1,4,5-P_3 and DAG.

stomach smooth muscle phosphatidic acid (PA) (Salmon and Honeyman, 1979). Also, decreases in the mass of PI (Jafferji and Michell, 1977; Egawa et al., 1982), increases in the mass of PA paralleling tension development (Salmon and Honeyman, 1980), increases in ^{32}P labeling of PI paralleling tension development (Villalobos-Molin et al., 1982; Baron et al., 1984), and PI decreases and PA increases paralleling tension development (Baron et al., 1984) have been reported.

The aim of this review is to describe, in detail, the present status of what is known of the agonist-stimulated pathways for inositol phospholipids, as well as the effect of phorbol esters and lipid-activated protein kinase C. A review of arachidonic acid and its metabolites is not included. Earlier (before 1984) reports of the material I will discuss are cited in the review articles. The specific aspects of what is known in smooth muscle will be noted within each section. In addition, a section will be devoted to airway smooth muscle.

II. CONTROL OF SECOND-MESSENGER PRODUCTION BY GUANINE NUCLEOTIDE-BINDING PROTEINS

A number of enzymes and channels activated upon interaction of an agonist with a membrane-bound receptor appear to be under the primary control of a family of guanine nucleotide-binding proteins (G proteins) (Gilman et al., 1985; Rodbell, 1985; Naccache and Sha'afi, 1986; Stryer and Bourne, 1986). The control of second-messenger production is dependent on the state of the G protein: inactive GDP state and active GTP state. The conversion from the GDP to the GTP state is dependent on receptor occupancy. Three of the systems which have been shown to be under control of such G proteins are the hydrolysis of PI-4,5-P$_2$ (Cockcroft and Gomperts, 1985; Enyedi *et al.*, 1986; Jackowski *et al.*, 1986; Lucus *et al.*, 1985; Sasaguri *et al.*, 1985; Huque, 1986; Uhing *et al.*, 1986), the formation of cyclic adenosine monophosphate (cAMP) (Strulovici *et al.*, 1985; Gilman *et al.*, 1985; Rodbell, 1985), and the formation of cyclic GMP (only for the visual transduction system) (Liebman, 1986, 1987; Stryer and Bourne, 1986).

III. INOSITOL PHOSPHOLIPID METABOLISM

A. Hydrolysis of PI-4,5-P$_2$ by Inositol Phospholipid-Specific Phospholipase C and Ca^{2+} Release

1. Formation of Inositol 1,4,5-Trisphosphate (I-1,4,5-P$_3$) and Phospholipase C

Agonist stimulation of a class of receptors appears to increase the hydrolysis of the phospholipid PI-4,5-P$_2$ into two moieties, water-soluble inositol 1,4,5-trisphosphate (I-1,4,5-P$_3$) and lipid-soluble diacylglycerol (DAG) (see Section III.B.1), whose fatty acid composition resembles the polyphosphoinositides (Cockcroft *et al.*, 1984). The pathway for inositol phospholipid metabolism is illustrated in Fig. 2. There have been numerous recent reviews on this subject (Berridge and Irvine, 1984; Berridge, 1985; Exton, 1985; Hawthorne, 1985; Hirasawa and Mishizuka, 1985; Hokin, 1985; Michell, 1985; Berridge, 1986a,b; Majerus *et al.*, 1986; Naccahe and Sha'afi, 1986; Rasmussen *et al.*, 1986; Sekar and Hokin, 1986; Troyer and Schwartz, 1986; Williamson, 1986; Williamson *et al.*, 1986). Hydrolysis of PI-4,5-P$_2$ has been reported in numerous smooth muscles (Akhtar and Abdel-Latiff, 1984, 1986; Sekar and Roufogalis, 1984; Derian and Moskowitz, 1986) as well as in airway smooth muscle (Takuwa *et al.*, 1986; Baron and Coburn, 1987). Subsequent to agonist stimulation, increases in the production of I-1,4,5-P$_3$ and other inositol phosphates have been observed in airway (Baron and Coburn, 1987; Grandordy *et al.*, 1986), other smooth muscles (Akhtar and Abdel-Latiff, 1980, 1986; Sekar and Roufogalis, 1984; Donaldson and Hill, 1985; Fox *et al.*, 1985; Legan *et al.*, 1985; Berta *et al.*, 1986; Derian and Moskowitz, 1986) and other tissues (Berridge *et al.*, 1984; Irvine *et al.*, 1985; Montague *et al.*, 1985; Rittenhouse and Sasson, 1985; Farese *et al.*, 1986a,b; Nanberg and Putney, 1986; Palmer *et al.*, 1986). These measurements are indications of inositol phospholipid turnover. In most instances, the inositol phospholipids were not radiolabeled to equilibrium; thus, counts appearing in the inositol phosphates represent flux and not mass (pool size) (see Section III.I). Increases in pool size of the inositol phosphates have been reported in few studies: one in which mass measurements were made by gas chromatography

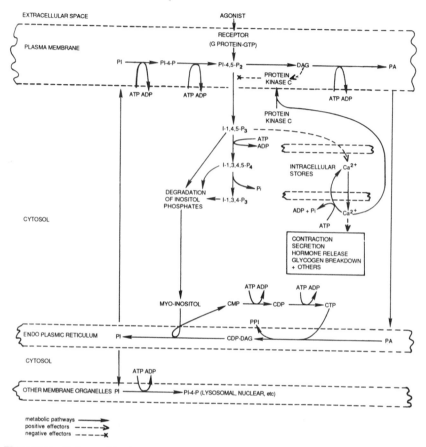

FIGURE 2. Pathway of inositol phospholipid metabolism. Certain aspects have been omitted from this scheme: (1) direct hydrolysis of PI and PI-4-P to their respective inositol phosphates; (2) formation of cyclic inositol phosphates; (3) indication that I-1,3,4,5-P_4 appears to affect the Ca^{2+} permeability of the plasma membrane. Also, the site of phosphorylation of DAG to PA may be in the plasma membrane, the cytosol, of microsomal membranes.

(Rittenhouse and Sasson, 1985) and others in which inositol phospholipids were labeled to equilibrium (Berridge *et al.*, 1984; J. B. Smith *et al.*, 1984; Baron and Coburn, 1987).

Studies on the hydrolytic activity of inositol phospholipid-specific phospholipase C indicate that, at low ($\sim$0.1 μM) [Ca^{2+}], PI-4,5-P_2 and PI-4-P are preferential substrates, while PI is hydrolyzed at much lower rates, from 10-fold less to undetectable (Wilson *et al.*, 1984; C. D. Smith *et al.*, 1985, 1986; Chung *et al.*, 1985; Litosch and Fain, 1985; Lucas *et al.*, 1985; Nakanishi *et al.*, 1985; Banno *et al.*, 1986; Deckmyn *et al.*, 1986; Jackowski *et al.*, 1986; Kelly and Ingraham, 1986; Melin *et al.*, 1986). Cholinergic stimulation of isolated membranes also causes hydrolysis of PI, PI-4-P, and PI-4,5-P_2 (Dunlop and Malaisse, 1986).

2. Correlation of I-1,4,5-P_3 Production and Ca^{2+} Release from Internal
   Stores

The parallel production of I-1,4,5-P_3 and the increase in cytosolic $[Ca^{2+}]$ have been demonstrated in numerous systems (Streb *et al.*, 1983; Thomas *et al.*, 1984; Campbell *et al.*, 1985; Bradford and Rubin, 1986; Jean and Klee, 1986; Portilla and Morrison, 1986; Ramsdell and Tashjian, 1986) as well as in smooth muscle (J. B. Smith *et al.*, 1984; Nabika *et al.*, 1985; Aiyar *et al.*, 1986; Bitar *et al.*, 1986) and has been reviewed recently (Berridge, 1985; Exton, 1985; Hokin, 1985; Berridge, 1986a,b; Sekar and Hokin, 1986; Williamson *et al.*, 1986). While there are no reports demonstrating that I-1,4,5-P_3 production precedes Ca^{2+} release, two kinetic studies in vascular smooth muscle (Alexander *et al.*, 1985) and hepatocytes (Charest *et al.*, 1985) demonstrated that the rate of elevation of I-1,4,5-P_3 following agonist stimulation is fast enough to account for the release of Ca^{2+} into the cytosol.

To date, the most convincing studies coupling I-1,4,5-P_3 to Ca^{2+} release from internal stores are those using cells or tissues permeabilized with saponin and Ca^{2+}-loaded microsomes. (Note: The detergent, saponin, appears only to affect the permeability of cholesterol-containing membranes and therefore would affect the plasma membrane, but not the endoplasmic or sarcoplasmic reticulum.) The methods employed generally involve the addition of I-1,4,5-P_3 to permeabilized cells, followed by increases in fluorescence of the Ca^{2+}-sensitive dyes, quin-2 or fura-2, or efflux of $^{45}Ca^{2+}$, monitored in smooth muscle (Suematsu *et al.*, 1984; Hashimoto *et al.*, 1985; J. B. Smith *et al.*, 1985; Somlyo *et al.*, 1985; Suematsu *et al.*, 1985a,b; Goldman *et al.*, 1986; Hashimoto *et al.*, 1986) or other tissues (Streb *et al.*, 1983; Berridge *et al.*, 1984; Burgess *et al.*, 1984; Dawson and Irvine, 1984; Joseph *et al.*, 1984; Prentki *et al.*, 1984; Clapper and Lee, 1985; O'Rourke *et al.*, 1985; Dawson *et al.*, 1986; Joseph *et al.*, 1986; Ueda *et al.*, 1986; Velasco *et al.*, 1986). In one case, not only I-1,4,5-P_3, but also inositol 1 : 2-cyclic 4,5-trisphosphate, were reported to release Ca^{2+} from permeabilized platelets and when injected into *Limulus* ventral photoreceptor cells (Wilson *et al.*, 1985b).

In work related to replenishment of the internal Ca^{2+} stores, Ca^{2+}–ATPase activity in vascular smooth muscle was inhibited by I-1,4,5-P_3 (Popescu *et al.*, 1986).

B. Formation of Diacylglycerol, Activation of Protein Kinase C, and Modulation of PI-4,5-P$_2$ Hydrolysis

1. Diacylglycerol and Protein Kinase C

Increases in the mass of diacylglycerol (DAG), the lipid-soluble product of the hydrolysis of PI-4,5-P$_2$, have been observed in a few cases. These have been rapid and transient increases (Rebecchi et al., 1983; Chaffoy et al., 1985; Hirasawa and Nishizuka, 1985; Sekar and Hokin, 1986; see also Berridge, 1985, and Sekar and Hokin, 1986, and references cited therein), as well as smaller sustained increases (Baron et al., 1984; Dixon and Hokin, 1984; Griendling et al., 1986; Takuwa et al., 1986). DAG has been demonstrated to stimulate the activity of the lipid-activated enzyme protein kinase C, which in turn phosphorylates cellular proteins of unknown function.

The role of protein kinase C in biological systems was recently reviewed (Berridge, 1986a; Kikkawa and Nishizuka, 1986a,b; Kikkawa et al., 1986). Activation of protein kinase C has also been shown to be associated with a wide variety of secretory processes and hormone-release mechanisms. Studies on protein kinase C generally employ various phorbol esters (4-β are active and 4-α are inactive); the DAG analogue 1-oleoyl-2-acetylglycerol or 1,2-dioctanoylglycerol to activate enzymatic activity, which in turn is associated with the phosphorylation of endogenous proteins and model proteins, e.g., histones. [Note: The nomenclature of phorbol esters can be confusing: phorbol 12-myristate 13-acetate (PMA) is the same chemical as 12-O-tetradecanoylphorbol 13-acetate (TPA).] All active diacylglycerols have 1,2-sn configuration (other stereoisomers are inactive) (Nomura et al., 1986).

Kikkawa and Nishizuka (1986a) noted that while the rapid and transient increases in DAG stimulate protein kinase C activity for only a short time, inactivated rapidly after the disappearance of DAG, not only by lowering activity, but by proteolytic degradation as well, the consequences of protein kinase C activity (i.e., phosphorylation of some cellular proteins) may persist for long periods. However, a recent study reports that an 80,000-dalton protein in Swiss 3T3 cells, which is phosphorylated in the presence of phorbol 12,13-dibutyrate (PDB) by protein kinase C, is rapidly dephosphorylated upon removal of the phorbol ester (Rodriquez-Pena et al., 1986). Thus, whether short-term activation of protein kinase C causes long-term or short-term effects and what the specific effects are is still to be resolved.

The intracellular location of protein kinase C appears to be dependent on [Ca^{2+}]. At lower (<2 nM) [Ca^{2+}] (in the presence of 3 mM Mg^{2+}), it is either in the cytoplasm or is loosely bound to membranes while at high (>4 nM) [Ca^{2+}], it is apparently translocated to membranes (Wolf et al., 1985). In the absence of Mg^{2+}, these values are shifted to slightly higher concentrations, 4 and 5 nM, respectively. In PYS cells, Ca^{2+} can enhance fourfold the amount of protein kinase C to membranes in the presence of TPA (Gopalakrishna *et al.*, 1986). Separately, Ca^{2+} and TPA can cause association of protein kinase C to membranes, although the properties of the interactions are quite different (e.g., Ca^{2+} is temperature independent and rapid ($<$1 min); TPA is temperature dependent and relatively slower).

2. Modulation of PI-4,5-P_2 Hydrolysis by Phorbol Esters and Protein Kinase C

In agonist-stimulated cells and tissues, phorbol esters appear to inhibit the stimulated response. Specifically, phorbol esters appear to inhibit the hydrolysis of PI-4,5-P_2 as monitored by decreases in I-1,4,5-P_3 formation and in the transient cytosolic [Ca^{2+}] elevation in smooth muscle (contraction can also inhibited) (Brock *et al.*, 1985; Cotecchia *et al.*, 1986; McMillan *et al.*, 1986; Mendoza *et al.*, 1986; Roth *et al.*, 1986) and other cells (Leeb-Lundberg *et al.*, 1985; Orellana *et al.*, 1985; Vicentini *et al.*, 1985; Watson and Lapetina, 1985; Zavoico *et al.*, 1985; Bianca *et al.*, 1986; Chuang, 1986; Garcia-Sainz *et al.*, 1986; Johnson *et al.*, 1986; Kikuchi *et al.*, 1986; Mizuguchi *et al.*, 1986; Molina y Vedia and Lapetina, 1986; Poll and Westwick, 1986; Sasakawa *et al.*, 1986; Tohmatsu *et al.*, 1986). The source of Ca^{2+} in this instance is from internal stores. Either phorbol esters and/or protein kinase C appears to inhibit hydrolysis of PI-4,5-P_2 by affecting the activity of inositol phospholipid-specific phospholipase C or by altering the state of the G protein associated with phospholipase C from active to inactive, halting the production of I-1,4,5-P_3, subsequent Ca^{2+} release, and the physiological response.

In smooth muscle, the addition of phorbol esters (presumably through activation of protein kinase C) can cause a contractile response (Danthuluri and Deth, 1984; Rasmussen *et al.*, 1984; Dale and Obianime, 1985; Baraban *et al.*, 1985; Nakaki *et al.*, 1985; Forder *et al.*, 1985; Park and Rasmussen, 1985; Menkes *et al.*, 1986; Wagner *et al.*, 1986). However, these contractile responses can be inhibited by Ca^{2+}-channel blockers and are dependent on external Ca^{2+} (Forder *et al.*, 1985; Menkes *et al.*, 1986). The contractile response to phorbol esters are much slower than to muscarinic and α_1-adrenergic agonists, and so forth. In view of the observations that (1) phorbol esters can increase the permeability of the plasma membrane to Ca^{2+} (Colucci *et al.*, 1986); (2) protein kinase C binding to membranes and activity is sensitive to Ca^{2+} (Wolf *et al.*, 1985); and (3) the source of Ca^{2+} for the contractile response is from an external source, one may hypothesize that the contraction is slow because upon addition of a phorbol ester there is not much protein kinase C bound to the membrane and thus available for activation to cause increases in the Ca^{2+} permeability of the plasma membrane. As the intracellular [Ca^{2+}] increases more protein kinase C becomes activated, thereby increasing the level of tension.

C. Metabolism of Inositol Phosphates

The transient increases in the cytosolic [I-1,4,5-P$_3$] and its return to a level near basal appears to be responsible for large transient increases in [Ca^{2+}], followed by a return to levels near basal even while the stimulus is maintained. The reduction of cytosolic [I-1,4,5-P$_3$] appears to occur by either of two enzymatic reactions. One is the action of a phosphatase, which converts I-1,4,5-P$_3$ to inositol 1,4-bisphosphate (I-1,4-P$_2$). The other is phosphorylation of I-1,4,5-P$_3$ to I-1,3,4,5-P$_4$ (Batty et al., 1985; Helsop et al., 1985; Hawkins et al., 1986; Irvine et al., 1986), which in turn is dephosphorylated to the inactive inositol trisphosphate, I-1,3,4-P$_3$ (Burgess et al., 1985; Irvine et al., 1985; Downes et al., 1986; Merritt et al., 1986; Rossier et al., 1986). (For Ca^{2+} release from internal stores only 1,4,5 species of inositol trisphosphate (I-1,4,5-P$_3$) is active; I-1,3,4-P$_3$ is inactive.) The isomerization of I-1,4,5-P$_3$ to I-1,3,4-P$_3$ is apparently a two-step phosphorylation that is calcium sensitive and ATP dependent (Irvine et al., 1986; Rossier et al., 1986). The percentage of I-1,4,5-P$_3$ metabolized by each of these reactions is yet to be determined, although preliminary indications are that the conversion of I-1,4,5-P$_3$ by direct dephosphorylation to I-1,4-P$_2$ occurs at low [Ca^{2+}], while at elevated [Ca^{2+}] the phosphorylation/dephosphorylation pathway may be more active. Recent reports have indicated that I-1,3,4,5-P$_4$ may play a role in regulating the entry of Ca^{2+} across the plasma membrane (Irvine and Moor, 1986; Imboden, 1987). Thus, cells are apparently able to regulate the cytosolic [Ca^{2+}] by regulating the cytosolic concentration and/or production of I-1,4,5-P$_3$ and I-1,3,4,5-P$_4$.

Further degradation of the inositol phosphates is complex and is not discussed further except to note that Li$^+$ can inhibit some of the subsequent phosphatase reactions; [a recent review (Williamson et al., 1988) partially clarifies inositol phosphate degradation].

D. Synthesis of Phosphatidylinositol

Continued production of I-1,4,5-P$_3$ requires stimulated synthesis of the inositol phospholipids (Carnessa de Scarnati and Lapetina, 1974; Villalobos-Molina et al., 1982; Baron et al., 1984; Sadler et al., 1984; J. B. Smith et al., 1984; Takuwa et al., 1986; Baron and Coburn, 1987). The metabolic reactions that result in the formation of PI, PI-4-P, and PI-4,5-P$_2$ are also presented in Fig. 2. Not only are enzymatic reactions involved, but the transfer of certain intermediates between the endoplasmic reticulum (site of PI synthesis) and the plasma membrane (site of agonist-stimulated hydrolysis of PI-4,5-P$_2$), as well as equilibration of newly synthesized PI among other organelle membranes.

The supply of lipid intermediates in the pathway appears come from an ATP-dependent phosphorylation of DAG to PA and its subsequent transfer to the endoplasmic reticulum. It is unclear whether phosphorylation preceeds lipid transfer. DAG kinase activity is both cytosolic and microsomal (Esko and Raetz, 1983; Coleman and Bell, 1983). PA then reacts with cytidine triphosphate (CTP) to yield the intermediate cytidine diphosphate–diacylglycerol (CDP–DAG), also known as phosphatidyl–CMP (PA–CMP), discussed in a recent review by Esko and Raetz (1983). Then, by action of the enzyme CDP-1,2-diacyl-*sn*-glycerol : myo-inositol 3-phosphatidyltransferase (PI synthetase), myo-inositol interacts with CDP–DAG to form PI with the release of CMP (Esko and Raetz, 1983; Parries and Hokin-Neaverson, 1984; Ghalayini and Eichberg, 1985; Iujvidin and Mordoh, 1986). This enzyme can also catalyze the CMP-dependent exchange of myo-inositol and PI (Berry *et al.*, 1983). PI is then transferred from the endoplasmic reticulum to the plasma membrane and other organelle membranes throughout the cell.

E. Rates of Synthesis of Inositol Phospholipids

Two phosphorylation steps in the plasma membrane replenish the pool of PI-4,5-P_2 to maintain its concentration and availability to provide a source for continued production of I-1,4,5-P_3. The rate of labeling of the inositol phospholipids has been shown to be stimulated by the action of a cholinergic agonist in trachealis smooth muscle (Baron and Coburn, 1987) and thrombin in human platelets (Wilson *et al.*, 1985c). The rates of synthesis of the inositol phospholipids have been reported in these two systems.

During agonist-stimulated incorporation of [^{3}H]-myo-inositol into trachealis smooth muscle inositol phospholipids (which occurred only during maintenance of tension and not during development of tension), it was observed that the rates of radiolabel incorporation into PI, PI-4-P, and PI-4,5-P_2 were almost identical. The K_i values (determined from radioactive flux of myo-inositol and pool sizes) were estimated to be 0.047 min^{-1} for PI, 0.63 min^{-1} for PI-4-P, and 0.25 min^{-1} for PI-4,5-P_2. By comparison, the rate of decrease in the pool size of PI (and parallel increase in PA) during the development of tension was estimated to be 0.43 nmoles/(100 nmoles total lipid P_i)(min). This is about three times faster than the rate of flux during maintenance of tension, suggesting that flux through the polyphosphoinositides, PI-4-P and PI-4,5-P_2, may be faster during tension development than during maintenance (Baron *et al.*, 1989).

The rates of synthesis that occurred in platelets subsequent to thrombin stimulation (Wilson *et al.*, 1985c) are more difficult to evaluate because no time variable is included. The value of the flux through PI-4-P was estimated to be 0.33–0.53 nmoles/10^9 platelets.

F. Compartmentation of Inositol Phospholipids

The distribution of inositol phospholipids in organelle cell membrane varies. PI appears to be ubiquitous. In hepatocytes (Seyfred and Wells, 1984), PI-4-

P appears to a significant extent in both the plasma (67%) and lysosomal (13%) membranes. PI-4,5-P_2 appears to reside mainly in the plasma membrane (90%). Lower levels of the polyphosphoinositides, PI-4-P (7%) and PI-4,5-P_2 (4%) are reported to be present in other individual organelle membranes. Also, in ^{32}P-labeled platelets, there appear to exist pools of phosphoinositides with different specific radioactivities (Vickers and Mustard, 1986). In carbamoylcholine-stimulated canine trachealis smooth muscle, at extended times (3–4 hr) when the specific radioactivities of the inositol phospholipids were apparently no longer increasing, estimates of maximum labeling (expressed as percentage of specific radioactivity of [^{3}H]-myo-inositol in the tissue) were 98% for PI, 87% for PI-4-P, and 66% for PI-4,5-P_2. These fractions apparently represent the compartments participating in cholinergic-stimulated metabolism. The unlabeled portions were speculated to reside mainly in pools either associated with other receptors or in plasma membranes that were internalized during the development of tension when no labeling of the inositol phospholipids was observed.

G. Other Reactions of Inositol Phospholipid Metabolism

Some other reactions, not included in Fig. 2, appear to exist. These have been demonstrated in broken cell preparations. One involves the hydrolysis of PI and PI-4-P directly to their corresponding inositol phosphates by phospholipase C. In a number of recent studies, however, action of inositol phospholipid-specific phospholipase C appears to prefer PI-4,5-P_2 and PI-4-P to PI by at least 10-fold (see Section III.A.1). The formation of the cyclic inositol phosphate species, myo-inositol 1,2-cyclic phosphate (cyclic I-1,2-P), and other inositol cyclic phosphates has also been demonstrated (Wilson *et al.*, 1985a,b; Dixon and Hokin, 1985; Connolly *et al.*, 1986; Majerus *et al.*, 1986).

H. ATP Utilization by Inositol Phospholipid Metabolism

The total energy cost, in terms of ATP hydrolysis, is relatively small but significant. Estimates were obtained from flux measurements of myo-inositol and rates of total ATP utilization in canine tracheal smooth muscle (Baron and Coburn, 1987). During the development of tension ATP flux used by the portion of the cycle that resides in the plasma membrane (PI to PA; three phosphorylation steps) (see Fig. 2) represents about 4% of the incremental increase in ATP flux (~1.5 times over resting muscle). However, during maintenance of tension, when the incremental increase is ~1.16 times that of resting muscle, the portion used by the entire cycle (five phosphorylation steps) (see Fig. 2) is ~7% of the incremental increase. These estimates do not include the phosphorylation of I-1,4,5-P_3 to I-1,3,4,5-P_4 and a number of ATP-utilizing reactions that may also be stimulated at the same time, including the formation of cyclic nucleotides, Ca^{2+} reuptake into internal stores, phosphorylation of proteins and phosphorylation of DAG to PA, which is then incorporated into noninositol-containing lipids.

I. Considerations for Labeling Methods in Assessment of Inositol Phospholipid Metabolism

There are two methods of labeling to equilibrium with [^{3}H]-myo-inositol: (1) labeling in the presence of agonist stimulation with subsequent removal of agonist and allowing tissues to return to at rest conditions, and (2) labeling of cells in culture for 24–48 hr. In nonequilibrium labeling conditions (short times of labeling, e.g., 30–90 min), the flux of radioactivity (radioactivity/tissue mass) can be measured with pool sizes measured independently. In equilibrium-labeled tissues, radioactivity is a measure of pool size and fluxes cannot be determined. In many studies, while the labeling conditions are given, the assessment of whether the state (nonequilibrium or equilibrium labeled) of the lipids was not made. This assessment is particularly critical when measurements of inositol phosphates are made, since radioactivity appearing in these species will represent pool sizes when the inositol phospholipids are equilibrium labeled and fluxes (not pool sizes) when the lipids are nonequilibrium labeled.

Incubation of tissues and cells with ^{32}P results in labeling of the γ-phosphate of ATP. There appears to be a fairly rapid exchange between the γ-phosphate of ATP and the 4- and 5-phosphates of PI-4-P and PI-4,5-P$_2$ resulting in equilibrium labeling of these positions.

IV. STATE OF KNOWLEDGE IN AIRWAY SMOOTH MUSCLE

Two reviews have addressed the areas of smooth muscle (Takenawa, 1982) and tracheal smooth muscle, in particular (Russell, 1986). Information on airway smooth muscle can be found in a conference summary (Nadel et al., 1985).

Investigators examining inositol phospholipid turnover in airway smooth muscle have reported effects of stimulation on inositol phospholipid and PA pool sizes (Baron et al., 1984, 1987; Takuwa et al., 1986; Baron and Coburn, 1987; Baron et al., 1989). These effects are a fall in the PI pool paralleled by an increase in the PA pool and, in some instances, fall in the PIP and PIP$_2$ pools. In general, the pools then remain at their new levels as long as the stimulus is maintained. This may be viewed as an effect of stimulation on the steady state levels of these lipids.

Labeling of inositol phospholipids following stimulation with carbamoyl-choline has been studied in canine (Baron et al., 1989) and bovine (Takuwa et al., 1986) trachealis smooth muscle. In case of canine, after incubation of the tissue for 30 min with [^{3}H]myo-inositol, the lipids are only minimally labeled. There was a slow increase up to 4 hr when the tissue was maintained at rest.

Stimulation by 5.5 μM carbamoylcholine, caused a rapid flux of radiolabel into all three lipids, linear up to $\sim$15 min and reaching maximal levels at $\sim$3–4 hr (levels at these late times were 2.5–3 times that observed in tissue at rest). The extent to which the pools were labeled, compared with the specific radioactivity of [^{3}H]myo-inositol, were 98% for PI, 87% for PIP and 66% for PIP$_2$. The interpretation of this finding is that these compartments represent the levels of rapidly turned-over pools. By contrast, bovine trachealis muscle, labeled for 3 hr with [^{3}H]myo-inositol, showed changes in radioactivity in the inositol phospholipids that paralleled the pool-size measurements for PIP and PIP$_2$. This would appear to indicate that the pools were labeled to, at, or near equilibrium prior to stimulation. The difference in [^{3}H]myo-inositol labeling in the canine and bovine trachealis muscle may be due to a difference in basal (in the absence of agonist) inositol phospholipid metabolism in the two tissues.

The I-1,4,5-P$_3$-mediated release of Ca^{2+} from internal stores has been demonstrated in canine trachealis smooth muscle (Hashimoto et al., 1985). The production of inositol phosphates has been observed in smooth muscle from bovine and canine trachea (Grandordy et al., 1986; Takuwa et al., 1986; Baron and Coburn, 1987; Baron et al., 1989).

The effect of the β-phorbol ester, PDB, on canine trachealis muscle strips contracted with carbamoylcholine has also been studied (Baron and Coburn, 1989). These experiments demonstrated that addition of PDB 20 min prior to stimulation, resulted in a general inhibition of inositol phospholipid turnover (changes in pool size of PI and PA, inositol phospholipid synthesis, production of inositol phosphates). When PDB was added 3 min after the onset of a car-bamoylcholine-stimulated contraction, several effects were observed: (I) pool sizes of PI and PA tended to return toward basal levels; (2) inositol phosphate production was inhibited at times beyond 12 min after the addition of PDB; and (3) inositol phospholipid synthesis was not inhibited. In addition, tissue maintained at rest, showed a twofold increase in the PIP pool size due to the presence of PDB.

Phorbol ester-induced contractions guinea pig lung parenchymal (Dale and Obianime, 1985) and tracheal (Menkes et al., 1986) strips have been reported. These contractions could be inhibited by the Ca^{2+} channel blockers verapamil, nifedipine, and diltiazem, and phorbol contracted strips could be relaxed about 22% by isoprenaline and forskalin (agents which cause elevation of cAMP levels).

V. SUMMARY

Airway smooth muscle contraction appears to be under the control of the metabolic processes involved in inositol phospholipid turnover. The metabolic events in inositol phospholipid metabolism have been elucidated to a much greater extent in numerous other tissues and cells than in the airway, although the very early events in agonist-stimulated metabolism have yet to be clearly demonstrated. Airway smooth muscle may prove useful in answering some of these questions about the early events, now that systems have been developed to measure metabolism as early as 250 msec (Kamm and Stull, 1986).

ACKNOWLEDGMENT. This work was supported by grant HL 19737 from the National Heart, Lung and Blood Institute.

VI. REFERENCES

Abdel-Latiff, A. A., 1974, Effects of neurotransmitters and other pharmacological agents on ^{32}Pi incorporation into phospholipids of the iris muscle of the rabbit, *Life Sci.* **15**:961–973.

Abdel-Latiff, A. A., Akhtar, R. A., and Hawthorne, J. N., 1977, Acetylcholine increases the breakdown of triphosphoinositide of rabbit iris muscle prelabelled with [^{32}P]phosphate, *Biochem. J.* **162**:61–73.

Aiyar, N., Nambi, P., Stassen, F. L., and Crooke, S. T., 1986, Vascular vasopressin receptors mediate phosphatidylinositol turnover and calcium efflux in an established smooth muscle cell line, *Life Sci.* **39**:37–45.

Akhtar, R. A., and Abdel-Latiff, A. A., 1980, Requirement for calcium ions in acetylcholine-stimulated phophodiesteratic cleavage of phosphatidyl-myo-inositol 4,5-bisphosphate in rabbit iris smooth muscle, *Biochem. J.* **192**:783–791.

Akhtar, R. A., and Abdel-Latiff, A. A., 1984, Carbachol causes rapid phophodiesteratic cleavage of phosphatidylinositol 4,5-bisphosphate and accumulation of inositol phosphates in rabbit iris smooth muscle; prazosin inhibits noradrenaline- and inophore A23187-stimulated accumulation of inositol phosphates, *Biochem. J.* **224**:291–300.

Akhtar, R. A., and Abdel-Latiff, A. A., 1986, Surgical sympathetic denervation increases alpha$_1$-adrenoceptor-mediated accumulation of myo-inositol trisphosphate and muscle contraction in rabbit iris dilator smooth muscle, *J. Neurochem.* **46**:96–104.

Alexander, R. W., Brock, T. A., Gimbrone, M. A., Jr., and Rittenhouse, S., 1985, Angiotensin increases inositol trisphosphate and calcium in vascular smooth muscle, *Hypertension* **7**:447–451.

Banno, Y., Nakashima, S., and Nozawa, Y., 1986, Partial purification of phosphoinositide phospholipase C from human platelet cytosol; characterization of its three forms, *Biochem. Biophys. Res. Commun.* **136**:713–721.

Baraban, J. M., Gould, R. J., Peroutka, S. J., and Snyder, S. H., 1985, Phorbol ester effects on neurotransmission: Interaction with neurotransmitters and calcium in smooth muscle, *Proc. Natl. Acad. Sci. USA* **82**:604–607.

Barnes, P., and Cuss, F. M., 1986, Biochemistry of airway smooth muscle, *Bull. Eur. Physiopathol. Respir.* **22**(Suppl. 7):191–200.

Baron, C. B., and Coburn, R. F., 1987, Inositol phospholipid turnover during contraction of canine trachealis muscle, *Ann. NY Acad. Sci.* **494**:80–83.

Baron, C. B., and Coburn, R. F., 1989, Phorbol 12,13-dibutyrate inhibits and reverses phosphatidylinositol and phosphatidic acid pool size changes and inhibits [^{3}H]-myo-inositol flux in canine trachealis smooth muscle, (in preparation).

Baron, C. B., Cunningham, M., Strauss, J. F. III, and Coburn, R. F., 1984, Pharmacomechanical coupling in smooth muscle may involve phosphatidylinositol metabolism, *Proc. Natl. Acad. Sci. USA* **81**:6899–6903.

Baron, C. B., Pring, M., and Coburn, R. F., 1989, Synthesis and compartmentation of inositol phospholipids in unstimulated and carbamoylcholine-stimulated smooth muscle, *Am. J. Physiol.* **256:** (in press).

Batty, I. R., Nahorski, S. R., and Irvine, R. F., 1985, Rapid formation of inositol 1,3,4,5-tetrakisphosphate following muscarinic receptor stimulation of rat cerebral cortical slices, *Biochem. J.* **232:**211–215.

Berridge, M. J., 1985, Inositol trisphosphate and diacylglycerol as intracellular second messengers, in: *Mechanisms of Receptor Regulation* (G. Poste and S. T. Crooke, eds.), pp. 111–130, Plenum, New York.

Berridge, M. J., 1986a, Agonist-dependent phosphoinositide metabolism: A bifurcating signal pathway, in: *New Insights into Cell and Membrane Transport Processes* (G. Poste and S. T. Crooke, eds.), pp. 201–216, Plenum, New York.

Berridge, M. J., 1986b, Inositol trisphosphate and calcium mobilization, in: *Calcium and the Cell,* Ciba Foundation Symposium No. 122 (D. Evered and J. Whelan, eds.), pp. 39–57, Wiley, Chichester.

Berridge, M. J., and Irvine, R. F., 1984, Inositol trisphosphate, a novel second messenger in cellular signal transduction, *Nature (Lond.)* **312:**315–321.

Berridge, M. J., Heslop, J. P., Irvine, R. F., and Brown, K. D., 1984, Inositol trisphosphate formation and calcium mobilization in Swiss 3T3 cells in response to platelet-derived growth factor, *Biochem. J.* **222:**195–201.

Berry, G., Yandrastiz, J. R., and Segal, S., 1983, CMP-dependent phosphatidylinositol: Myo-inositol exchange activity in isolated nerve-endings, *Biochem. Biophys. Res. Commun.* **112:** 817–821.

Berta, P., Sladeczek, F., Travo, P., Bockaert, J., and Haiech, J., 1986, Activation of phosphatidylinositol synthesis by different agonists in a primary culture of smooth muscle cells grown on collagen microcarriers, *FEBS Lett.* **200:**27–31.

Bianca, V. D., Grzeskowiak, M., Cassatella, M. A., Zeni, L., and Rossi, F., 1986, Phorbol 12, myristate 13, acetate protentiates the respiratory burst while inhibits phosphoinositide hydrolysis and calcium mobilization by formyl-methionyl-leucyl-phenylalanine in human neutrophils, *Biochem. Biophys. Res. Commun.* **135:**556–565.

Bitar, K. N., Bradford, P. G., Putney, J. W., Jr., and Makhlou, G. M., 1986, Stoichiometry of contraction and Ca^{2+} mobilization by inositol 1,4,5-trisphosphate in isolated gastric smooth muscle cells, *J. Biol. Chem.* **261:**16591–16596.

Bradford, P. G., and Rubin, R. P., 1986, Quantitative changes in inositol 1,4,5-trisphosphate in chemoattractant-stimulated neurophils, *J. Biol. Chem.* **261:**15644–15647.

Brock, T. A., Rittenhouse, S. E., Powers, C. W., Ekstein, L. S., Gimbrone, M. A., Jr., and Alexander, R. W., 1985, Phorbol ester and 1-oleoyl-2-acetylglycerol inhibit angiotensin activation of phospholipase C in cultured vascular smooth muscle cells, *J. Biol. Chem.* **260:**15158–14162.

Burgess, G. M., Godfrey, P. P., McKinney, J. S., Berridge, M. J., Irvine, R. F., and Putney, J. W., Jr., 1984, The second messenger linking receptor activation to internal Ca^{2+} in liver, *Nature (Lond.)* **309:**63–66.

Burgess, G. M., McKinney, J. S., Irvine, R. F., and Putney, J. W., Jr., 1985, Inositol 1,4,5-trisphosphate and inositol 1,3,4-trisphosphate formation in Ca^{2+}-mobilizing-hormone-activated cells, *Biochem. J.* **232:**237–243.

Campbell, M. D., Deth, R. C., Payne, R. A., and Honeyman, T. W., 1985, Phosphoinositide hydrolysis is correlated with agonist-induced calcium flux and contraction in the rabbit aorta, *Eur. J. Pharmacol.* **116:**129–136.

Canessa de Scarnati, O., and Lapetina, E. G., 1974, Adrenergic stimulation of phosphatidylinositol labelling in rat vas deferens, *Biochim. Biophys. Acta* **360:**298–305.

Cauvin, C., and Van Breemen, C., 1984, Regulation of Ca^{2+} levels in smooth-muscle cells, *Biochem. Soc. Trans.* **12:**939–941.

Charest, R., Prpic, V., Exton, J. H., and Blackmore, P. F., 1985, Stimulation of inositol trisphosphate formation in hepatocytes by vasopressin, adrenalin and angiotensin II and its relationship to changes in cytosolic free Ca^{2+}, *Biochem. J.* **227**:79–90.

Chuang, D.-M., 1986, Carbachol-induced accumulation of inositol-1-phosphate in neurohybridoma NCB-20 cells: Effects of lithium and phorbol esters, *Biochem. Biophys. Res. Commun.* **136**: 622–629.

Chung, S. M., Proia, A. D., Klintworth, G. K., Watson, S. P., and Lapetina, E. G., 1985, Deoxycholate induces the preferential hydrolysis of polyphosphoinositides by human platelet and rat corneal phospholipase C, *Biochem. Biophys. Res. Commun.* **129**:411–416.

Clapper, D. L., and Lee, H. C., 1985, Inositol trisphosphate induces calcium release from non-mitochondrial stores in sea urchin egg homogenates, *J. Biol. Chem.* **260**:13947–13964.

Cockcroft, S., Baldwin, J. M., and Allan, D., 1984, The Ca^{2+}-activated polyphosphoinositide phosphodiesterase of human and rabbit neutrophil membranes, *Biochem. J.* **221**:477–482.

Cockcroft, S., and Gomperts, B. D., 1985, Role of guanine nucleotide binding protein in the activation of polyphosphoinositide phosphodiesterase, *Nature (Lond.)* **314**:534–536.

Coleman, R. A., and Bell, R. M., 1983, Topography of membrane-bound enzymes that metabolize complex lipids, in: *The Enzymes*, Vol. XVI (P. D. Boyer, ed.), pp. 605–625, Academic, New York.

Colucci, W. S., Gimbrone, M. A., Jr., and Alexander, R. W., 1986, Phorbol diester modulates alpha-adrenergic receptor-coupled calcium efflux and alpha-adrenergic receptor number in culture vascular smooth muscle cells, *Circ. Res.* **58**:393–398.

Connolly, T. M., Wilson, D.B., Bross, T. E., and Majerus, P. W., 1986, Isoation and characterization of the inositol cyclic phosphate products of phosphoinositide cleavage by phospholipase C, *J. Biol. Chem.* **261**:122–126.

Cotecchia, S., Leeb-Lundberg, L. M. F., Hagen, P-O., Lefkowitz, R. J., and Caron, M. G., 1985, Phorbol ester effects on $alpha_1$-adrenoceptor binding and phosphatidylinositol metabolism in cultured vascular smooth muscle cells, *Life Sci.* **37**:2389–2398.

Dale, M. M., and Obianime, W., 1985, Phorbol myristate acetate causes in guinea-pig lung parenchymal strip a maintained spasm which is relatively resistant to isoprenaline, *FEBS Lett.* **190**: 6–10.

Danthuluri, N. R., and Deth, R. C., 1984, Phorbol ester-induced contraction of arterial smooth muscle and inhibition of alpha-adrenergic response, *Biochem. Biophys. Res. Commun.* **125**: 1103–1109.

Dawson, A. P., and Irvine, R. F., 1984, Inositol 1,4,5-trisphosphate-promoted Ca^{2+} release from microsomal fraction of rat liver, *Biochem. Biophys. Res. Commun.* **120**:858–864.

Dawson, A. P., Comerford, J. G., and Fulton, D. V., 1986, The effect of GTP on inositol 1,4,5-trisphosphate-stimulation of Ca^{2+} efflux from a rat liver microsomal fraction: Is a GTP-dependent protein phosphorylation involved?, *Biochem. J.* **234**:311–315.

De Chaffoy de Courcelles, D., Leysen, J. E., De Clerck, F., Van Belle, H., and Janssen, P. A. J., 1985, Evidence that phospholipid turnover is the signal transducing system coupled to serotonin-S_2 receptor sites, *J. Biol. Chem.* **260**:7603–7608.

Deckmyn, H., Tu, S-M., and Majerus, P. W., 1986, Guanine nucleotides stimulate soluble phosphoinositide-specific phospholipase C in the absence of membranes, *J. Biol. Chem.* **261**:16553–16558.

Derian, C. K., and Moskowitz, M. A., 1986, Polyphosphoinositide hydrolysis in endothelial cells and carotid artery segments: Bradykinin-2 receptor stimulation is calcium-independent, *J. Biol. Chem.* **261**:3831–3837.

Dixon, J. F., and Hokin, L. E., 1984, Secretogogue-stimulated phosphatidylinositol breakdown in the exocrine pancreas literates arachidonic acid, steraic acid, and glycerol by sequential action of phospholipase C and diglyceride lipase, *J. Biol. Chem.* **259**:14418–14425.

Dixon, J. F., and Hokin, L. E., 1985, The formation of inositol 1,2-cyclic phosphate on agonist stimulation of phosphoinositide breakdown in mouse pancreatic minilobules: Evidence for direct phosphodiesteratic cleavage of phosphatidylinositol, *J. Biol. Chem.* **260**:16068–16071.

Donaldson, J., and Hill, S. J., 1985, Histamine-induced inositol phospholipid breakdown in the longitudinal smooth muscle of guinea-pig ileum, *Br., J. Pharmacol.* **85**:499–512.

Downes, C. P., Hawkins, P. T., and Irvine, R. F., Inositol-1,3,4,5-tetrakisphosphate and not phosphatidylinositol 3,4-bisphosphate is the probable precursor of inositol 1,3,4-trisphosphate in agonist-stimulated parotid glands, *Biochem. J.* **238**:501–506.

Dunlop, M. E., and Malaisse, W. J., 1986, Phosphoinositide phosphorylation and hydrolysis in pancreatic islet cell membrane, *Arch. Biochem. Biophys.* **244**:421–429.

Egawa, K., Sacktor, B., and Takenawa, T., 1981, Ca^{2+}-dependent and Ca^{2+}-independent degradation of phosphatidylinositol in rabbit vas deferens, *Biochem. J.* **194**:129–136.

Enyedi, P., Mucsi, I., Hunyady, L., Catt, K. J., and Spat, A., 1986, The role of guanine nucleotide binding proteins in the formation of inositol phosphates in adrenal glomerulosa cells, *Biochem. Biophys. Res. Commun.* **140**:941–947.

Esko, J. D., and Raetz, C. R., 1983, Synthesis of phospholipids in animal cells, in: *The Enzymes,* Vol. XVI (P. D. Boyer, ed.), pp. 207–253, Academic, New York.

Everd, D., and Whelan, J. (eds.), 1986, *Calcium and the Cell,* Ciba Foundation Symposium No. 122, Wiley, Chichester.

Exton, J. H., 1985, Role of calcium and phosphoinositides in the action of certain hormones and neurotransmitters, *J. Clin. Invest.* **75**:1753–1757.

Farese, R. V., Kuo, J. Y., Babischkin, J. S., and Davis, J. S., 1986a, Insulin provokes a transient activation of phospholipase C in the rat epididymal fat pad, *J. Biol. Chem.* **261**:8589–8592.

Farese, R. V., Rosic, N., Babischkin, J., Farese, M. G., Foster, R., and Davis, J. S., 1986b, Dual activation of the inositol–trisphosphate–calcium and cyclic nucleotide intracellular signaling systems by adrenocorticotropin in rat adrenal cells, *Biochem. Biophys. Res. Commun.* **135**:742–748.

Forder, J., Scriabine, A., and Rasmussen, H., 1985, Plasma membrane calcium flux, protein kinase C activation and smooth muscle contraction, *J. Pharmacol. Exp. Ther.* **235**:267–273.

Fox, A. W., Abel, P. W., and Minneman, K. P., 1985, Activation of alpha$_1$-adrenoceptors increases [^{3}H]inositol metabolism in rat vas deferens and caudal artery, *Eur. J. Pharmacol.* **116**:145–152.

Garcia-Sainz, J. A., Tussie-Luna, M. I., and Hernandez-Sotomayor, S. M. T., 1986, Phorbol esters, vasopressin and angiotensin II block alpha$_1$-adrenergic action in rat hepatocytes. Possible role of protein kinase C, *Biochim. Biophys. Acta* **887**:69–72.

Ghalayini, A., and Eichberg, J., 1985, Purification of phosphatidylinositol synthetase from rat brain by CDP-diacylglycerol affinity chromatography and properties of the purified enzyme, *J. Neurochem.* **44**:175–182.

Gilman, A. G., Smigel, M. D., Bokoch, G. M., and Robishaw, J. D., 1985, Guanine-nucleotide-binding regulatory proteins: membrane-bound information transducers, in: *Mechanisms of Receptor Regulation* (G. Poste and S. T. Crooke, eds.), pp. 149–158, Plenum, New York.

Goldman, Y. E., Reid, G. P., Somlyo, A. P., Somlyo, A. V., Trentham, D. R., and Water, J. W., 1986, Activation of skinned vascular smooth muscle by photolysis of ''caged inositol trisphosphate'' to inositol 1,4,5-trisphosphate (InsP$_3$), *J. Physiol. (Lond.)* **377**:100P.

Gonzales, R. A., and Crews, F. T., 1985, Guanine nucleotides stimulate production of inositol trisphosphate in rat cortical membranes, *Biochem. J.* **232**:799–804.

Gopalakrishna, R., Barsky, S. H., Thomas, T. P., and Anderson, W. B., 1986, Factors influencing chelator-stable, detergent-extractable, phorbol diester-induced membrane association of protein kinase C: Differences between Ca^{2+}-induced and phorbol ester-stabilized membrane bindings of protein kinase C, *J. Biol. Chem.* **261**:16438–16445.

Grandordy, B. M., Cuss, F. M., Sampson, A. S., Palmer, J. H. B., and Barnes, P. J., 1986, Phosphatidylinositol response to cholinergic agonists in airway smooth muscle: Relationship to contraction and muscarinic receptor occupancy, *J. Pharmacol. Exp. Ther.* **238**:273–279.

Griendling, K. K., Rittenhouse, W. E., Brock, T. A., Ekstein, L. S., Gimbrone, M. A., Jr., and Alexander, R. A., 1986, Sustained diacylglycerol formation form inositol phospholipids in angiotensin II-stimulated vascular smooth muscle cells, *J. Biol. Chem.* **261:**5901–5906.

Hashimoto, T., Hirata, M., and Itoh, Y., 1985, A role for inositol 1,4,5-trisphosphate in the initiation of agonist-induced contractions of dog tracheal smooth muscle, *Br. J. Pharmacol.* **86:** 191–199.

Hashimoto, T., Hirata, M., Itoh, Y., Kanmura, Y., and Kuriyama, H., 1986, Inositol 1,4,5-trisphosphate activates pharmacomechanical coupling in smooth muscle of the rabbit mesenteric artery, *J. Physiol. (Lond.)* **370:**605–618.

Hawkins, P. T., Stephens, L., and Downes, C. P., 1986, Rapid formation of inositol-1,3,4,5-tetrakisphosphate and inositol 1,3,4-trisphosphate in rat parotid glands may both result indirectly from receptor-stimulated release of inositol 1,4,5-trisphosphate from phosphatidylinositol 4,5-bisphosphate, *Biochem. J.* **238:**507–516.

Hawthorne, J. N., 1985, Inositol phospholipid and phosphatidic acid metabolism in response to membrane receptor activation, *Proc. Nutr. Soc.* **44:**167–172.

Helsop, J. P., Irvine, R. F., Tashjian, A. H., and Berridge, M. J., 1985, Inositol tetrakis- and pentakisphosphates in GH_4 cells, *J. Exp. Biol.* **119:**395–401.

Hirasawa, K., and Nishizuka, Y., 1985, Phosphatidylinositol turnover in receptor mechanism and signal transduction, *Annu. Rev. Pharmacol. Toxicol.* **25:**147–170.

Hokin, L. E., 1985, Receptors and phosphoinositide-generated second messengers, *Annu. Rev. Biochem.* **54:**205–235.

Hokin, L. E., and Hokin, M. R., 1955, Effects of acetylcholine on the turnover of phosphoryl units in individual phospholipids of pancreas slices and brain cortex slices, *Biochim. Biophys. Acta* **18:**102–110.

Hokin, M. R., and Hokin, L. E., 1953, Enzyme secretion and the incorporation of ^{32}P into phospholipids of pancreatic slices, *J. Biol. Chem.* **203:**967–977.

Huque, T., and Bruch, R. C., 1986, Odorant- and guanine nucleotide-stimulated phosphoinositide turnover in olfactory cilia, *Biochem. Biophys. Res. Commun.* **137:**36–42.

Imboden, J., 1987, Regulation of the inositol tris/tetrakisphosphate pathway during T cell activation, Presented at the 71st Meeting of the Federation of American societies for experimental biology, Washington, D.C.

Irvine, R. F., and Moor, R. M., 1986, Micro-injection of inositol-1,3,4,5-tetrakisphosphate activates sea urchin eggs by a mechanism dependent on external Ca^{2+}, *Biochem. J.* **240:** 917–920.

Irvine, R. F., Anggard, E. E., Letcher, A. J., and Downs, C. P., 1985, Metabolism of inositol 1,4,5-trisphosphate and inositol 1,3,4-trisphosphate in rat parotid glands, *Biochem. J.* **229:**505–511.

Irvine, R. F., Letcher, A. J., Helsop, J. P., and Berridge, M. J., 1986, The inositol tris/tetrakisphosphate pathway—Demonstration of Ins(1,4,5)P$_3$ 3-kinase activity in animal tissues, *Nature (Lond.)* **320:**631–634.

Iujvidin, S., and Mordoh, J., 1986, Metabolism of phosphatidyl-dCMP in sarcoma 180 cells: Effect of chlorpromazine, phosphatidic acid and inositol, *Eur. J. Biochem.* **154:**187–192.

Jackowski, S., Rettenmier, C. W., Sherr, C. J., and Rock, C. O., 1986, A guanine nucleotide-dependent phosphatidylinositol 4,5-diphosphate phospholipase C in cells transformed by the v-fms and v-fes oncogenes, *J. Biol. Chem.* **261:**4878–4985.

Jafferji, S. S., and Michell, R. H., 1976, Muscarinic cholinergic stimulation of phosphatidylinositol turnover in the longitudinal smooth muscle of guinea-pig ileum, *Biochem. J.* **154:**653–657.

Jean, T., and Klee, C. B., 1986, Calcium modulation of inositol 1,4,5-trisphosphate-induced calcium release from neuroblastoma × glioma hybrid (NG108-15) microsomes, *J. Biol. Chem.* **261:**16414–16420.

Johnson, R. M., Connelly, P. A., Sisk, R. B., Pobiner, B. F., Hewlett, E. L., and Garrison, J. C., 1986, Pertussis toxin or phorbol 12-myristate 13-acetate can distinguish between epidermal growth factor- and angiotensin-stimulated signals in hepatocytes, *Proc. Natl. Acad. Sci. USA* **83**:2032–2036.

Joseph, S. K., and Williamson, J. R., 1986, Characteristics of inositol trisphosphate-mediated Ca^{2+} release from permeabilized hepatocytes, *J. Biol. Chem.* **261**:14658–14664.

Joseph, S. K., Thomas, A. P., Williams, R. J., Irvine, R. F., and Williamson, J. R., 1984, Myo-inositol 1,4,5-trisphosphate: A second messenger for the hormonal mobilization of intracellular Ca^{2+} in liver, *J. Biol. Chem.* **259**:3077–3081.

Kamm, K. E., and Stull, J. T., 1986, Activation of smooth muscle contraction: Relation between myosin phosphorylation and stiffness, *Science* **232**:80–82.

Kelly, S. M., and Ingraham, L. M., 1986, Action of granulocyte phospholipase C on inositol-containing phospholipids, *Fed. Proc.* **45**:960.

Kikkawa, U., and Nishizuka, Y., 1986a, The role of protein kinase C in transmembrane signalling, *Annu. Rev. Cell Biol.* **2**:149–178.

Kikkawa, U., and Nishizuka, Y., 1986b, Protein kinase C, in: *The Enzymes*, Vol. XVII (P. D. Boyer, ed.), pp. 167–189, Academic, New York.

Kikkawa, U., Kitano, T., Saito, N., Kishimoto, A., Tankyama, K., Tanaka, C., and Nishizuka, Y., 1986, Role of protein kinase C in calcium-mediated signal transduction, in: *Calcium and the Cell*, Ciba Foundation Symposium No. 122 (D. Evered and J. Whelan, eds.), pp. 197–211, Wiley, Chichester.

Kikuchi, A., Kozawa, O., Hamamori, Y., Kaibuchi, K., and Takai, Y., 1986, Inhibition of chemotactic peptide-induced phosphoinositide hydrolysis by phorbol esters through the activation of protein kinase C in differentiated human leukemia (HL-60) cells, *Cancer Res.* **46**:3401–3406.

Leeb-Lundberg, L. M. R., Cotecchia, S., Lomasney, J. W., DeBernardis, J. F., Lefkowitz, R. J., and Caron, M. G., 1985, Phorbol esters promote alpha$_1$-adrenergic receptor phosphorylation and receptor uncoupling from inositol phospholipid metabolism, *Proc. Natl. Acad. Sci. USA* **82**: 5651–5655.

Legan, E., Chernow, B., Parrillo, J., and Roth, B. L., 1985, Activation of phosphatidylinositol turnover in rat aorta by alpha$_1$-adrenergic receptor stimulation, *Eur. J. Pharmacol.* **110**:389–390.

Liebman, P. A., 1986, The role of cGMP control in visual receptor transduction, *Neurosci. Res.* **4**(Suppl.):S35–S43.

Liebman, P. A., 1987, Visual receptor transduction, *Ann. NY Acad. Sci.* **494**:65–73.

Litosch, I., and Fain, J. N., 1985, 5-Methyltryptamine stimulates phospholipase C-mediated breakdown of exogenous phosphoinositides by blowfly salivary gland membranes, *J. Biol. Chem.* **260**:16052–16055.

Lucas, D. O., Bajjalieh, S. M., Kowalchyk, J. A., and Martin, T. F. J., 1985, Direct stimulation of thyrotropin-releasing hormone (TRH) of polyphosphoinositide hydrolysis in GH$_3$ cell membranes by a guanine nucleotide-modulated mechanism, *Biochem. Biophys. Res. Commun.* **132**: 721–728.

Majerus, P. W., Connolly, T. M., Deckmyn, H., Ross, T. A., Bross, T. E., Ishii, H., Bansal, V. S., and Wilson, D. B., 1986, The metabolism of phosphoinositide-derived messenger molecules, *Science* **234**:1519–1526.

McMillan, M., Chernow, B., and Roth, B. L., 1986, Phorbol esters inhibit alpha$_1$-adrenergic receptor-stimulated phosphoinositide hydrolysis and contraction in rat aorta: Evidence for a link between vascular contraction and phosphoinositide turnover, *Biochem. Biophys. Res. Commun.* **134**:970–974.

Melin, P-M., Sundler, R., and Jergil, B., 1986, Phospholipase C in rat liver plasma membranes, *FEBS Lett.* **198**:85–88.

Mendoza, S. A., Lopez-Rivas, A., Sinnett-Smith, J. W., and Rozengurt, E., 1986, Phorbol esters and diacylglycerol inhibit vasopressin-induced increases in cytoplasmic-free Ca^{2+} and $^{45}Ca^{2+}$ efflux in Swiss 3T3 cells, *Exp. Cell Res.* **164**:536–545.

Menkes, H., Barban, J. M., and Snyder, S. H., 1986, Protein kinase C regulates smooth muscle tension in guinea-pig trachea and ileum, *Eur. J. Pharmacol.* **122:**19–28.

Merritt, J. E., Taylor, C. W., Rubin, R. P., and Putney, J. W., Jr., 1986, Isomers of inositol trisphosphate in exocrine pancreas, *Biochem. J.* **238:**825–829.

Michell, R. H., 1985, Receptor-controlled phosphatidylinositol 4,5-bisphosphate hydrolysis in the control of rapid receptor-mediated cellular responses and of cell proliferation, in: *Mechanisms of Receptor Regulation* (G. Poste and S. T. Crooke, eds.), pp. 75–94, Plenum, New York.

Mizuguchi, J., Beaven, M. A., Li, J. H., and Paul, W. E., 1986, Phorbol myristate acetate inhibits anti-IgM-mediated signaling in resting B cells, *Proc. Natl. Acad. Sci. USA* **83:**4474–4478.

Molina y Vedia, L., and Lapetina, E. G., 1986, Phorbol 12,13-dibutyrate and 1-oleyl-2-acetyldiacylglycerol stimulate inositol trisphosphate dephosphorylation in human platelets, *J. Biol. Chem.* **261:**10493–10495.

Montague, W., Morgan, N. G., Rumford, G. M., and Prince, C. A., 1985, Effect of glucose on polyphosphoinositide metabolism in isolated rat islets of Langerhans, *Biochem. J.* **227:**483–489.

Nabika, T., Velletri, P. A., Lovenberg, W., and Beavens, M. A., 1985, Increase in cytosolic calcium and phosphoinositide metabolism induced by angiotensin II and [Arg]vasopressin in vascular smooth muscle cells, *J. Biol. Chem.* **260:**4661–4670.

Naccache, P. H., and Sha'afi, R. I., 1986, Neutrophil activation, polyphosphoinositide hydrolysis, and the guanine nucleotide regulatory proteins, in: *New Insights into Cell and Membrane Transport Processes* (G. Poste and S. T. Crooke, eds.), pp. 175–198, Plenum, New York.

Nadel, J., Coburn, R., Murphy, R., Szurszewski, J., and Gail, D., 1985, Workshop on Airway Smooth Muscle. Summary of a Conference held September 25–27, 1983, *Am. Rev. Respir. Dis.* **131:**159–162.

Nahorski, S. R., and Batty, I., 1986, Inositol tetrakisphosphate: Recent developments in phosphoinositide metabolism and receptor function, *Trends Pharmacol. Sci.* **7:**83–85.

Nakaki, T., Roth, B. L., Chuang, D.-M., and Costa, E., 1985, Phasic and tonic components in 5-HT$_2$ receptor-mediated rat aorta contraction: Participation of Ca^{2+} channels and protein kinase C, *J. Pharmacol. Exp. Ther.* **234:**442–446.

Nakanishi, H., Nomura, H., Kikkawa, U., Kishioto, A., and Nishizuka, Y., 1985, Rat brain and liver soluble phospholipase C: Resolution of two forms with different requirements for calcium, *Biochem. Biophys. Res. Commun.* **132:**582–590.

Nakashima, S., Tohmatsu, T., Hattori, H., Okano, Y., and Nozawa, Y., 1986, Inhibitory action of cyclic GMP on secretion, polyphosphoinositide hydrolysis and calcium mobilization in thrombin-stimulated human platelets, *Biochem. Biophys. Res. Commun.* **135:**1099–1104.

Nanberg, E., and Putney, J., Jr., 1986, Alpha$_1$-adrenergic activation of brown adipocytes leads to an increased formation of inositol polyphosphates, *FEBS Lett.* **195:**319–322.

Nomura, H., Ase, K., Sekiguchi, K., Kikkawa, U., and Nishizuka, Y., 1986, Stereospecificity of diacylglycerol for stimulus–response coupling in platelets, *Biochem. Biophys. Res. Commun.* **140:**1143–1151.

Orellana, S. A., Solski, P. A., and Brown, J. H., 1985, Phorbol ester inhibits phosphoinositide hydrolysis and calcium mobilization in cultured astrocytoma cells, *J. Biol. Chem.* **260:**5236–5239.

O'Rourke, F. A., Helenda, S. P., Zavoico, G. B., and Feinstein, M. B., 1985, Inositol 1,4,5-trisphosphate releases Ca^{2+} for a Ca^{2+}-transporting membrane vesicle fraction derived from human platelets, *J. Biol. Chem.* **260:**956–962.

Palmer, S., Hawkins, P. T., Michell, R. H., and Kirk, C. J., 1986, The labelling of polyphosphoinositides with [^{32}P]Pi and the accumulation of inositol phosphates in vasopressin-stimulated hepatocytes, *Biochem. J.* **238:**491–499.

Park, S., and Rasmussen, H., 1985, Activation of tracheal smooth muscle contraction: synergism between Ca^{2+} and activators of protein kinase C, *Proc. Natl. Acad. Sci. USA* **82:**8835–8839.

Parries, G. S., and Hokin-Neaverson, M., 1984, Phosphatidylinositol synthase from canine pancreas: Solubilization by *n*-octyl glucopyranoside and stabilization by manganese, *Biochemistry* **23**:4785–4791.

Poll, C., and Westwick, J., 1986, Phorbol esters modulate thrombin-operated calcium mobilization and dense granule release in human platelets, *Biochim. Biophys. Acta* **886**:434–440.

Popescu, L. M., Hinescu, M. E., Musat, S., Ionescu, M., and Pistritzu, F., 1986, Inositol trisphosphate and the contraction of vascular smooth muscle cells, *Eur. J. Pharmacol.* **123**:167–169.

Portilla, D., and Morrison, A. R., 1986, Bradykinin-induced changes in inositol trisphosphate mass in MDCK cells, *Biochem. Biophys. Res. Commun.* **140**:644–649.

Poste, G., and Crooke, S. T. (eds.), 1985, *Mechanisms of Receptor Regulation,* Plenum, New York.

Poste, G., and Crooke, S. T. (eds.), 1986, *New Insights into Cell and Membrane Transport Processes,* Plenum, New York.

Prentki, M., Janjic, D., Irvine, R. F., Berridge, M. J., and Wollheim, C. B., 1984, Rapid mobilization of Ca^{2+} from rat insulinoma microsomes by inositol-1,4,5-trisphosphate, *Nature (Lond.)* **309**:562–564.

Ramsdell, J. S., and Tashjian, A. H., Jr., 1986, Thyrotropin-releasing hormone (TRH) elevation of inositol trisphosphate and cytosolic free calcium is dependent on receptor number: Evidence for multiple rapid interactions between TRH and its receptor, *J. Biol. Chem.* **261**:5301–5306.

Rasmussen, H., Forder, J., Kojima, I., and Scriabine, A., 1984, TPA-induced contraction of isolated rabbit vascular muscle, *Biochem. Biophys. Res. Commun.* **122**:776–784.

Rasmussen, H., Kojima, I., and Barrett, P., 1986, Information flow in the calcium messenger system, in: *New Insights into Cell and Membrane Transport Processes* (G. Poste and S. T. Crooke, eds.), pp. 145–174, Plenum, New York.

Rebecchi, M. J., Kolesnick, R. N., and Gershengorn, M. C., 1983, Thyrotropin-releasing hormone stimulates rapid loss of phosphatidylinositol and its conversion to 1,2-diacylglycerol and phosphatidic acid in rat mammotropic pituitary cells: Association with calcium mobilization and prolactin secretion, *J. Biol. Chem.* **258**:227–234.

Rittenhouse, S. E., and Sasson, J. P., 1985, Mass change in myoinositol trisphosphate in human platelets stimulated by thrombin: Inhibitory effects of phorbol ester, *J. Biol. Chem.* **260**:8657–8660.

Rodbell, M., 1985, Signal transduction in biological membranes, in: *Mechanisms of Receptor Regulation* (G. Poste and S. T. Crooke, eds.), pp. 65–74, Plenum, New York.

Rodriguez-Pena, A., Zachary, I., and Rozengurt, E., 1986, Rapid dephosphorylation of a 80000 protein, a specific substrate of protein kinase C upon removal of phorbol esters, bombesin and vasopressin, *Biochem. Biophys. Res. Commun.* **140**:379–285.

Rosier, M. F., Dentand, A., Lew, P. D., Capponi, A. M., Vallotton, M. B., 1986, Interconversion of inositol (1,4,5)-trisphosphate to inositol (1,3,4,5)-tetrakisphosphate and (1,3,4)-trisphosphate in permeabilized adrenal glomerulosa cells is calcium-sensitive and ATP-dependent, *Biochem. Biophys. Res. Commun.* **139**:259–265.

Roth, B. L., Nakake, T., Chaung, D-M., and Costa, E., 1986, 5-Hydroxytryptamine$_2$ receptors coupled to phospholipase C in rat aorta: Modulation of phosphoinositide turnover by phorbol ester, *J. Pharmacol. Exp. Ther.* **238**:480–485.

Russell, J. A., 1986, Tracheal smooth muscle, *Clin. Chest Med.* **7**:189–200.

Sadler, K., Litosch, I., and Fain, J. N., 1984, Phosphoinositide synthesis and Ca^{2+} gating in blowfly salivary glands exposed to 5-hydroxytryptamine, *Biochem. J.* **222**:327–334.

Salmon, D. M., and Honeyman, T. W., 1979, Increased phosphatidate accumulation during single contractions of isolated smooth-muscle cells, *Biochem. Soc. Trans.* **7**:986–988.

Salmon, D. M., and Honeyman, T. W., 1980, Proposed mechanism of cholinergic action in smooth muscle, *Nature (Lond.)* **284**:344–345.

Sasaguri, T., Hirata, M., and Kuriyama, H., 1985, Dependence on Ca^{2+} of the activities of phosphatidylinositol 4,5-bisphosphate phosphodiesterase and inositol 1,4,5-trisphosphatase in smooth muscles in the procine coronary artery, *Biochem. J.* **231**:497–503.

Sasakawa, N., Ishii, K., Yamamoto, S., and Kato, R., 1986, Differential effects on protein kinase C activators on carbamylcholine- and high K$^+$-induced rises in intracellular free calcium concentration in cultured adrenal chromaffin cells, *Biochem. Biophys. Res. Commun.* **139**:903–909.

Sekar, M. C., and Hokin, L. E., 1986, The role of phosphoinositides in signal transduction, *J. Membr. Biol.* **89**:193–210.

Sekar, M. C., and Roufogalis, M. D., 1984, Muscarinic-receptor stimulation enhances polyphosphoinositide breakdown in guinea-pig ileum smooth muscle, *Biochem. J.* **223**:527–531.

Seyfred, M. A., and Wells, W. W., 1984, Subcellular site and mechanism of vasopressin-stimulated hydrolysis of phosphoinositides in rat hepatocytes, *J. Biol. Chem.* **259**:7666–7672.

Sha'afi, R. I., Molski, T. F. P., Huang, C-H., and Naccache, P. H., 1986, The inhibition of neutrophil responsiveness caused by phorbol esters is blocked by the protein kinase C inhibitor H7, *Biochem. Biophys. Res. Commun.* **137**:50–60.

Smith, C. D., Lane, B. C., Kusaka, I., Verghese, M. W., and Snyderman, R., 1985, Chemoattractant receptor-induced hydrolysis of phosphatidylinositol 4,5-bisphosphate in human polymorphonuclear leukocyte membranes: Requirement for guanine nucleotide regulatory protein, *J. Biol. Chem.* **260**:5875–5878.

Smith, C. D., Cox, C. C., and Snyderman, R., 1986, Receptor-coupled activation of polyphosphoinositide-specific phospholipase C by an N protein, *Science* **232**:97–100.

Smith, J. B., Smith, L. S., Brown, E. R., Barnes, D., Sabir, M. A., Davis, J. S., and Farese, R. V., 1984, Angiotensin II rapidly increases phosphatidate–phosphoinositide synthesis and phosphoinositide hydrolysis and mobilizes intracellular calcium in cultured arterial muscle cells, *Proc. Natl. Acad. Sci. USA* **81**:7812–7816.

Smith, J. B., Smith, L., and Higgins, B. L., 1985, Temperature and nucleotide dependence of calcium release by myo-inositol 1,4,5-trisphosphate in cultured vascular smooth muscle cells, *J. Biol. Chem.* **260**:14413–14416.

Somlyo, A. V., Bond, M., Somlyo, A. P., and Scarpa, A., 1985, Inositol trisphosphate-induced calcium release and contraction in vascular smooth muscle, *Proc. Natl. Acad. Sci. USA* **82**:5231–5235.

Strand, F. L. (ed.), 1987, *Third Colloquium in Biological Sciences: Cellular Signal Transduction,* Vol. 494, New York Academy of Sciences, New York.

Streb, H., Irvine, R. F., Berridge, M. J., and Schulz, I., 1983, Release of Ca^{2+} from a nonmitochondrial intracellular store in pancreatic acinar cells by inositol 1,4,5-trisphosphate, *Nature (Lond.)* **306**:67–69.

Stephens, N. L. (ed.), 1977, *Biochemistry of Smooth Muscle,* University Park Press, Baltimore.

Strulovici, B., Stadel, J. M., and Lefkowitz, R. J., 1985, Adenylate-cyclase-coupled beta-adrenergic receptors: Biochemical mechanisms of desensitization, in: *Mechanisms of Receptor Regulation* (G. Poste and S. T. Crooke, eds.), pp. 279–294, Plenum, New York.

Stryer, L., and Bourne, H. R., 1986, G proteins: A family of signal transducers, *Annu. Rev. Cell Biol.* **2**:391–419.

Suematsu, E., Hirata, M., Hashimoto, T., and Kuriyama, H., 1984, Inositol 1,4,5-trisphosphate releases Ca^{2+} from intracellular store sites in skinned single cells of porcine coronary artery, *Biochem. Biophys. Res. Commun.* **120**:481–485.

Suematsu, E., Hirata, M., Sasguri, T., Hashimoto, T., and Kuriyama, H., 1985a, Roles of Ca^{2+} on the Inositol 1,4,5-trisphosphate-induced release of Ca^{2+} from saponin-permeabilized single cells of the porcine coronary artery, *Comp. Biochem. Physiol.* **82A**:645–649.

Suematsu, E., Nishimura, J., Hirata, M., Inamitsu, T., and Ibayashi, H., 1985b, Inositol 1,4,5-trisphosphate and intracellular Ca^{2+} store sites in human periperal lymphocytes, *Biomed. Res.* **6**:279–286.

Takenawa, T., 1982, Inositol phospholipids in stimulated smooth muscles, *Cell Calcium* **3**:359–368.

Takuwa, Y., Takuwa, N., and Rasmussen, H., 1986, Carbachol induces a rapid and sustained hydrolysis of polyphosphoinositide in bovine tracheal smooth muscle: Measurements of the mass of polyphosphoinositides, 1,2-diacylglycerol, and phosphatidic acid, *J. Biol. Chem.* **261:** 14670–14675.

Thomas, A. P., Alexander, J., and Williamson, J. R., 1984, Relationship between inositol polyphosphoinositide production and the increase of cytosolic free Ca^{2+} induced by vasopressin in isolated hepatocytes, *J. Biol. Chem.* **259:**5574–5584.

Tohmatsu, T., Hattori, H., Nagao, S., Ohki, K., and Nozawa, Y., 1986, Reversal by protein kinase C inhibitor of suppressive actions of phorbol 12-myristate 13-acetate on polyphosphoinositide metabolism and cytosolic Ca^{2+} mobilization in thrombin-stimulated human platelets, *Biochem. Biophys. Res. Commun.* **134:**868–875.

Troyer, D. A., Schwartz, D. W., Kreisberg, J. I., and Venkatachalam, M. A., 1986, Inositol phospholipid metabolism in the kidney, *Annu. Rev. Physiol.* **48:**51–71.

Ueda, T., Chueh, S-H., Noel, M. W., and Gill, D. L., 1986, Influence of inositol 1,4,5-trisphosphate and guanine nucleotides on intracellular calcium release within the N1E-115 neuronal cell line, *J. Biol. Chem.* **261:**3184–3192.

Uhing, R. J., Prpic, V., Jiang, H., and Exton, J. H., 1986, Hormone-stimulated polyphosphoinositide breakdown in rat liver plasma membranes: Role of guanine nucleotides and calcium, *J. Biol. Chem.* **261:**2140–2146.

Velasco, G., Shears, S. B., Michell, R. H., and Lazo, P. S., 1986, Calcium uptake by intracellular compartments in permeabilised enterocytes: Effect of inositol 1,4,5-trisphosphate, *Biochem. Biophys. Res. Commun.* **139:**612–618.

Vicentini, L. M., Di Virgilio, F., Ambrosini, A., Pozzan, T., and Meldolesi, J., 1985, Tumor promoter phorbol 12-myristate, 13-acetate inhibits phosphoinositide hydrolysis and cytosolic Ca^{2+} rise induced by the activation of muscarinic receptors in PC12 cells, *Biochem. Biophys. Res. Commun.* **127:**310–317.

Vickers, J. D., and Mustard, J. F., 1986, The phosphoinositides exist in multiple pools in rabbit platelets, *Biochem. J.* **238:**411–417.

Villalobos-Molina, R., Hong, E., and Garcia-Sainz, J. A., 1982, Correlation between phosphatidylinositol labeling and contraction in rabbit aorta: Effect of alpha-1 adrenergic activation, *J. Pharmacol. Exp. Ther.* **222:**258–261.

Wagner, B., Fugner, M-L., Schachtele, C., Marme, D., and Osswald, H., 1986, Phorbolester-induced contractions of vascular smooth muscles, *Pflugers Arch.* **406:**R41.

Watson, S. P., and Lapetina, E. G., 1985, 1,2-Diacylglycerol and phorbol ester inhibit agonist-induced formation of inositol phosphates in human platelets: Possible implications for negative feedback regulation of inositol phospholipid hydrolysis, *Proc. Natl. Acad. Sci. USA* **82:**2623–2626.

Williamson, J. R., 1986, Role of inositol lipid breakdown in the generation of intracellular signals: State of the art lecture, *Hypertension* **8** [Suppl II]:II-140–II-156.

Williamson, J. R., Joseph, S. K., Coll, K. E., Thomas, A. P., Verhoeven, A., and Prentki, M., 1986, Hormone-induced inositol lipid breakdown and calcium-mediated cellular responses in liver, in: *New Insights into Cell and Membrane Transport Processes* (G. Poste and S. T. Crooke, eds.), pp. 217–248, Plenum, New York.

Williamson, J. R., Hansen, C. A., Johanson, R. A., Cooll, K. E., and Williamson, M., 1988, Formation and metabolism of inositol phosphates: The inositol tris/tetrakisphosphate pathway, *Adv. Expt. Med. Biol.* **232:**183–195.

Wilson, D. B., Bross, T. E., Hofmann, S. L., and Majerus, P. W., 1984, Hydrolysis of polyphosphoinositides by purified sheep seminal vesicle phospholipase C enzymes, *J. Biol. Chem.* **259:**11718–11724.

Wilson, D. B., Bross, T. E., Sherman, W. R., Berger, R. A., and Majerus, P. W., 1985a, Inositol cyclic phosphates are produced by cleavage of phosphatidylphosphoinositols (polyphosphoinositides) with purified sheep seminal vesicle phospholipase C enzymes, *Proc. Natl. Acad. Sci. USA* **82:**4013–4017.

Wilson, D. B., Connolly, T. M., Bross, T. E., Majerus, P. W., Sherman, W. R., Tyler, A. N., Rubin, L. J., and Brown, J. E., 1985b, Isolation and characterization of the inositol cyclic phosphate products of polyphosphoinositide cleavage by phospholipase C: Physiological effects in permeabilized platelets and *Limulus* photoreceptor cells, *J. Biol. Chem.* **260**:13496–13501.

Wilson, D. B., Neufeld, E. J., and Majerus, P. W., 1985c, Phosphoinositide interconversion in thrombin-stimulated human platelets, *J. Biol. Chem.* **260**:1046–1051.

Wolf, M., LeVine, H. III, May, W. S., Jr., Cuatrecasas, P., and Sahyoun, N., 1985, A model for intracellular translocation of protein kinase C involving synergism between Ca^{2+} and phorbol esters, *Nature (Lond.)* **317**:546–549.

Zavoico, G. B., Helenda, S. P., Sha'afi, R. I., and Feinstein, M. B., 1985, Phorbol myristate acetate inhibits thrombin-stimulated Ca^{2+} mobilization and phosphatidylinositol 4,5-bisphosphate hydrolysis in human platelets, *Proc. Natl. Acad. Sci. USA* **82**:3859–3862.

Chapter 8

Electrical Properties of Airway Smooth Muscle

Tadao Tomita

Department of Physiology
School of Medicine
Nagoya University
Nagoya 466, Japan

I. INTRODUCTION

Smooth muscle can be classified according to electrical properties, which are closely related to contractile behavior. One group is highly excitable, having a spikelike action potential and phasic contraction. This includes the spontaneously active muscles (e.g., intestinal and uterine muscles) and quiescent muscles (e.g., vas deferens). The second group is moderately excitable, responding in gradations to electrical stimulation. Smooth muscle in this group may exhibit oscillatory electrical activities. The third group is normally electrically inexcitable. Electrophysiological experiments on airway muscle have been carried out thus far only on the tracheal muscle. These results show that tracheal muscle belongs to either the second or third group, depending on the animal species, indicating a relatively low excitability. It is very likely, however, that the properties of airway smooth muscle differ according to the location of the muscle in the respiratory tract.

The Ca permeability of the plasma membrane is one of the most important steps in modifying the intracellular Ca concentration, thereby controlling the contraction of smooth muscle; this is generally thought to be regulated by two different systems: voltage-operated Ca channel (VOC) and receptor-operated Ca channel (ROC) present in the membrane (Bolton, 1979; Van Breemen et al., 1979). The VOC is responsible for Ca influx during the action potential or the membrane depolarization caused by excess K, while the ROC is involved in the agonist-induced contraction independent of membrane depolarization. Since the open state of the VOC is strongly influenced by membrane potential, it is important to understand the mechanism determining the membrane potential. However, an understanding of not only the underlying mechanism of determining the membrane potential but also ionic mechanism of membrane excitation is still rather limited in smooth muscle, particularly in airway muscle. Although a low sensitivity of agonist-induced contraction to Ca entry blockers, such as verapamil, compared with K-induced contraction suggests that the ROC is a different entity from the VOC, direct electrophysiological evidence has not yet been obtained.

II. TECHNICAL PROBLEMS

Electrical properties of smooth muscle can be studied mainly using three different techniques: the sucrose-gap, intracellular microelectrode, and patch-clamp methods. Intracellular potential (transmembrane, or resting, potential) is usually measured with intracellular microelectrodes. However, there are several drawbacks to this method. A stable prolonged recording is often difficult, owing to small fiber diameter and the movement of preparations. Insertion of an electrode into a cell may produce electrical leakage in the plasma membrane around the electrode, resulting in alteration of transmembrane potential. This factor is expected to be larger in smaller cells, such as smooth muscle (2–4 μm in diameter). Therefore, it is important to use electrodes of a very fine tip. By contrast, as the tip is made finer, the tip potential tends to increase more (Adrian, 1956). This introduces faulty estimation of the absolute value of resting potential (fortunately, not of changes of membrane potential). Therefore, in smooth muscles, reported values of transmembrane potential should be interpreted with some reservation.

Another way of measuring transmembrane potential is to use so-called whole-cell recording, using the patch electrode (cf. Chapter 9, this volume). However, there are also some problems with this method, such as imperfect shielding between the electrode and the plasma membrane as well as possible alteration of intracellular milieu.

In order to analyze the electrical properties, it is essential to measure the membrane resistance (or conductance). Since muscle fibers are electrically interconnected in many smooth muscles, the spatial spread of current applied through intracellular electrode is complicated and cannot be used for analysis of the membrane parameters (Tomita, 1975). Extracellular polarization of the tissue in a longitudinal direction of fibers using an insulated partition method is suitable for this purpose to some extent (Abe and Tomita, 1968), as applied for the bovine and canine tracheal muscles (Kirkpatrick, 1975; Suzuki et al., 1976).

The sucrose-gap method can be used for simultaneous recording of electrical and mechanical activities of smooth muscles and is often used for their pharmacological studies. However, since a reasonable length (~5 mm) is necessary for a conventional sucrose-gap recording, this method has been applied only to the airway muscle of large animals, such as bovine (Kirkpatrick, 1975) and canine tracheal muscle (Coburn and Yamaguchi, 1977). The applicability of sucrose-gap method is interpreted to indicate that these muscles have sufficient cell–cell electrical coupling. However, at the moment, the structural basis (e.g., the gap junction) for electrical coupling has not, at least quantitatively, been established.

The membrane potential can be displaced by passing current through the membrane using the double sucrose-gap method in order to measure the membrane resistance and active electrical response. However, this approach has also inherent problems, such as electrical shunting by sucrose solution, spatial inhomogeneity of current spread within the tissue, changes in the internal tissue resistance with time of exposure to sucrose solution, and the possible effects of sucrose solution on the plasma membrane (Coburn et al., 1975). Therefore, this method is suitable for qualitative, rather than quantitative, analysis of drug action.

Recently, patch-clamp experiments are becoming a main stream of electrophysiology of smooth muscle. In isolated single cells, the transmembrane current flow can be precisely measured with the patch-clamp technique. (This topic is separately dealt with in Chapter 9, this volume).

III. MEMBRANE POTENTIAL AND SPONTANEOUS ELECTRICAL ACTIVITY

There is general agreement that in the smooth muscle the main factor determining the membrane potential is a concentration gradient of K ion across the membrane. Although the membrane potential differs in different smooth muscles, the reason for the difference has not been properly explained in terms of either a relative ionic permeability of other ions or of an electrogenic contribution of ionic pump, or both.

Some smooth muscle is quiescent, while other types show various degrees of spontaneous electrical activity. The pattern of the activity is not apparently correlated with the resting membrane potential. The spontaneous activity is believed to be determined not only by the property of the Ca channel, but by that of K-channel as well, because the excitation of the membrane is determined by the net inward current through the membrane, i.e., the difference between the inward current (mainly Ca current in smooth muscle) and the outward (mainly K) current. Therefore, it is possible that the effect of many drugs is exerted through modification of the K permeability of the plasma membrane.

Canine tracheal muscle, removed from the muscosal layer, has a stable membrane potential of -54 to -60 mV (Farley and Miles, 1977; Kannan et al., 1983; Kroeger and Stephens, 1975; Stephens and Kroeger, 1970; Suzuki et al.,

1976). The resting potential was the same (about −60 mV) in upper, middle, and lower portions of trachea, but this was slightly (∼3 mV) lower than that in the bronchial muscle (Souhrada et al., 1983). In the bronchiole, a much higher membrane potential (about −70 mV) was reported (Inoue and Ito, 1986). However, the physiological significance of the regional difference is unclear.

The resting potential recorded from bovine tracheal muscle was essentially the same as that of canine muscle. The average membrane potential was reported to be −47.6 mV (Kirkpatrick, 1975) and −60 mV (Souhrada et al., 1981). However, ∼50% of the preparations showed the spontaneous slow waves during the first 3 hr after mounting in the organ bath (Fettes et al., 1981; Kirkpatrick, 1981). In other experiments, 36.6% of impalements in 112 fibers showed regular slow waves and 26.8% irregular oscillation of membrane potential (Souhrada et al., 1981).

The isolated guinea pig tracheal muscle has muscle tone and produces slow rhythmic electrical activity (slow waves), without regenerative spikes, in normal solution. The presence of spontaneous electrical activity was first noted by Clark and Small (1979). This is easily detected with extracellular recordings; and the properties of electrical activity have been analyzed with extracellular as well as intracellular electrodes in later studies by Small and colleagues (Ahmed et al., 1984, 1985; Allen et al., 1985a,b; Foster et al., 1983a,b, 1984; Small, 1982) and by Souhrada and colleagues (McCaig and Souhrada, 1980; Souhrada and Souhrada, 1981a,b; Souhrada et al., 1981). Impalement of the microelectrode is usually made from the mucosal surface, after removal of the mucosal layer. The pattern of the slow wave varies in different preparations, probably related to differences in stretch or degrees of muscle tone development. The slow waves usually appear continuously, but in some preparations they stop periodically or their amplitude is modulated. According to McCaig and Souhrada (1980) and Souhrada et al. (1981b), 27–33.6% of impalements showed regular activity, 26.7–44% showed irregular activity, and the remaining showed no activity. No correlation was found between the pattern of activity and the membrane potential. By contrast, a close correlation was found between the slow-wave amplitude and the membrane potential (Honda et al., 1986). The mean amplitude and frequency of typical slow waves are reported to be 11–13 mV and 44–65/min, showing reasonable agreement in separate experiments (Ahmed et al., 1984, 1985; Honda et al., 1986; McCaig, 1986; McCaig and Souhrada, 1980; Souhrada et al., 1981b). Since muscle tone can be maintained even in a preparation lacking electrical activity, it is likely that the individual slow wave has apparently disappeared because of either high frequency or desynchronization in such a preparation.

Spontaneously generated muscle tone is eliminated by indomethacin or aspirin in guinea pig tracheal muscle, suggesting involvement of prostaglandins (Brink et al., 1981; Farmer et al., 1974; Ito et al., 1985; Mansour and Daniel, 1986; Ono et al., 1977; Orehek et al., 1975; Saad and Burka, 1983). The slow waves and tension development, which appeared periodically in this preparation, were slowly abolished by indomethacin (Fig. 1). The electrical and mechanical activity reappeared by prostaglandin E_2 (PGE_2) in the presence of indomethacin. Therefore, it is an attractive idea that spontaneous electrical and mechanical activities are caused by endogenous production of arachidonic products. It has also been considered in guinea pig intestinal smooth muscles that prostaglandin may play an important role in the generation and/or modification of spontaneous activity or muscle tone (Coburn, 1980; Ferreira *et al.*, 1976; Frankhuijzen and Bonta, 1975; Yamaguchi *et al.*, 1976).

Human tracheal muscle strips obtained from autopsies showed electrical and mechanical activity similar to that in guinea pig tracheal muscle (Honda and Tomita, 1987). The membrane potential was slightly lower (about -45 mV) than that in the guinea pig. The mean amplitude and frequency of slow waves were 7.7 mV and 19.7 waves/min when observed in nine preparations obtained from five autopsies (51–73 years old). In contrast to guinea pig, the electrical and mechanical activity of the human trachea was not inhibited but was slightly potentiated by indomethacin. This result agrees with previous reports showing that the active tension in human tracheal and bronchial muscle is not blocked by indomethacin (Brink *et al.*, 1980; Davis *et al.*, 1982; Hutás *et al.*, 1981; Ito *et al.*, 1985). Since FPL 55712, a leukotrine antagonist, inhibits muscle tone, it is likely in the human trachea that endogenous leukotrine, rather than prostaglandin, is involved in spontaneously generated contraction (Ito *et al.*, 1985). This finding is consistent with observations on human tracheal and bronchial muscles that leukotrienes C_4 and D_4 are 500 times more potent than prostaglandin $F_{2\alpha}$ ($PGF_{2\alpha}$) and that the leukotriene-induced contraction is antagonized by FPL 55712 (Dahlén *et al.*, 1980; Jones *et al.*, 1982).

In dogs treated with indomethacin (1 mg/kg) sc daily for 5 days, 6 of 12 dogs developed cough and wheeze. The tracheal muscle obtained from the coughing dog had significantly low membrane potential (-52.4 ± 1.8 versus -59.0 ± 2.1 mV in control dog trachea, $N = 35$ for both) and showed spontaneous rhythmic electrical and mechanical activity. The electrical activity consisted of slow oscillation of membrane potential, with an amplitude of 1–5 mV (3.8 on average) and a frequency of 6–18/min. This activity was abolished by atropine (1 μM). Therefore, it is considered that indomethacin removed the inhibitory effect of prostaglandins on acetylcholine (ACh) release from the nerve terminal and that an increase in the spontaneous release of ACh may be responsible for the rhythmic activity (Ito and Tajima, 1981).

The addition of ouabain (10 μM) or removal of external K caused the membrane to depolarize both in guinea pig and bovine tracheal muscles (Souhrada *et al.*, 1981). This has been interpreted to be the result of inhibition of an electrogenic Na pump. However, it may also be partly attributable to an increase in prostaglandins, as demonstrated for guinea pig taenia coli (Coburn and Soltoff, 1977).

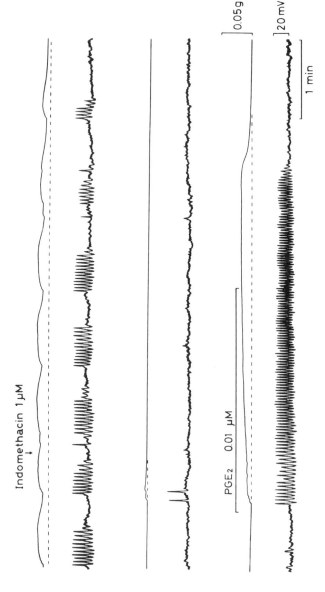

FIGURE 1. Effects of indomethacin (1 μM) and prostaglandin E₂ (PGE₂, 0.01 μM) on mechanical (upper) and electrical activity (lower trace) of guinea pig tracheal muscle. Continuous recording from the same cell with an intracellular microelectrode.

As shown in guinea pig taenia (Casteels *et al.*, 1971; Tomita and Yamamoto, 1971), the membrane of the tracheal muscles was transiently hyperpolarized because of activation of the electrogenic Na pump by readmitting K following exposure to K-free solution (Souhrada *et al.*, 1981). However, the degree of contribution by the electrogenic pump to the steady state of membrane potential is still unclear, as discussed for rabbit small intestinal muscle (El-Sharkawy and Daniel, 1975). It is quite possible, however, that the electrogenic Na pump contributes significantly to the membrane potential under some pathological conditions, such as the hyperreactivity displayed in immunological reactions or alteration of the temperature (McCaig and Souhrada, 1980; Souhrada and Souhrada, 1981a,b, 1982, 1985).

IV. EVOKED ACTION POTENTIALS

When the cell membrane of bovine tracheal muscle was depolarized by outward current pulses using the partition method, no action potential of an all-or-none type could be evoked and the membrane resistance was markedly reduced, indicating outward-going rectification (Kirkpatrick, 1975). Thus, the lack of action potential generation protects the tracheal muscle from brisk contraction. The poor excitability can be explained by a significant contribution of outward current because of the increase in K conductance of the membrane by depolarization that antagonizes a depolarizing action of inward current. In the presence of tetraethylammonium (TEA, 30 mM), which is known to reduce the K conductance of the membrane, the membrane was depolarized from -50.9 to -37.6 mM, rectification was reduced, and spontaneous electrical activity was produced accompanied by rhythmic contraction (Kirkpatrick, 1975). The electrical activity was of a slow-wave type, with an amplitude of 16.1 mV ($N = 23$) and a frequency of 20.9/min ($N = 21$) on average.

In canine tracheal muscle, analysis of membrane potential changes with the partition method gives a membrane time constant of 282–449 msec and a space constant of 1.6–3.2 mm (Kannan *et al.*, 1983; Kroeger and Stephens, 1975; Suzuki *et al.*, 1976). As in bovine muscle, the action potential could not be evoked by electrical stimulation; this is likely to be related to the fact that membrane resistance was reduced when the membrane was depolarized more than 5 mV.

Tetraethylammonium (5 mM) depolarized the membrane of canine tracheal muscle by 8 mV and increased the membrane resistance (Suzuki *et al.*, 1976). In the presence of TEA, depolarizing current pulses evoked graded action potentials. TEA (33 mM) depolarized the membrane from -47 to -34 mV, increased membrane resistance, and produced spontaneous oscillation in membrane potential (15–20 mV) at frequency of 15–20/min (Kroeger and Stephens, 1975). In later experiments, it was shown that the membrane potential was decreased from

-55 to -37 mV with 30 mM TEA (Kannan *et al.*, 1983) or from -58 to -44 mV with 20 mM TEA (Richards *et al.*, 1986). The electrical activity, including evoked action potentials, caused by TEA were abolished by 1 μM D-600 (Kannan *et al.*, 1983) and by 0.1 mM 8-bromo-cyclic guanosine monophosphate (cGMP) (Richards *et al.*, 1986).

In guinea pig tracheal muscle, TEA also depolarized the membrane, increased slow wave amplitude, and induced spike-type action potential on the upstroke of slow wave (Ahmed *et al.*, 1985; Allen *et al.*, 1985b; Foster *et al.*, 1984; McCaig and Souhrada, 1980; see also Fig. 3).

V. RESPONSE TO EXCITATORY AGENTS

In canine tracheal muscle, 5-hydroxytryptamine (5-HT, 0.5–1 μg/ml) depolarized the membrane by ~10 mV and produced sustained contraction (Coburn and Yamaguchi, 1977). In the presence of 5-HT, depolarizing current pulses occasionally evoked action potentials. The contraction caused by 5-HT was accompanied by much smaller depolarization compared with the same degree of tension response caused by excess K. Furthermore, the contraction caused by excess K could be abolished by repolarizing the membrane to the original level with applying inward current, whereas the contraction produced by 5-HT was only partially reduced by the same procedure. These results suggest that a potential-independent process, i.e., the pharmacomechanical coupling (Somlyo and Somlyo, 1968) is involved in the 5-HT-induced contraction (Coburn, 1977; Coburn and Yamaguchi, 1977).

On the basis of the similar experiments also on the canine tracheal muscle, contractions induced by ACh (20 μM) are largely attributed to the pharmacomechanical coupling and that, due to this mechanism, ACh contraction is resistant to Ca-channel blockers, such as verapamil, D-600, or La ions, compared with K-contracture (Coburn, 1977, 1979).

The membrane potential decreased with increasing the external K concentration, the slope being ~37 mV per 10-fold change in the concentration. The contraction appeared at the membrane potential less negative than -50 mV (normal resting membrane potential: -60 mV) with 15 mM K, and a good correlation was obtained between tension development and membrane potential. By contrast, the ACh-induced depolarization reached about -40 mM at 1 μM, and no further depolarization was observed by increasing the concentrations to >1 μM, whereas the contraction started at a membrane potential of -55 mV with 0.1 μM ACh, and it increased to a concentration of 1 mM (Farley and Miles, 1977). Thus, this result is consistent with the conclusion, reached with the sucrose-gap method, that pharmacomechanical coupling is involved in the ACh contraction.

Pharmacomechanical coupling can partly explained by intracellular Ca release, likely to be mediated by inositol 1,4,5-trisphosphate, which is produced by hydrolysis of phosphatidylinositol in the plasma membrane following receptor activation with agonists (Berridge, 1986; Hashimoto *et al.*, 1985). However, transmembrane Ca influx is necessary for a sustained contraction. Since the contraction caused by receptor activation is resistant to Ca-channel blocker at concentrations that readily suppress K contracture, the ROC is generally considered responsible for the Ca influx. However, direct evidence for the ROC is still lacking. In guinea pig tracheal muscle, as in the dog, the K contracture is readily blocked by verapamil, while carbachol-induced contraction is only weakly inhibited. When Ca is replaced with Ba, which is known to pass the VOC in the cardiac muscle, carbachol-induced contraction becomes susceptible to verapamil (Baba *et al.*, 1985). On the basis of the pharmacological analysis of verapamil effects on Ca (or Ba) concentration, it was suggested that the VOC is modified by carbachol to various degrees depending on the presence of Ca or Ba and that the affinity of the channel to verapamil changes due to this modification (Baba *et al.*, 1985). In recent studies, we have obtained the results showing that carbachol increases the affinity of the channel to Ca dose dependently but that it does not alter the dissociation constant of verapamil at the channel, when estimated from the shift of Ca concentration–tension curves taking their Hill coefficients into consideration (Gonda *et al.*, 1988). It appears that the channels responsible for K contracture and carbachol-induced contraction are fundamentally the same and that the low sensitivity of carbachol-induced contraction to verapamil is not attributable to involvement of special channels (the ROC) but to alteration of the kinetics of channel gating by carbachol.

During depolarization by ACh, no action potential was generated, when studied with intracellular microelectrodes. ACh produced contraction with a depolarization smaller than that of excess K; when the membrane was depolarized to the same level, the contraction was larger for ACh than for K (Suzuki *et al.*, 1976).

In bovine tracheal muscle, histamine (1 μg/ml) depolarized the membrane from −46.7 to −34.3 mV and produced slow oscillation of membrane potential (10.4 mV in amplitude and 11.8/min in frequency), accompanied by rhythmic contraction (Kirkpatrick, 1975). By contrast, ACh (1 μg/ml) produced only smooth depolarization from −42.8 to −29.9 mV. When the effect of ACh (1 μg/ml) was studied with the sucrose-gap method, the membrane was slowly depolarized, and small oscillatory potentials were superimposed on the sustained depolarization. However, in contrast to electrical oscillatory activity, ACh produced a smooth contraction without oscillation (Cameron and Kirkpatrick, 1977).

In guinea pig tracheal muscle, ACh (1 μM) depolarized the membrane and increased the slow-wave amplitude and frequency; at high concentrations (10–100 μM), it reduced or abolished the slow wave with a sustained large depolarization (Ahmed *et al.*, 1984). Electrical membrane noise was often observed at 100 μM, at a random oscillation of few millivolts at relatively high frequency. The mechanical response to ACh was smooth contraction. Histamine (1–200 μM) produced a similar electrical and mechanical response.

The excitatory responses to TEA and carbachol in the human tracheal muscle were essentially similar to those observed in the guinea pig muscle (Honda and Tomita, 1967). The properties of ionic channels involved in the response to receptor activation should be investigated further, using more precise methods, such as the patch-clamp technique.

VI. RESPONSE TO INHIBITORY AGENTS

In the canine tracheal muscle, isoprenaline at a higher concentration than 0.5 μM hyperpolarized the membrane with a reduction of membrane resistance (Ito and Tajima, 1982). The phasic contraction evoked by membrane depolarization in the presence of TEA (5 mM) was inhibited by isoproterenol (5 μM) without affecting the action potential, suggesting that an intracellular mechanism is mainly responsible for inhibition of the tension development.

In guinea pig tracheal muscle, isoproterenol hyperpolarized the membrane dose dependently, and the hyperpolarization was ~20 mV at 1 μM. Since this muscle has slow waves, the hyperpolarization was accompanied by a decrease in frequency or abolition at high concentrations of the slow wave, and the relaxation was usually correlated well with the electrical changes (Allen et al., 1985b; Honda et al., 1986). A typical example of the isoproterenol effect is shown in Fig. 2.

Tetraethylammonium reduced isoproterenol-induced hyperpolarization (Allen et al., 1985b). However, higher concentrations of isoproterenol still inhibit the electrical activity and cause membrane hyperpolarization (Fig. 3). The link between the slow wave and the maintenance of muscle tone is considered relatively weak, because nifedipine, verapamil, or methoxyverapamil (D-600) at a concentration of 1 μM abolishes the slow wave without reducing the muscle tone significantly (Ahmed et al., 1985; Allen et al., 1985a; Foster et al., 1984; Small, 1982). When the electrical activity and muscle tone were increased by TEA (1–8 mM), nifedipine (1 μM), or verapamil (10 μM) suppressed the electrical activity but decreased the muscle tone only to the level before the TEA application (Ahmed et al., 1985; Foster et al., 1984).

Dissociation between electrical and mechanical activity was also observed in the effects of nicorandil (Allen et al., 1986a) and aminophylline (Allen et al., 1986b). A small relaxation caused by nicorandil (1 μM) appeared without any change in slow waves or membrane potential, although at 10 μM it abolished the slow wave. Similarly, 10 μM aminophylline produced relaxation in the absence of electrical changes. Inhibition of slow-wave and membrane hyperpolarization appeared only at >100 μM.

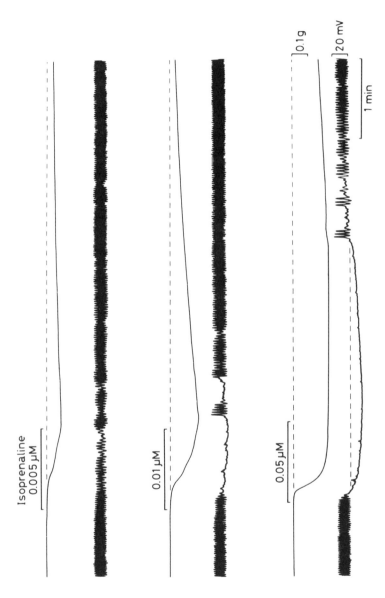

FIGURE 2. Effects of isoproterenol on mechanical (upper trace) and electrical activity (lower trace) of guinea pig tracheal muscle (continuous intracellular recording). Isoproterenol was applied for 1 min at 20-min intervals, by increasing the concentration successively as indicated.

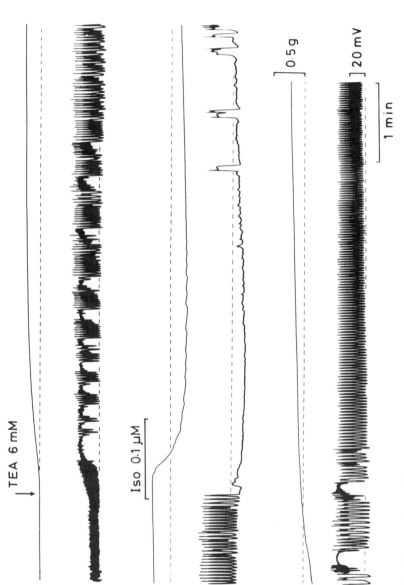

FIGURE 3. Effects of TEA (6 mM) on tension (upper) and intracellular membrane potential and effects of isoproterenol (0.1 μM) in the presence of TEA. Isoproterenol was applied for 1 min, 10 min after TEA.

The inhibitory responses to isoproterenol in the human tracheal muscle were essentially similar to those observed in guinea pig muscle (Honda and Tomita, 1987). There was a reasonable correlation between the electrical and mechanical activity.

VII. SUMMARY

Tracheal smooth muscle may be quiescent or spontaneously active, depending on the animal species and experimental conditions. The isolated canine muscle has little muscle tone and no spontaneous electrical activity. Some isolated bovine muscles show spontaneous mechanical and electrical activity. By contrast, most guinea pig and human tracheal muscles produce high muscle tone and rhythmic electrical activity. Muscle tone and electrical activity appear to be causally related and both are likely to be produced by endogenous arachidonic products, prostaglandins in guinea pig tracheal muscle, and leukotrienes in the human tracheal muscle.

Electrical activity consists of oscillatory fluctuation of membrane potential, i.e., a slow-wave type. When TEA is applied, spontaneous action potentials of spike type are generated even in quiescent muscles, or spike potentials appear on top of the augmented slow wave in spontaneously active muscles. It is likely that the low excitability of airway smooth muscle is related to a relatively high K conductance and that TEA facilitates generation of spontaneous activity and spike activity by reducing the K conductance of the membrane. Although the ionic mechanism underlying the slow-wave or spike activity has not been properly analyzed, an increase in the Ca conductance seems to be mainly responsible for the depolarizing phase. However, the property of Ca channels has to be properly studied using the patch-clamp technique, before we can understand the VOC and ROC in smooth muscle.

There is often reasonable correlation between changes in membrane potential and mechanical response during contraction or relaxation caused by receptor activation. However, poor correlation also exists, particularly in the contraction caused by ACh or carbachol. The potential-independent component of the contraction can be explained by the assumption that the contraction is mediated by pharmacomechanical coupling (Somlyo and Somlyo, 1968), either because Ca influx through the ROC (Bolton, 1979; Van Breemen et al., 1979) or because of the release of Ca from intracellular store, or both. This assumption will be clarified, together with the ionic mechanism of electrical activity. The electrical properties of airway smooth muscle have been analyzed mainly in the tracheal muscle. The studies should be extended to the electrophysiology of smooth muscle in smaller airways.

VIII. REFERENCES

Abe, Y., and Tomita, T., 1968, Cable properties of smooth muscle, *J. Physiol. (Lond.)* **196**:87–100.

Adrian, R. H., 1956, The effect of internal and external potassium concentration on the membrane potential of frog muscle, *J. Physiol. (Lond.)* **133**:631–658.

Ahmed, F., Foster, R. W., Small, R. C., and Weston, A. H., 1984, Some features of the spasmogenic actions of acetylcholine and histamine in guinea-pig isolated trachealis, *Br. J. Pharmacol.* **83**:227–233.

Ahmed, F., Foster, R. W., and Small, R. C., 1985, Some effects of nifedipine in guinea-pig isolated trachealis, *Br. J. Pharmacol.* **84**:861–869.

Allen, S. L., Foster, R. W., Small, R. C., and Towart, R., 1985a, The effects of the dihydropyridine Bay K 8644 in guinea-pig isolated trachealis, *Br. J. Pharmacol.* **86**:171–180.

Allen, S. L., Beech, D. J., Foster, R. W., Morgan, G. P., and Small, R. C., 1985b, Electrophysiological and other aspects of relaxant action of isoprenaline in guinea-pig isolated trachealis, *Br. J. Pharmacol.* **86**:843–854.

Allen, S. L., Foster, R. W., Morgan, G. P., and Small, R. C., 1986a, The relaxant action of nicorandil in guinea-pig isolated trachealis, *Br. J. Pharmacol.* **87**:117–127.

Allen, S. L., Cortijo, J., Foster, R. W., Morgan, G. P., Small, R. C., and Weston, A. H., 1986b, Mechanical and electrical aspects of the relaxant action of aminophylline in guinea-pig isolated trachealis, *Br. J. Pharmacol.* **88**:473–483.

Baba, K., Wawanishi, M., Satake, T., and Tomita, T., 1985, Effects of verapamil on the contractions of guinea-pig tracheal muscle induced by Ca, Sr, and Ba, *Br. J. Pharmacol.* **84**:203–211.

Baba, K., Satake, T., Takagi, K., and Tomita, T., 1986, Effects of verapamil on the response of the guinea-pig tracheal muscle to carbachol, *Br. J. Pharmacol.* **88**:441–449.

Berridge, M. J., 1986, Regulation of ion channels by inositol trisphosphate and diacylglycerol, *J. Exp. Biol.* **124**:323–335.

Bolton, T. B., 1979, Mechanisms of action of transmitters and other substances on smooth muscle, *Physiol. Rev.* **59**:606–718.

Brink, C., Grimaud, C., Guillot, C., and Orehek, J., 1980, The interaction between indomethacin and contractile agents on human isolated airway muscle, *Br. J. Pharmacol.* **69**:383–388.

Brink, C., Duncan, P. G., and Douglas, J. S., 1981, Histamine, endogenous prostaglandins and cyclic nucleotides in the regulation of airway muscle responses in the guinea pig, *Prostaglandins* **22**:729–738.

Cameron, A. R., and Kirkpatrick, C. T., 1977, A study of excitatory neuromuscular transmission in the bovine trachea, *J. Physiol. (Lond.)* **270**:733–745.

Casteels, R., Droogmans, G., and Hendrickx, H., 1971, Electrogenic sodium pump in smooth muscle cells of the guinea-pig's taenia coli, *J. Physiol. (Lond.)* **217**:297–313.

Clark, L. A., and Small, R. C., 1979, Simultaneous recording of electrical and mechanical activity from smooth muscle of the guinea pig isolated trachea, *J. Physiol. (Lond.)* **300**:5P.

Coburn, R. F., 1977, The airway smooth muscle cell, *Fed. Proc.* **36**:2692–2697.

Coburn, R. F., 1979, Electromechanical coupling in canine trachealis muscle: Acetylcholine contractions, *Am. J. Physiol.* **236**:C177–C184.

Coburn, R. F., 1980, Evidence for oscillation of prostaglandin release from spontaneously contracting guinea pig taenia coli, *Am. J. Physiol.* **239**:G53–G58.

Coburn, R. F., and Soltoff, S., 1977, Na^+-K^+-ATPase inhibition stimulates PGE release in guinea pig taenia coli, *Am. J. Physiol.* **232**:C191–C195.

Coburn, R. F., and Yamaguchi, T., 1977, Membrane potential-dependent and -independent tension in the canine tracheal muscle, *J. Pharmacol. Exp. Ther.* **201**:276–284.

Coburn, R. F., Ohba, M., and Tomita, T., 1975, Recording of intracellular electrical activity with the sucrose-gap method, in: *Methods in Pharmacology*, Vol. 3 (E. E. Daniel and D. M. Paton, eds.), pp. 231–245, Plenum, New York.

Dahlén, S. E., Hedqvist, P., Hammarström, S., and Samuelsson, B., 1980, Leukotrienes are potent constrictors of human bronchi, *Nature (Lond.)* **288**:484–486.

Davis, C., Kannan, M. S., Jones, T. R., and Daniel, E. E., 1982, Control of human airway smooth muscle: In vitro studies, *J. Appl. Physiol.* **53**:1080–1087.

El-Sharkawy, T.Y., and Daniel, E. E., 1975, Electrogenic sodium pumping in rabbit small intestinal smooth muscle, *Am. J. Physiol.* **229**:1277–1286.

Farley, J. M., and Miles, P. R., 1977, Role of depolarization in acetylcholine-induced contractions of dog trachealis muscle, *J. Pharmacol. Exp. Ther.* **201**:199–205.

Farmer, J. B., Farrar, D. G., and Wilson, J., 1974, Antagonism of tone and prostaglandin mediated responses in a tracheal preparation by indomethacin and SC-19220, *Br. J. Pharmacol.* **52**:559–565.

Ferreira, S. H., Herman, A. G., and Vane, J. R., 1976, Prostaglandin production by rabbit isolated jejunum and its relationship to the inherent tone of the preparation, *Br. J. Pharmacol.* **56**:469–477.

Fettes, J., Kirkpatrick, C. T., Morrow, R. J., and Tomita, T., 1981, Spontaneous and induced slow wave activity in bovine tracheal smooth muscle, *J. Physiol. (Lond.)* **312**:54–55P.

Foster, R. W., Small, R. C., and Weston, A. H., 1983a, Evidence that the spasmogenic action of tetraethylammonium in guinea-pig trachealis is both direct and dependent on the cellular influx of calcium ion, *Br. J. Pharmacol.* **79**:255–263.

Foster, R. W., Small, R. C., and Weston, A. H., 1983b, The spasmogenic action of potassium chloride in guinea-pig trachealis, *Br. J. Pharmacol.* **80**:553–559.

Foster, R. W., Okpalugo, B. I., and Small, R. C., 1984, Antagonism of Ca^{2+} and other actions of verapamil in guinea-pig isolated trachealis, *Br. J. Pharmacol.* **81**:499–507.

Frankhuijzen, A. L., and Bonta, I. L., 1975, Role of prostaglandins in tone and effector reactivity of the isolated rat stomach preparation, *Eur. J. Pharmacol.* **31**:44–52.

Gonda, H., Baba, K., Satake, T., Takagi, K., and Tomita, T., 1988, The dissociation constant of verapamil estimated from its effect on Ca concentration–tension curves in guinea-pig tracheal muscle, *Pulm. Pharmacol.* **1**:7–13.

Hashimoto, T., Hirata, M., and Ito, Y., 1985, A role of for inositol 1,4,5-trisphosphate in the initiation of agonist-induced contractions of dog tracheal smooth muscle, *Br. J. Pharmacol.* **86**:191–199.

Honda, K., Satake, T., Takagi, K., and Tomita, T., 1986, Effects of relaxants on electrical and mechanical activities in the guinea-pig tracheal muscle, *Br. J. Pharmacol.* **87**:665–671.

Honda, K., and Tomita, T., 1987, Electrical activity in isolated human tracheal muscle, *Jpn. J. Physiol.* **37**:333–336.

Hutás, I., Hadházy, P., Debreczeni, L., and Vizi, E. S., 1981, Relaxation of human isolated bronchial smooth muscle. *Lung* **159**:153–161.

Inoue, T., and Ito, Y., 1986, Characteristics of neuroeffector transmission in the smooth muscle layer of dog bronchiole and modifications by autacoids, *J. Physiol. (Lond.)* **370**:551–565.

Ito, M., Baba, K., Takagi, K., Stake, T., and Tomita, T., 1985, Some properties of calcium-induced contraction in the isolated human and guinea-pig tracheal smooth muscle, *Respir. Physiol.* **59**:143–153.

Ito, Y., and Tajima, K., 1981, Spontaneous activity in the trachea of dogs treated with indomethacin: An experimental model for aspirin-related asthma, *Br. J. Pharmacol.* **73**:563–571.

Ito, Y., and Tajima, K., 1982, Dual effects of catecholamines on pre- and post-junctional membranes in the dog trachea, *Br. J. Pharmacol.* **75**:433–440.

Jones, T. R., Davis, C., and Daniel, E. E., 1982, Pharmacological study of the contractile activity of leukotriene C_4 and D_4 on isolated human airway smooth muscle, *Can. J. Physiol. Pharmacol.* **60**:638–643.

Kannan, M. S., Jager, L. P., Daniel, E. E., and Garfield, R. E., 1983, Effects of 4-aminopyridine and tetraethylammonium chloride on the electrical activity and cable properties of canine tracheal smooth muscle, *J. Pharmacol. Exp. Ther.* **227**:706–715.

Kirkpatrick, C. T., 1975, Excitation and contraction in bovine tracheal smooth muscle, *J. Physiol. (Lond.)* **244**:263–281.

Kirkpatrick, C. T., 1981, Tracheobronchial smooth muscle, in: *Smooth Muscle: An Assessment of Current Knowledge* (E. Bülbring, A. F. Brading, A. W. Jones, and T. Tomita, eds.), pp. 385–395, Edward Arnold, London.

Kroeger, E. A., and Stephens, N. L., 1975, Effect of tetraethylammonium on tonic airway smooth muscle: Initiation of phasic electrical activity, *Am. J. Physiol.* **228**:633–636.

Mansour, S., and Daniel, E. E., 1986, Maintenance of tone, role of arachidonate metabolites, and effects of sensitization in guinea pig trachea, *Can. J. Physiol. Pharmacol.* **64**:1096–1103.

McCaig, D. J., 1986, Electrophysiology of neuroeffector transmission in the isolated, innervated trachea of the guinea-pig, *Br. J. Pharmacol.* **89**:793–801.

McCaig, D. J., and Souhrada, J. F., 1980, Alteration of electrophysiological properties of airway smooth muscle from sensitized guinea-pigs, *Respir. Physiol.* **41**:49–60.

Ono, T., Ohtsuka, M., Sakai, S., Ohno, S., and Kumada, S., 1977, Relaxant effect of aspirin-like drugs on isolated guinea-pig tracheal chain, *Jpn. J. Pharmacol.* **27**:887–898.

Orehek, J., Douglas, J. S., and Bouhuy, A., 1975, Contractile responses of the guinea-pig trachea in vitro: Modification by prostaglandin synthesis-inhibiting drugs, *J. Pharmacol. Exp. Ther.* **194**: 554–564.

Richards, I. S., Murlas, C., Ousterhout, J. M., and Sperelakis, N., 1986, 8-Bromo-cyclic GMP abolishes TEA-induced slow action potentials in canine trachealis muscle, *Eur. J. Pharmacol.* **128**:299–302.

Saad, M. H., and Burka, J. F., 1983, Effects of immunological sensitization on the responses and sensitivity of guinea pig airways to bronchoconstrictors. Modulation by selective inhibition by arachidonic acid metabolism, *Can. J. Physiol. Pharmacol.* **61**:876–887.

Small, R. C., 1982, Electrical slow waves and tone of guinea-pig isolated trachealis muscle: Effects of drugs and temperature changes, *Br. J. Pharmacol.* **77**:45–54.

Somlyo, A. V., and Somlyo, A. P., 1968, Electromechanical and pharmacomechanical coupling in vascular smooth muscle, *J. Pharmacol. Exp. Ther.* **159**:129–145.

Souhrada, M., and Souhrada, J. F., 1981a, The direct effect of temperature on airway smooth muscle, *Respir. Physiol.* **44**:311–323.

Souhrada, M., and Souhrada, J. F., 1981b, Reassessment of electrophysiological and contractile characteristics of sensitized airway smooth muscle, *Respir. Physiol.* **46**:17–27.

Souhrada, M., and Souhrada, J. F., 1982, Potentiation of Na^+-electrogenic pump of airway smooth muscle by sensitization, *Respir. Physiol.* **47**:69–81.

Souhrada, M., and Souhrada, J. F., 1985, Alterations of airway smooth muscle cell membrane by sensitization, *Pediatr. Pulmonol.* **1**:207–214.

Souhrada, M., Souhrada, J. F., and Cherniak, R. M., 1981, Evidence for a sodium electrogenic pump in airway smooth muscle, *J. Appl. Physiol.* **51**:346–352.

Souhrada, M., Klein, J. J., Berend, N., and Souhrada, J. F., 1983, Topographical differences in the physiological response of canine airway smooth muscle, *Respir. Physiol.* **52**:245–258.

Stephens, N. L., and Kroeger, E., 1970, Effect of hypoxia on airway smooth muscle mechanics and electrophysiology, *J. Appl. Physiol.* **28**:630–635.

Suzuki, H., Morita, K., and Kuriyama, H., 1976, Innervation and properties of the smooth muscle of the dog trachea, *Jpn. J. Physiol.* **26**:303–320.

Tomita, T., 1975, Electrophysiology of mammalian smooth muscle, *Prog. Biophys. Mol. Biol.* **30**: 185–203.

Tomita, T., and Yamamoto, T., 1971, Effects of removing the external potassium on the smooth muscle of the guinea-pig taenia coli, *J. Physiol. (Lond.)* **212**:851–868.

Van Breemen, C., Aaronson, P., and Loutzenhiser, R., 1979, Sodium-calcium interactions in mammalian smooth muscle, *Pharmacol. Rev.* **30:**167–208.

Yamaguchi, T., Hitzig, B., and Coburn, R. F., 1976, Endogenous prostaglandin in guinea pig taenia coli, *Am. J. Physiol.* **230:**149–157.

Chapter 9

Ion Channels in Airway Smooth Muscle

Michael I. Kotlikoff

Department of Animal Biology
School of Veterinary Medicine
and Cardiovascular–Pulmonary Division
Department of Medicine,
University of Pennsylvania School of Medicine
Philadelphia, Pennsylvania 19104-6046

I. INTRODUCTION

Recent advances in the study of membrane ion channels have been made through the use of cell disaggregation and single-cell patch-clamp recording techniques. This approach has been responsible for the rapidly advancing knowledge about specific ion-channel proteins and their control processes in numerous tissues. Ion channels have been studied either by isolating a patch of membrane and recording single-channel events or by examination of the aggregate activity of a single type of channel by precisely controlling the ionic species available on both sides of the cell membrane. These techniques have led to the characterization of channels in terms of surface receptor coupling, ion selectivity, voltage activation, voltage-dependent inactivation, and modulation by cellular second messengers. Recent parallel advances in the isolation and sequencing of channel proteins, their reconstitution into lipid bilayers, and the expression of channel proteins in *xenopus* oocytes by injection of RNA have led to specific hypotheses about their three-dimensional structure, interaction with other membrane proteins, and regulation of synthesis.

Although mammalian smooth muscle presents particular difficulties with respect to cell isolation and maintenance of phenotypic differentiation in tissue culture, progress has been made toward the characterization and study of ion channels in these tissues. This chapter briefly summarizes the basic biophysics of ion channels, reviews recent experimental data identifying specific membrane ion channels in airway smooth muscle, and provides some discussion of the function of these channel in cellular excitation processes.

II. ION CHANNELS

Ion channels are integral membrane proteins that provide a pathway for ions to permeate the membrane in which they are inserted. As such, they provide a regulated pore through which ions are able to cross the hydrophobic lipid bilayer. A characteristic feature of these membrane proteins is the ability to change the configuration from a closed, or ion-impermeant, to an open, or ion-permeant, configuration. In the open configuration, most channels are ion selective. That is, resistance to permeation is dependent on the particular ionic species, most channels showing a marked selectivity for a particular cytosolic or extracellular ion. Ions move passively through the channel in a direction determined by the transmembrane potential and the equilibrium potential of the ion. This driving force and the conductance of the channel in the open state determine the ion flux. For any given ion, conductance through the channel is fixed, the channel being either completely open or completely closed. The process of moving from the closed to the open configuration, termed channel gating, is the pivotal step in the regulation of channel function.

The specific types of channels present in a particular cell membrane enable the cell to respond stereotypically to electrical depolarization or neurotransmitter binding. Many ion channels are voltage gated; i.e., the probability that a given ion channel is in the open or closed state is a function of the transmembrane potential. By altering membrane conductance, the gating of specific ion channels alters membrane potential, thereby activating or inactivating other voltage-gated channels. In addition to activation by the electrical field surrounding the channel, binding of various substances can result in altered channel function by altering either the voltage-gating characteristics, the mean open time, or the number of functional channels. An example of the regulation of channel gating by binding of a substance to the channel is the action of calcium in activating a potassium channel, the ubiquitous calcium-activated potassium, K(Ca), channel. This voltage-gated channel is activated by calcium binding to the internal aspect of the channel protein (Meech, 1978). Thus, at any given transmembrane potential, the probability of a K(Ca) channel being open is dependent on the cytosolic free calcium concentration. A more general example of this process is the modification of channel gating by hormone binding to cell-surface receptors. This binding triggers other intracellular events that ultimately modify the channel; two prominent examples of this process are (1) phosphorylation of a channel protein domain subsequent to activation of protein kinases (Levitan, 1985), and (2) binding of a G protein subunit to the channel following dissociation of the receptor–G protein complex (Codina et al., 1987). Hormonal modulation of potassium and calcium channel function by these mechanisms have been demonstrated (e.g., Reuter et al., 1982; Siegelbaum et al., 1982; Codina et al., 1987). The implications of the modulation of ion-channel function by external ligand binding are widespread. The function of membrane ion channels may be regulated by external hormones that subsequently release cellular second messengers or activate protein kinases. Ion channels may thus play important functional roles in cellular events that are not characterized by large changes in transmembrane potential.

The molecular features underlying the functional properties of ion channels are beginning to be clarified. While initial characteristics of ion channels were inferred from electrophysiological characteristics of whole membranes, the ability to resolve individual ion fluxes associated with single-channel fluctuations of state provided impressive confirmation of these early hypotheses. Several ion channels have now been isolated and sequenced (e.g., Noda *et al.*, 1983; Tanabe *et al.*, 1987). Further physical information about the tertiary structure of ion channels will undoubtedly enhance our understanding of channel selectivity and channel gating.

III. PATCH-CLAMP TECHNIQUES

The study of individual smooth muscle cells eliminates many of the disadvantages and experimental limitations associated with multicellular tissue strips and permits the use of patch-clamp techniques to resolve small ionic currents. To date, studies on mammalian smooth muscle have been limited due to difficulties encountered during isolation of these long thin cells from the dense extracellular matrix. Protocols have been developed, however, that allow for the preparation of single dissociated vascular (Defeo and Morgan, 1985; Warshaw *et al.*, 1986) and gastric (Mitra and Morad, 1985) smooth muscle cells. Similar techniques have been used to obtain individual relaxed airway myocytes from the dog trachealis muscle that can be studied acutely (McCann and Welsh, 1985; Kotlikoff, 1988b) or after they have been plated and maintained in tissue culture media (Panettieri and Kotlikoff, 1987; Kotlikoff *et al.*, 1987). The dissociation, using enzyme solutions containing mixtures of proteolytic enzymes in nominally calcium-free balanced salt solutions, is preceded by careful dissection of the mucosal and serosal layers so that contamination by nonsmooth muscle cell types is minimized. Particular care must be exercised to limit digestion in order to retain the cells in a relaxed and viable state. Dissociated airway smooth muscle cells can be grown in tissue culture using standard smooth muscle culture methods (Chamley-Campbell *et al.*, 1979). Monoclonal antibody staining has indicated no evidence of contamination of primary cultures of airway smooth muscle cells with nonsmooth muscle cell types (Panettieri and Kotlikoff, 1987).

These single cells are accessible to study using patch-clamp techniques, since the enzymatic dissociation has effectively removed the basement membrane substance surrounding the cell membrane. The reader is referred to several complete reviews of patch-clamp techniques for detailed information (Sakmann and Neher, 1983; Hamill *et al.*, 1981). Briefly, single-channel recording is accomplished by the initial formation of a gigohm resistance seal (gigaseal) between the glass recording pipette and the cell membrane. For single-channel recording, the electrically isolated patch of membrane under the pipette is usually excised from the cell giving access to both sides of the cell patch. The small fluxes of current associated with the opening of an individual ion channel in the roughly 1-μm^2 membrane patch can then be resolved and studied. A second patch technique, termed whole cell recording (Hamill *et al.*, 1981), entails the formation of a gigaseal similar to single-channel techniques, but the patch of membrane is then sucked out of the pipette tip, leaving the pipette seal intact. In this fashion, a 1-μm^2 hole is created in the membrane, and the contents of the patch pipette diffuse into the cell. This technique is especially useful for recording from channels that are vulnerable to excision from the cell membrane, such as calcium channels, and for examining hormonal gating effects, such as those mediated by second messengers, which require an intact cell.

Patch-clamp techniques have recently been employed in the characterization of a number of specific ion channels in smooth muscle. As expected, there appears to be considerable variation in ion channel make-up between different smooth muscle tissues. Of the channels identified in smooth muscle, only potassium and calcium channels have been reported in airway smooth muscle to date.

IV. POTASSIUM CHANNELS IN AIRWAY SMOOTH MUSCLE

In certain types of smooth muscle cells, the net effect of the coordinated activation of ion channels is a spike depolarization. Other smooth muscle cell types, of which airway smooth muscle is a prominent example, do not produce spike depolarizations, but respond to the application of neurotransmitter substances or depolarization by KCl with a graded decrease in membrane potential (Kirkpatrick, 1975; Farley and Miles, 1977; Coburn, 1979). Furthermore, spontaneous electrical activity is not recorded under normal conditions. This electrical stability appears to be the result of the outwardly rectifying potassium channels in the airway smooth muscle cell membrane. That is, as the airway smooth muscle cell depolarizes, potassium channels open increasing membrane potassium conductance and limiting further depolarization. Under conditions in which potassium channels are blocked by tetraethylammonium (TEA), spike potentials, slow-wave activity, and spontaneous contractions are observed (Kirkpatrick, 1975; Stephens *et al.*, 1975; Kroeger and Stephens, 1975; Suzuki *et al.*, 1976). Membrane potassium conductance therefore appears to account for electrical stability of the airway smooth muscle cell. The specific types of potassium channels responsible for this potassium conductance are beginning to be described.

McCann and Welsh (1986) reported results of patch-clamp experiments that identified a specific potassium channel in airway smooth muscle. These workers

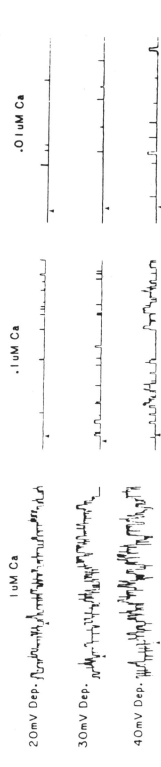

FIGURE 1. Single-channel recordings of calcium-activated potassium currents in antibody smooth muscle cells. Effect of internal Ca concentration and voltage on channel activity in excised inside-out patches. The internal solution contained Ca concentrations as indicated. The effect of voltage on channel kinetics can be observed by comparing tracings in the same row. The arrowheads mark the current level when no channels are open. (From McCann and Welsh, 1986.)

recorded the single-channel activity of a large conductance calcium-activated potassium K(Ca) channel in inside-out patches from dissociated canine tracheal myocytes. In this technique, the cytosolic side of the membrane patch is exposed to the recording bath solution, and the external membrane surface is bathed in the pipette solution. When the calcium concentration was raised on the cytoplasmic side of the membrane patch or when the patch was depolarized, both the frequency of channel openings and the duration of the open state increased (Fig. 1). The channel was shown to be highly selective for potassium and very sensitive to blockage by TEA applied externally or internally or for cesium applied internally. The presence of a K(Ca) channel has been reported in other smooth muscle cells (Benham *et al.*, 1986; Mitra and Morad, 1985; Singer and Walsh, 1984), and nonmammalian smooth muscle (Singer and Walsh, 1984) and appears to be widespread in other cell types.

Using whole-cell patch-clamp recording techniques, the macroscopic currents elicited by depolarization can be analyzed (Kotlikoff, 1988a). Figure 2a shows the currents evoked by voltage-clamp depolarizations of an airway smooth muscle cell. Inward currents (downward current deflections) precede outward currents that are progressively larger in magnitude with depolarizations to more positive potentials. Longer clamp steps (Fig. 2b) indicate that the outward currents are activated by voltage-clamp depolarizations positive to -20 mV and that depolarizations to potentials positive to 0 mV elicit inactivating and noninactivating current components. The reversal potential of the current, as well as current blockade with high concentrations of TEA applied at the external surface of the cell, or cesium substituted for potassium in the cytosolic solution, identify the charge carrier as K^+ ions. The two current components can be separated on the basis of both voltage-dependent inactivation and pharmacological sensitivity to potassium-channel blockers (Kotlikoff, 1988a). To identify the component of this current that is attributable to the K(Ca) current, experiments were conducted in which the whole-cell configuration was achieved in the presence of external calcium; the bath solution was then changed to one containing Mn^{2+} substituted for calcium. The traces shown in Fig. 3 were obtained in the same cell with calcium in the bath and after substitution. The K^+ currents were not altered by substitution of the calcium-channel impermeant ion Mn^{2+}, even though the inward calcium currents were clearly blocked (compare small inward currents in current families in Fig. 3). In addition, charybdotoxin, a naturally occurring toxin that specifically blocks large conductance K(Ca) channels (Smith *et al.*, 1986), did not block these potassium currents.

These experiments indicate that several voltage sensitive potassium channels are present in the canine airway smooth muscle cell. The lack of an effect on the potassium current by calcium-channel blockade or application of the K(Ca)-channel antagonist charybdotoxin suggests that the current is carried by potassium passing through non-K(Ca) channels; the current resembles voltage-activated delayed rectifier channels found in numerous neuronal cells (Hille,

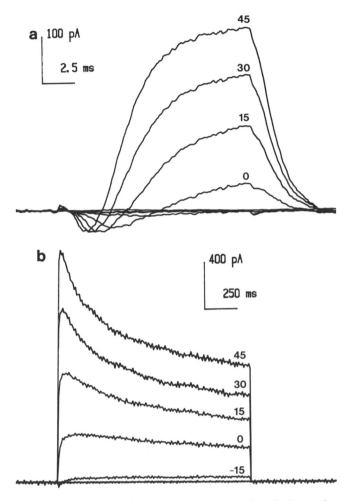

FIGURE 2. Whole-cell recordings of potassium channel currents in antibody smooth muscle cells. Current recordings during voltage clamp steps from −60 to 45 mV in steps of 15 mV from a holding voltage of −70 mV. The external solution was Krebs buffer, containing 10 mM calcium. The internal solution was calcium-buffered with 11 mM EGTA and contained 140 mM K$^+$. (a) Traces showing the first 15 msec following depolarization show inward currents preceding outward currents and the delayed activation of outward currents. (b) Longer clamp steps demonstrate the time course of inward currents relative to outward currents and the inactivating component of the K$^+$ currents.

1984). K(Ca) channels seen at the single-channel level may not be readily observed in whole-cell preparations because of calcium buffering in the internal solution, although a K(Ca) current is still not observed in experiments using low calcium-buffering capacity intracellular solutions. It is possible that cytosolic calcium levels must be elevated by the release of intracellular calcium stores

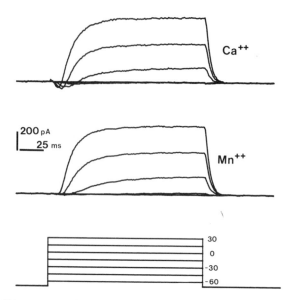

FIGURE 3. K$^+$ currents are delayed rectifier currents. Voltage-clamp experiment in the presence of 10 mM Ca^{2+} in the external bath (above), and with 10 mM Mn^{2+} substituted for Ca^{2+} (lower current traces). K$^+$ currents are slightly larger in the absence of inward currents, consistent with a slight cancellation of the outward current by the inward current. Internal solutions were as in Fig. 2; however, identical results have been obtained with 1 mM EGTA in the pipette solution. Bars show time course of voltage-clamp steps.

(Kotlikoff *et al.*, 1987) in order to activate calcium-activated membrane conductances. Presumably, delayed rectifier potassium channels activate when the cell is depolarized, accounting for the outward rectification observed in tissue strips (Kirkpatrick, 1975; Farley and Miles, 1977; Coburn, 1979). Of note is the fact that outward rectification in these studies is only partially blocked by high concentrations of TEA (Kirkpatrick, 1975). These data are consistent with the action of delayed rectifier channels that are more resistant to TEA blockade than K(Ca) channels. Thus, electrical depolarization of the airway smooth muscle cell results in the opening of voltage-sensitive channels that act to stabilize membrane potential. The degree to which calcium-activated potassium channels and delayed rectifier potassium channels are activated during neurotransmitter- or autacoid-induced cellular activation remains to be determined.

V. CALCIUM CHANNELS IN AIRWAY SMOOTH MUSCLE

Two types of calcium channels were recently reported in vascular smooth muscle, based on whole-cell calcium current recordings (Sturek and Hermsmeyer, 1986). These channels have been termed the transient (T) channel and long-lasting (L) channel, by analogy to similar currents recorded in cardiac myocytes (Bean, 1985) and neuronal cells (Nowycky *et al.*, 1985). In cardiac and neuronal cells, the T channel activates at more negative voltages than the L channel and inactivates with a relatively rapid time course. The L current activates only with stronger depolarizations, inactivates slowly, and is highly sensitive to dihydropyridine blocking agents. Most published accounts of calcium currents in smooth muscle cells indicate that the principal current component is a current that inactivates rapidly (time constant <100 msec) (Mitra and Morad, 1985; Klockner and Isenberg, 1985; Bean *et al.*, 1986; Yatani *et al.*, 1987; Clapp *et al.*, 1987; Kotlikoff, 1988b), which would seem to suggest that the underlying channels are T type. However, the voltage activation of the current and the single-channel conductance (Yatani *et al.*, 1987; Benham *et al.*, 1987) are more compatible with the L-type dihydropyridine-sensitive calcium channel. The sensitivity to blockade by dihydropyridines does not unambiguously identify the current either, since the main current component is blocked by dihydropyridines, but usually at concentrations substantially higher than those necessary to block L channels in skeletal muscle, cardiac muscle, or neuronal cells. Thus, it appears that the major calcium channel protein in smooth muscle cells displays significant functional differences when compared with the L channel that has been characterized in other tissues.

The electrophysiological and pharmacological characteristics of the calcium current in canine tracheal myocytes were recently reported (Kotlikoff, 1988b). Under normal conditions of high K^+ internal solutions and high Na^+ external solutions, small inward currents are often seen proceeding the outward currents described above (see Figs. 2a and 3a). Calcium currents, which are smaller in amplitude than K^+ currents, can be pharmacologically isolated by blocking all outward currents. Figure 4a shows a typical experiment in which potassium channels were blocked by replacement of cytosolic potassium with the impermeant cation cesium (Cs^+) and by the replacement of NaCl with TEA chloride in the external bath. Under these conditions, the activation kinetics and voltage dependence of inward currents may be studied in isolation. Inward currents were identified as calcium currents on the basis of their presence under conditions in which calcium was the only permeant cation (as in Fig. 4a) and by their blockage by the substitution of cobalt or manganese for calcium (see Fig. 3). Calcium currents were elicited by voltage-clamp steps positive to −45 mV. The current-voltage plot shown in Fig. 4b demonstrates that the current amplitude increases with more positive depolarizations, reaching a maximum at 15 mV. Inactivation of the current occurred with a time constant of approximately 30 msec and at holding potentials more positive than −40 mV. The calcium current was not sensitive to nanomolar concentrations of dihydropyridines but was blocked by nifedipine and augmented by BAY K8644 in micromolar concentrations. It is worth noting that a current almost identical in voltage activation and inactivation characteristics is increased by acetylcholine (ACh) in smooth muscle cells isolated from the toad stomach (Clapp *et al.*, 1987). Similar modulation of the

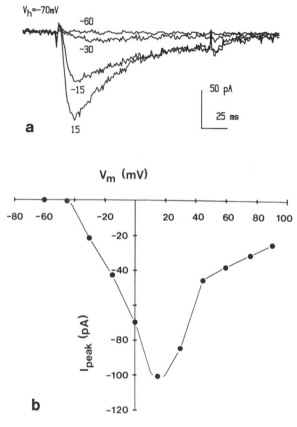

FIGURE 4. Whole-cell recordings of calcium-channel currents in airway smooth muscle cells. (a) Current recordings during voltage-clamp steps to -60 mV, -30 mV, -15 mV, and 15 mV. Internal solution contained CsCl (130 mM), and external solution contained TEACl (140 mM) to block potassium currents. The external solution contained calcium (10 mM) as the only permeant cation. V_h indicates the holding potential. (b) Current–voltage relationship for same cell with additional voltage-clamp steps. Peak current is plotted against command potential.

calcium current in airway smooth muscle cells, however, has not been observed to date.

Unlike reports in some smooth muscle cells, no sustained noninactivating calcium current could be clearly recorded in airway smooth muscle cells. If present, such a noninactivating current would account for very little of the macroscopic inward current. Moreover, distinct current components could not be separated on the basis of holding potential or sensitivity to dihydropyridines (Kotlikoff, 1988b). In light of the fact that purified calcium-channel proteins from smooth muscle show different reactivity to antibodies than cardiac or neuronal cells (Schmid *et al.*, 1986), it is not surprising that differences in channel function at the whole-cell level are also observed. Experiments that relate the structural differences between the calcium-channel proteins in airway smooth muscle and those in other cell types will be of considerable interest in airway smooth muscle physiology, especially in light of the demonstration that the cardiac L channel is modulated by neurotransmitter or hormonal receptor binding, resulting in altered voltage/open-state channel characteristics. Equally important will be studies that determine the role of the calcium current in the integrated response of the airway myocyte to neurotransmitter substances.

VI. FUNCTIONAL CHARACTERISTICS OF ION CHANNELS IN AIRWAY SMOOTH MUSCLE

Although our knowledge of the total ion-channel makeup of airway smooth muscle cells is still formative, certain functional observations about the importance of these channels in excitation–contraction coupling can be made. One characteristic of airway smooth muscle that appears to distinguish it from most vascular and gastrointestinal (GI) smooth muscles is its electrical quiescence. This property can be ascribed to the presence of at least two types of voltage-sensitive delayed rectifier potassium channels in the airway myocyte membrane. The presence of these channels and the aggregate outward current that they conduct is sufficient to explain the rectifying nonspiking properties of the airway smooth muscle cell membrane (Kotlikoff, 1988a) (see Fig. 2). Furthermore, it is likely that K(Ca) channels are activated whenever cytosolic calcium rises. Thus, the generation of inositol trisphosphate by agonist binding (Baron *et al.*, 1984), and the subsequent release of intracellular calcium stores (Kotlikoff *et al.*, 1987), would be expected to activate K(Ca) channels (unless the rise of cytosolic calcium were tightly compartmentalized). The spike action potentials observed in double sucrose-gap current-clamp experiments in the presence of potassium channel-blocking agents (Kirkpatrick, 1975) are likely mediated by calcium influx, since experiments to date have failed to demonstrate the presence of sodium channels in airway smooth muscle, which have been reported in vascular smooth muscle cells (Sturek and Hermsmeyer, 1986). Voltage-activated calcium channels will open when the cell is modestly depolarized from its resting potential of approximately -50 mV (Kirkpatrick, 1975), thereby providing one pathway for cellular excitation. The depolarization that attends this inward current will be blunted by the almost simultaneous activation of outward currents, resulting in an isopotential influx of extracellular calcium.

Neurotransmitters or bronchoactive autacoids cause a depolarization of the airway smooth muscle cell following receptor binding (Kirkpatrick, 1975; Farley and Miles, 1977; Coburn, 1979). The ionic processes that mediate this depolarization are incompletely understood, although it appears that the opening of ion channels other than those described above is triggered by these agents. The demonstration of strictly voltage-activated calcium channels in airway smooth muscle is insufficient to explain the depolarization induced by agonist binding, unless these channels are activated by neurotransmitter/autacoid binding, or by subsequent postreceptor events. The demonstration of cholingergic modulation of voltage-dependent calcium channels would not explain this depolarization, since ACh does not activate an inward current in the range of the resting membrane potential; rather, it augments the current evoked by depolarization (Clapp et al., 1987). In nonmammalian smooth muscle, a potassium channel that is sensitive to muscarinic agents has been shown to account for ACh-induced depolarization (Sims et al., 1985). This channel, the M channel, is inactivated by ACh binding, resulting in a decrease in potassium conductance and subsequent depolarization. The M current has not been demonstrated in mammalian smooth muscle to date. Other candidate channels mediating agonist-induced depolarization are nonselective ligand-gated cation channels (Benham et al., 1987) and calcium-activated chloride channels (Byrne and Large, 1987) that would account for an inward current at resting potentials negative to the chloride equilibrium potential (approximately -20 mV in smooth muscle (Aickin and Brading, 1982)). Future studies directed at determining the presence of these or other channels will help elucidate the molecular nature of the ion channels that mediate agonist-induced depolarization, and define the extent of ion-channel participation in airway smooth muscle activation.

ACKNOWLEDGMENTS. This work was supported by the American Lung Association, the University of Pennsylvania Research Foundation, and by grants HL 36150 and HL 41084 from the National Institutes of Health. The author thanks Dr. R. K. Murray for helpful comments.

VII. REFERENCES

Aickin, C. C., and Brading, A. F., 1982, Measurement of intracellular chloride in guinea-pig vas deferens by ion analysis, [36]chloride efflux and micro-electrodes, *J. Physiol. (Lond.)* **326:**139–154.

Baron, C. B., Cunningham, M., Strauss, J. F. III, and Coburn, R. F., 1984, Pharmacomechanical coupling in smooth muscle may involve phosphatidylinositol metabolism, *Proc. Natl. Acad. Sci. USA* **81:**6899–6903.

Bean, B. P., 1985, Two types of calcium channels in canine atrial cells, *J. Gen. Physiol.* **86:**1–30.

Bean, B. P., Sturek, M., Puga, A., and Hermsmeyer, K., 1986, Calcium channels in muscle cells isolated from rat mesenteric arteries: modulation by dihydropyridine drugs, *Circ. Res.* **59:**229–235.

Benham, C. D., and Bolton, T. B., 1983, Patch-clamp studies of slow potential-sensitive potassium channels in longitudinal smooth muscle cells of rabbit jejunum, *J. Physiol. (Lond.)* **340:**469–486.

Benham, C. D., Bolton, T. B., Lang, R. J., and Takewaki, T., 1986, Calcium-activated potassium channels in single smooth muscle cells of rabbit jejunum and guinea-pig mesenteric artery, *J. Physiol. (Lond.)* **371:**45–67.

Benham, C. D., Bolton, T. B., Byrne, N. G., and Large, W. A., 1987, Action of externally applied adenosine triphosphate on single smooth muscle cells dispersed from rabbit ear artery, *J. Physiol. (Lond.)* **387**:473–488.

Byrne, N. G., and Large, W. A., 1987, Action of noradrenaline on single smooth muscle cells freshly dispersed from the rat anococcygeus muscle, *J. Physiol. (Lond.)* **389**:513–525.

Chamley-Campbell, J., Campbell, G. R., and Ross, R., 1979, The smooth muscle cell in culture, *Physiol. Rev.* **59**:606–718.

Clapp, L. H., Vivaudou, M. B., Walsh, J. V., Jr., and Singer, J. J., 1987, Acetylcholine increases voltage-activated Ca^{++} current in freshly dissociated smooth muscle cells, *Proc. Natl. Acad. Sci. USA* **84**:2092–2096.

Coburn, R. F., 1979, Electromechanical coupling in canine trachealis muscle: Acetylcholine contractions, *Am. J. Physiol.* **5**:C177–C184.

Codina, J., Yatani, A., Grenet, D., Brown, A. M., and Birnbaumer, L., 1987, *Science* **236**:442–445.

Defeo, T. T., and Morgan, K. G., 1985, Responses of enzymatically isolated mammalian vascular smooth muscle cells to pharmacological and electrical stimuli, *Pflugers Arch.* **404**:100–102.

Droogmans, G., and Callewaert, G., 1986, Ca^{2+}-channel current and its modification by the dihydropyridine agonist BAY K8644 in isolated smooth muscle cells, *Pflugers Arch.* **406**:259–265.

Farley, J. M., and Miles, P. R., 1977, Role of depolarization in acetylcholine-induced contractions of dog trachealis muscle, *J. Pharmacol. Exp. Ther.* **201**:199–205.

Friedman, M. E., Suarez-Kurtz, G., Kaczorowski, G. J., Katz, G. M., and Reuben, J. P., 1986, Two calcium currents in a smooth muscle cell line, *Am. J. Physiol.* **250**:H699–H703.

Hamill, O. P., Marty, A., Neher, A., Sakmann, B., and Sigworth, F. J., 1981, Improved patch-clamp techniques for high-resolution current recording from cells and cell-free membrane patches, *Pflugers Arch.* **391**:85–100.

Hille, B., 1984, *Ionic Channels of Excitable Membranes,* Sinauer, Sunderland, Massachusetts.

Kirkpatrick, C. T., 1975, Excitation and contraction in bovine tracheal smooth muscle, *J. Physiol. (Lond.)* **244**:263–281.

Kotlikoff, M. I., 1988a, Voltage-dependent potassium currents in airway smooth muscle cells: modulation by histamine and cholinergic agonists, *J. Physiology* (submitted).

Kotlikoff, M. I., 1988b, Calcium currents in isolated canine airway smooth muscle cells, *Am. J. Physiol.* **254**:C793–C801.

Kotlikoff, M. I., Murray, R. K., and Reynolds, E. E., 1987, Histamine-induced calcium release and phorbol antagonism in cultured airway smooth muscle cells, *Am. J. Physiol.* **253**:C561–C566.

Kroeger, E. A., and Stephens, N. L., 1975, Effect of tetraethylammonium on tonic airway smooth muscle: Initiation of phasic electrical activity, *Am. J. Physiol.* **228**:633–636.

Levitan, I. B., Phosphorylation of ion channels, 1985, *J. Membr. Biol.* **87**:177–190.

McCann, J. D., and Welsh, M. J., 1986, Voltage-gated Ca-activated K channels in isolated canine airway smooth muscle cells, *J. Physiol. (Lond.)* **372**:113–127.

Meech, R. W., 1978, Calcium-dependent potassium activation in nervous tissues, *Annu. Rev. Biophys. Bioeng.* **7**:1–18.

Mitra, R., and Morad, M., 1985, Ca^{++} and Ca^{++}-activated K^+ currents in mammalian gastric smooth muscle cells, *Science* **229**:269–272.

Noda, M., Furutani, Y. F., Takahashi, H., Toyosato, M., Tanabe, T., Shimizu, S., Kikyotani, S., Kayano, T., Hirose, T., Inayama, S., and Numa, S., Cloning and sequence analysis of calf cDNA and human genomic DNA encoding α-subunit precursor of muscle acetylcholine receptor, *Nature (Lond.)* **305**:818–823.

Nowycky, M. C., Fox, A. P., and Tsien, R. W., 1985, Three types of neuronal calcium channel with different calcium agonist sensitivity, *Nature (Lond.)* **316**:440–443.

Panettieri, R. A., and Kotlikoff, M. I., 1987, The isolation and growth of cultured airway smooth muscle, *Am. Rev. Respir. Dis.* **135**:A272.

Reuter, H., Stevens, C. F., Tsien, R. Q., and Yellen, G., 1982, Properties of single calcium channels in cardiac cell culture, *Nature (Lond.)* **297:**501–504.

Sakmann, B., and Neher, E., 1983, *Single-Channel Recording,* Plenum, New York.

Schmid, A., Barhanin, J., Coppola, T., Borsotto, M., and Lazdunski, M., 1986, *Biochemistry,* **25:** 3492–3495.

Siegelbaum, S. A., Camardo, J. S., and Kandel, E. R., 1982, Serotonin and cAMP close single K channels in Aplysia sensory neurons, *Nature (Lond.)* **299:**413–417.

Sims, S. M., Singer, J. J., and Walsh, J. V., Jr., 1985, Cholinergic agonists suppress a potassium current in freshly dissociated smooth muscle cells of the toad, *J. Physiol. (Lond.)* **367:**503–529.

Singer, J. J., and Walsh, J. V., Jr., 1984, Large conductance Ca^{++}-activated K^+ channels in smooth muscle cell membrane, *Biophys. J.* **45:**68–70.

Smith, C., Phillips, M., and Miller, C., 1986, Purification of charybdotoxin, a specific inhibitor of the high-conductance Ca^{++}-activated K^+ channel, *J. Biol. Chem.* **261:**14607–14613.

Stephens, N. L., Kroeger, E. A., and Kromer, U., 1975, Induction of a myogenic response in tonic airway smooth muscle by tetraethylammonium, *Am. J. Physiol.* **228:**628–632.

Sturek, M., and Hermsmeyer, K., 1986, Calcium and sodium channels in spontaneously contracting vascular muscle cells, *Science* **233:**475–478.

Suzuki, H., Morita, K., and Kuriyama, H., 1976, Innervation and properties of the smooth muscle of the dog trachea, *Jpn. J. Physiol.* **26:**303–320.

Tanabe, T., Takishima, H., Mikami, A., Flockerzi, V., Takahashi, H., Kangawa, K., Kojima, M., Matsuo, H., Hirose, T., and Numa, S., 1987, Primary structure of the receptor for calcium channel blockers from skeletal muscle, *Nature (Lond.)* **328:**313–318.

Warshaw, D. W., Szarek, J. L., Hubbard, M. S., and Evans, J. N., 1986, Pharmacology and force development of single freshly isolated bovine carotid artery smooth muscle cells, *Circ. Res.* **58:** 399–406.

Yatani, A. C., Seidel, C. L., Allen, J., and Brown, A. M., 1987, *Circ. Res.* **60:**523–533.

Chapter 10

Coupling Mechanisms in Airway Smooth Muscle

Ronald F. Coburn and Kenji Baba

Department of Physiology
University of Pennsylvania School of Medicine
Philadelphia, Pennsylvania 19104-6085

I. INTRODUCTION

Our goal in this chapter is to summarize knowledge about the coupling mechanisms that operate during contraction of airway smooth muscle and to coordinate the material presented in other chapters. As electrophysiologists, our orientation to this area is to classify coupling mechanisms as membrane potential (E_m) dependent and E_m independent. Early observations indicated that K^+-depolarized smooth muscle can be contracted by various agonists (Evans *et al.*, 1958). Agonist-evoked smooth muscle contractions can occur without detectable membrane depolarizations (Ito *et al.*, 1979). Agonist-induced contraction of canine trachealis muscle occurs with a smaller membrane depolarization than that seen at equivalent force during K^+-evoked contractions (Coburn, 1979). During ACh-induced depolarization and contraction, there is very little effect on force of repolarizing the membrane by injecting hyperpolarizing current, whereas during K^+-evoked contractions, repolarization of membrane depolarization using this technique, reverses contraction (Coburn, 1977, 1979). K^+-induced contractions of airway smooth muscle appear to be entirely dependent on extracellular Ca^{2+}, but ACh-induced contractions are resistant to removal of extracellular Ca^{2+} (Farley and Miles, 1977). Thus, ACh-evoked contractions are driven, at least in part, by the release of Ca^{2+} from intracellular stores. The concept of E_m-independent force development and maintenance was strengthened with the observation that Ca^{2+} antagonists that block voltage-dependent Ca^{2+} channels have only a small effect on ACh-induced contractions of airway smooth muscle (Coburn, 1977; Eberlin *et al.*, 1982), however, they completely inhibit K^+-evoked force. The value of this approach has been supported by the demonstration that smooth muscle voltage-gated L Ca^{2+} channels are blocked by low concentrations of these agents; T Ca^{2+} channels are also blocked by these agents, but at higher concentrations (Friedman *et al.*, 1986; Sturek and Hermsmeyer, 1986). Since the only known mechanism whereby membrane depolarization can activate smooth muscle is via opening of plasma membrane voltage-gated Ca^{2+} channels, the resistance of agonist-induced contractions to calcium antagonists and depletion of extracellular Ca^{2+} argue for the existence of E_m-independent contraction mechanisms. Membrane potential-independent coupling was labeled pharmacomechanical coupling (PMC) (Somlyo and Somlyo, 1968), a term that perhaps deserves to be dropped, however, we shall use this term in this present chapter. Electromechanical coupling (EMC) refers to coupling mechanisms dependent on membrane depolarization. PMC and EMC mechanisms involve cascades of events triggered by receptor occupancy. In the following discussion, we emphasize muscarinic activation of airway smooth muscle, an area of interest in our laboratory.

II. PHARMACOMECHANICAL COUPLING MECHANISMS

The major hypothesis in this area at the time of writing is that PMC can be explained on the basis of second messengers generated by inositol phospholipid metabolism (Berridge *et al.*, 1983; Streb *et al.*, 1983; Hashimoto *et al.*, 1985, 1986; Baron and Coburn, 1987a; Baba *et al.*, 1988). Inositol phospholipid metabolism is activated in airway smooth muscle, which predominantly uses PMC mechanisms, and is not caused by membrane depolarization (Baron *et al.*, 1984). The state of knowledge regarding inositol phospholipid metabolism in smooth muscle and evidence that inositol 1,4,5-trisphosphate [I(1,4,5)-P_3] releases Ca^{2+} from the SR are summarized in Chapter 7 (this volume). In canine trachealis muscle preincubated with [^{3}H]myo-inositol, IP_3 radioactivity increased maximally within 3–5 sec following the onset of field stimulation; following this, during maintained stimulation, IP_3 radioactivity decreased slightly (Baron and Coburn, 1987a). The metabolic flux rate in inositol phospholipids, determined from the rate of incorporation of myo-inositol into inositol phospholipids, increased during development at contraction and remained elevated during force maintenance, at 30–60 times baseline rate (Baron *et al.*, 1988). It appears that during the development of force, inositol phospholipid metabolism is driven by decrease in phosphatidylinositol (PI) content, whereas during maintained force, metabolic flux is driven by the rate of resynthesis of PI (Baron *et al.*, 1988). Although there is no method for the direct measurement of inositol phosphate concentrations, estimates of increases in IP_3 concentration obtained from IP_3 radioactivity and phosphatidylinositol bisphosphate (PIP_2) specific radioactivity indicated that carbachol-induced increases in IP_3 were large enough to induce the release of Ca^{2+} from the sarcoplasmic reticulum of skinned muscle fibers (Baron et al., 1988).

During muscarinic activation of canine or bovine trachealis muscle, tissue concentrations of diacylglycerol (DAG), another second messenger by-product of inositol phospholipid metabolism, become elevated (Baron and Coburn, 1984; Rasmussen et al., 1987). Diacylglycerol activates protein kinase C (PKC) (Kikkawa et al., 1983), and it is possible that PKC-mediated protein phosphorylation is involved in generating PMC-generated force. This possibility is supported by the findings that phorbol esters, which activate PKC (Kikkawa et al., 1983), cause smooth muscle contractions (Chatterjee and Tejada, 1986; Menkes et al., 1986; Jiang and Morgan, 1987; Rasmussen et al., 1987; Baba and Coburn, 1988) and that patterns of protein phosphorylations seen in canine trachealis muscle activated by carbachol are similar to those seen with phorbol ester-induced contractions (Rasmussen et al., 1987). Caldesmon, a component of a proposed actin filament Ca^{2+}-sensitive contraction mechanisms in smooth muscle (Smith et al., 1987), is apparently phosphorylated during these contractions, but this occurs very slowly during the time of maintained force (Rasmussen et al., 1987) and is therefore not involved in PMC mechanisms operating early during development and maintenance of force. In nonmuscle tissues, agonist-induced increases in tissue [DAG] occur as a result of the metabolism of phosphatidylcholine, as well as inositol phospholipids (Farese et al., 1987). It is not known for airway smooth muscle whether muscarinic-induced increases in [DAG] are completely explained by activation of inositol phospholipid metabolism. In our study (Baron et al., 1984), decreases in PI were equimolar with increases in phosphatidic acid, and there was no decrease in phosphatidylcholine or phosphatidylethanolamine pool sizes, suggesting that all or most of the increase in phosphatidic acid and DAG pool sizes was attributable to inositol phospholipid metabolism. Diacylglycerol can be formed by hydrolysis of PI and phosphatidylinositol phosphate (PIP), as well as by PIP_2 hydrolysis. Multiple phospholipase C (PLC) isoforms, membrane bound or cytosolic, have now been identified in various other tissues (Litosch and Fain, 1985; Rebecchi and Posen, 1987). Although all the isoforms have activity using PI, PIP, and PIP_2 as substrate, hydrolysis of PIP_2 occurs at near physiological $[Ca^{2+}]$, whereas effects on PI and on PIP require higher $[Ca^{2+}]$. This finding suggests that PIP_2 is the predominant substrate under physiological conditions. There is little evidence bearing on the issue of whether PMC can be explained entirely by second messengers produced by inositol phospholipid metabolism. The PMC component of carbachol-evoked contractions of swine trachealis muscle is practically eliminated by pretreatment with phorbol dibutyrate, which completely inhibits metabolic flux in the inositol phospholipid transduction system (Baron and Coburn, 1987a; Baba et al., 1988). Thus, PMC may be entirely attributable to the second messengers $I(1,4,5)\text{-}P_3$ and DAG generated by inositol phospholipid metabolism.

Evidence that inositol phospholipid metabolism is not voltage dependent is that there was no increase in rate of ^{32}P uptake into phosphatidylinositol during K^+-induced contractions, such as occurs during ACh-evoked contractions (Baron *et al.*, 1984). We note that in other tissues K^+ evoked contractions are associated with increases in IP_3 radioactivity, possibly as a result of activation of PLC by an increase in cytosolic free $[Ca^{2+}]$ (Sasakawa *et al.*, 1987).

Muscarinic contractions of airway smooth muscle are associated with activation of phospholipase A_2 (PLA_2), release of arachidonate, and formation of arachidonate by-products (Yamaguchi *et al.*, 1976). Arachidonate and/or arachidonate by-products may be involved in PMC, but this area is poorly developed in smooth muscle. Muscarinic contractions of canine trachealis muscle are associated with increases in [cGMP] (R. F. Coburn, unpublished data), which could evoke cyclic guanosine monophosphate (cGMP)-mediated phosphorylation involved in PMC, an area deserving further investigation.

It has been postulated that a voltage-independent mechanism for contraction involves activation of voltage-independent Ca^{2+} channels (Bolton, 1979). There seems to be little evidence for this in airway smooth muscle. No such Ca^{2+} channel has been observed in any tissue. Voltage-dependent channels can be modulated, either by GTP-binding proteins (G proteins) or other messengers, so that their voltage-current characteristics are altered. Thus, voltage-dependent Ca^{2+} channels could presumably by opened by G proteins or other messengers without a change in E_m.

III. ELECTROMECHANICAL COUPLING MECHANISMS DURING MUSCARINIC CONTRACTIONS

As is currently visualized, force developed using EMC mechanisms is caused by increases in free cytosolic $[Ca^{2+}]$ resulting from influx of Ca^{2+} across the plasma membrane via voltage-gated Ca^{2+} channels. The evidence that Ca^{2+} is not released from intracellular stores as a result of membrane depolarization is that K^+-evoked contractions are inhibited by Ca^{2+}-antagonists and are dependent on bathing solution $[Ca^{2+}]$. Thus, EMC depends on alterations in open times and on the frequency of opening of channels that determine membrane depolarization and the properties of plasma membrane voltage-gated Ca^{2+} channels. Muscles in the trachea and large bronchi are similar to other slow or tonic smooth muscles, which show agonist, or K^+-activated, graded, maintained depolarization without action potentials. Ionic currents activated during muscarinic-evoked membrane depolarization are still not well described. In the guinea pig longitudinal ileal smooth muscle, ACh-induced depolarization is caused by an inward current via a nonspecific channel (Benham *et al.*, 1985). (In this discussion, we use the convention that inward current carried by a cation depolarizes the membrane; that outward K^+ current, hyperpolarizes; and that outward current carried by an anion, i.e., Cl^-, results in membrane depolarization.) So far, patch-clamp studies have not identified channels that carry inward-depolarizing current. Carbachol-induced depolarization of swine trachealis muscle is inhibited by substitution of bathing solution Na^+ by impermeant ions, but not by Cl^- substitution or by amiloride (Baba et al., 1988), suggesting that inward Na^+ current is depolarizing the plasma membrane. This can not be proved using this approach, since Na^+ substitution has many effects, including inhibition of Na^+-H^+ and Na^+-Ca^{2+} exchange. Although these processes are seen as electrically neutral, changes in cytosolic Ca^{2+} are known to alter K^+ and Ca^{2+} currents (McCann and Welsh, 1986; Ohya et al., 1988). Carbachol and ACh-induced depolarization of swine or canine trachealis muscle are not influenced by agents that block voltage-gated Ca^{2+} channels, and it is unlikely that inward Ca^{2+} currents are large enough in these muscles to have much effect on membrane potential. It is possible that muscarinic-evoked depolarization of airway smooth muscle is caused by activation of outward Cl^- current (Byrne and Large, 1988).

Whole-cell patch-clamp studies have provided valuable data about the characteristics of smooth muscle plasma membrane voltage-gated Ca^{2+} channels (see Chapter 9, this volume). Two general subtypes of Ca^{2+} channels have been found in smooth muscle, type T, which rapidly inactivate, and type L, which do not inactivate (Friedman et al., 1986; Sturek and Hermsmeyer, 1986; Nakazawa et al., 1988; Ohya et al., 1988). Like other tissues (Nowycky et al., 1985), L channels in smooth muscle can be activated over the range -20 to $+40$ mV and are inhibited by dihydropyridine calcium antagonists. T channels are activated over the range -40 to -20 mV and are relatively resistant to dihydropyridine, Ca^{2+} antagonists. In smooth muscle cells from canine trachealis muscle, T Ca^{2+} have been characterized showing them to be similar to T channels found in other smooth muscle, but L Ca^{2+} channels are difficult to demonstrate (Kotlikoff, 1988). Trying to relate the properties of L and T Ca^{2+} channels found in smooth muscle to the operation of voltage-sensitive Ca^{2+} channels in intact airway smooth muscle is difficult. During maintained contractions, in which the membrane is depolarized, contractions are sensitive to Ca^{2+} antagonists, and L channels would be suspected to be operational. However, contraction occurs at membrane potentials considerably more negative than the threshold for opening of L channels. There are two possibilities to explain this conflict of data: (1) T channels are operating, perhaps using I_L (Nakazawa et al., 1988), a component of T-channel current that does not inactivate and is more sensitive to dihydropyridine Ca^{2+} antagonists than is the inactivating current component; or (2) characteristics of these channels determined using patch-clamp approaches are only qualitative.

Voltage-current plots from airway smooth muscle show marked rectification in that it is practically impossible to depolarize the plasma membrane by >20 mV, even with large cathodal current injections. It has long been known that this is attributable to the characteristics of voltage-gated K^+ channel-blocking agents. Single-channel experiments performed on canine tracheal muscle were able to identify only voltage-gated Ca^{2+}-activated K^+ channels, since rectification is inhibited by K^+-channel blocking agents. Single channel experiments performed on canine tracheal muscle were able to identify only voltage gated Ca^{2+}-activated K^+ currents (McCann and Welsh, 1986); however, recently whole-cell clamps in this tissue unmasked another voltage-gated K^+ channel that was Ca^{2+} independent (Kotlikoff, 1987). Thus, muscarinic agents activate at least two rectifying outward K^+ currents as a result of membrane depolarization and increases in cytosolic $[Ca^{2+}]$. The marked rectifying properties of the membrane due, at least in part, to activation of K^+ currents seems to explain the graded nature of membrane potential changes with airway smooth muscle in that the action potential threshold is never achieved under normal conditions. Muscarinic agents inactivate a late K^+ current (a M current) in some smooth muscles, but this effect does not explain muscarinic-evoked membrane depolarization (Sims *et al.*, 1988).

IV. COUPLING AMONG RECEPTORS, CHANNELS, AND SECOND-MESSENGER SYSTEMS

In the previous sections of this chapter we have discussed coupling mechanisms that appear to be important in muscarinic evoked contractions of airway smooth muscle. Evidence indicates that the inositol phospholipid transduction system and membrane-depolarization and activation of membrane-gated Ca^{2+} channels are important coupling mechanisms. Contractions of smooth muscle involve a cascade of events, some of which result in increases in cytosolic $[Ca^{2+}]$ and consequent activation of contraction by means of phosphorylation of myosin light-chain kinase. The mechanism for activation of contractile proteins during the latch state (during maintained force) is still poorly understood. These areas have been covered in Chapter 6 (this volume). We will now consider some of the initial steps in the cascade of events involved in coupling—events at receptors and how signals originating in receptors, when occupied by ACh, are delivered to channels or enzymes that control the inositol phospholipid transduction system and other transduction systems. There are gaps in our knowledge about how muscarinic receptors in airway smooth muscle direct the opening of channels that result in membrane depolarization. On the basis of studies in other systems, it is very likely that this occurs via G proteins. There are protein–protein interactions between liganded receptor proteins and G proteins with resultant G protein disaggregation into subunits, seen as interacting with channel proteins and altering their voltage–current properties (Hescheler *et al.*, 1987; Yatoni *et al.*, 1987). The first demonstration of an effect of a G protein on an mammalian ion channel (Yatoni *et al.*, 1987) was published in 1985, and we can expect that in the coming years additional knowledge in this area will become available. G proteins were discovered by investigators studying signals between the β-adrenergic receptor and adenylate cyclase (Lefkowitz et al., 1983). Ago-nist-induced PIP_2 hydrolysis is activated by a G protein acting on PLC, in a manner analogous to regulation of adenylate cyclase (Litosch and Fain, 1985; Taylor and Merritt, 1986). It is possible that activation of PLA_2 also involves transduction from receptor via G proteins and/or activation by increases in cyto-solic $[Ca^{2+}]$ due to influx via Ca^{2+} channels or release from the SR (Chang, 1987). The number of known G proteins obtained in non-smooth muscle prepara-tions is proliferating, with different G proteins having specificity for different actions, i.e., different channels, activation of adenylate cyclase, PLC, and so forth.

The use of molecular biological techniques has demonstrated the presence of multiple muscarinic receptor proteins. We are at a stage at which questions about the coupling roles of various types of muscarinic receptors are being approached. A single type of muscarinic receptor can activate at least two differ-ent G proteins (Ashkenazi et al., 1987), suggesting that single receptor classes do not have only one function in cell control.

V. CYTOSOLIC FREE CALCIUM CONCENTRATION

Cytosolic $[Ca^{2+}]$ is an important component of the cascade of events that operates during EMC and PMC in smooth muscle. It can be argued that despite the use of Fura-2 and aequorin to measure free cytosolic $[Ca^{2+}]$ in smooth muscle, there is still uncertainty regarding changes in $[Ca^{2+}]$ during K^+ and agonist-induced contractions. Our discussion about EMC and PMC mechanisms has emphasized different mechanisms of increasing cytosolic free $[Ca^{2+}]$. Ac-tivation of non-airway smooth muscle by various agonists is associated with an initial rapid increase in cytosolic $[Ca^{2+}]$ followed by a decline to a steady-state level (Kotlikoff et al., 1987; Takuwa et al., 1987; Himpens and Somlyo, 1988). Using activation of phosphorylase as an indicator of intracellular $[Ca^{2+}]$, Silver and Stull (1985) recorded a similar pattern of intracellular $[Ca^{2+}]$ changes in canine trachealis muscle. With K^+-induced contractions, the pattern is dis-similar to that seen with agonist activation in that $[Ca^{2+}]$ are maintained (Tak-uwa et al., 1987; Himpens and Somlyo, 1988). There are differences in patterns of increases of free cytosolic $[Ca^{2+}]$ in different smooth muscles and with the use of different indicators that need to be explained (Bruschi et al., 1988). Although there is evidence that during the onset of contraction activating Ca^{2+} originates in the SR, it still is not certain that inositol phospholipid metabolism can be activated this rapidly. During maintained force, activating Ca^{2+} appears to originate both in the SR and the extracellular space (Rasmussen et al., 1987; Himpens and Somlyo, 1988; Baba et al., 1988).

VI. MODULATION OF ELECTROMECHANICAL AND PHARMACOMECHANICAL COUPLING MECHANISMS

There is evidence that many of the events in the cascade of reactions that activate smooth muscle can be modulated by second messengers:

1. *Plasma membrane receptors—uncoupling by phosphorylation:* There is direct evidence that receptors can be phosphorylated via cAMP-dependent protein kinase (Lefkowitz *et al.*, 1983) as well as indirect evidence for PKC-mediated phosphorylation of muscarinic receptors (Lundberg *et al.*, 1985). Since it is likely that different classes of muscarinic receptors are coupled to different G proteins and control different cellular processes, receptor phosphorylation may control the turning on or off of these processes during muscarinic activation.

2. *Receptor–receptor modulation:* G proteins subunits coupled to one type of receptor, can modulate coupling in another type of receptor. The classic example is the inhibitory effect of G_i, which is activated by β-adrenergic receptor occupancy, on muscarinic receptor coupling (Lefkowitz *et al.*, 1983).

3. *Ion channels:* The concept that ion channels can be phosphorylated and that phosphorylation alters their voltage–current relationships is established (Levitan, 1985; Richards and Murlas, 1987). Phorbol esters cause membrane depolarization in swine trachealis muscle (Baba and Coburn, 1988), suggesting that the functions of channels controlling resting E_m are altered by PKC-mediated phosphorylations. There is considerable evidence that phorbol esters alter the operation of plasma membrane Ca^{2+} channels in vascular and airway smooth muscle and other cells (Osugi *et al.*, 1986; Galizzi *et al.*, 1987; Litten *et al.*, 1987; Baba *et al.*, 1988). Increasing [cAMP] or [cGMP] suppresses canine trachealis muscle TEA-evoked action potentials (Richards *et al.*, 1987; Richards and Murlas, 1987). Cytosolic [Ca^{2+}] modulates currents in smooth muscle plasma membrane Ca^{2+} channels (Ohya *et al.*, 1988). Inositol tetrakisphosphate may modulate plasma membrane Ca^{2+} transport (Irvine and Moor, 1986; Morris *et al.*, 1987).

4. *Inositol phospholipid metabolism:* Phorbol esters inhibit the rate of carbachol-activated inositol phospholipid metabolism, suggesting that PKC exerts feedback control, via phosphorylation, on the major excitatory transduction system in airway smooth muscle. cGMP may also exert control on inositol phospholipid transduction (Rapoport, 1986). Inositol phosphate metabolism can be modulated by a PKC-mediated phosphorylation (Williamson *et al.*, 1987).

5. *Effects of cytosolic [Ca^{2+}] on myofilaments:* The effects of cAMP on the Ca^{2+} affinity of myosin light-chain kinase are discussed in Chapter

6 (this volume). Phorbol esters cause smooth muscle contraction without a detectable increase in free cytosolic $[Ca^{2+}]$ (Jiang and Morgan, 1987; Rasmussen et al., 1987). This finding, as well as other data (Itoh et al., 1988), suggest that PKC-mediated phosphorylation may alter the Ca^{2+} affinity at regulatory sites or that PKC-mediated phosphorylation evokes a Ca^{2+}-independent contraction. The decrease in free cytosolic $[Ca^{2+}]$ that apparently occurs during maintained force induced by agonists suggests that Ca^{2+} has different actions during development and during the maintenance of force.

6. *Possible modulation of coupling mechanisms by pH:* The activation of smooth muscle by agonists that activate inositol phospholipid metabolism is associated with activation of $Na^+ - H^+$ exchange, with resulting alkalinization (Danthuluri et al., 1987). Whether this plays a physiological role is unknown. This area has been reviewed recently (Wray, 1988).

The use of phorbol esters in the study of coupling mechanisms in airway smooth muscle has uncovered data that show the potential of PKC-mediated phosphorylation for modulation of both PMC and EMC (Baron et al., 1987b; Baba et al., 1988). The phorbol ester-induced contraction is associated with membrane depolarization (Baba and Coburn, 1988), suggesting that channels involved in depolarization can be controlled by PKC-evoked phosphorylation. Carbachol-evoked contractions in muscle preincubated with phorbol ester occur with a smaller membrane depolarization and become similar to K^+-evoked contractions in that they are dependent on external Ca^{2+} and are inhibited by verapamil and nifedipine. Metabolic flux in inositol phospholipids is markedly or completely inhibited by treatment with phorbol esters, and it appears that PMC mechanisms are not operating under this condition. Thus, PKC-related reactions are able to modulate coupling mechanisms such that PMC is inhibited and EMC mechanisms are responsible for contractions. In addition, there is evidence PKC-mediated phosphorylation alters the sensitivity of contractile filaments to Ca^{2+} (Jiang and Morgan, 1987; Itoh et al., 1988). Figure 1 compares K^+-evoked contractions and carbachol-evoked contractions under conditions in which only EMC mechanisms are operative. Figure 1 emphasizes the extent of modulation that can occur, as shown with E_m–force plots.

VII. SUMMARY

Figure 2 summarizes many of the points made in this chapter. It is shown that the two coupling mechanisms are not entirely separate, in that muscarinic receptors and inositol phospholipid-generated signals appear to influence both coupling mechanisms. We have drawn multiple muscarinic receptors and multi-

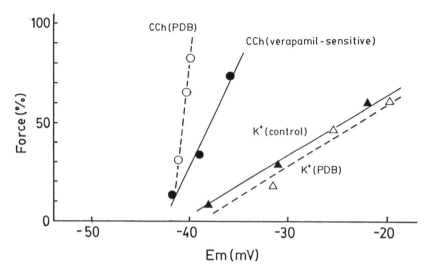

FIGURE 1. Different membrane potential–force relationships in different electromechanical coupling type of contractions of swine trachealis muscle: 40 mM K, 40 mM K in phorbol dibutyrate preincubated tissue, verapamil-sensitive carbachol-evoked force, and carbachol-evoked force in the presence of PDB. As reviewed in the text, the carbachol-induced contraction in phorbol ester-preincubated tissue has been converted to a EMC-type contraction. The marked shift in E_m–force to the left with carbachol, compared with that seen with K^+-evoked contractions, is probably due to an effect of G proteins on voltage-dependent Ca^{2+} channels. Phorbol ester preincubation had no effect on K^+-evoked contractions but causes a shift in the carbachol membrane potential–force plot to the left, compared with verapamil-sensitive data. (From Baba *et al.*, 1988.)

→

FIGURE 2. Schematized representation of EMC and PMC mechanisms. (A) EMC following binding of a muscarinic agent to muscarinic receptors; there is a (1) G protein subunit mediated activation of channels that depolarize the plasma membrane to values less than the voltage-gated Ca^{2+} channel threshold, (2) influx of Ca^{2+} via these channels, (3) increase in cytosolic $[Ca^{2+}]$, (4) activation of myosin light-chain kinase, (5) phosphorylation of the myosin light chain, and (6) activation of actomyosin ATPase. (B) The PMC cascade involves (1) G-subunit activation of phospholipase C, (2) activation of the inositol phospholipid transduction system with resultant synthesis of I(1,4,5)-P$_3$, (3) release of Ca^{2+} from the sarcoplasmic reticulum, (4) activation of myosin light-chain kinase and myosin light-chain phosphorylation, (5) possible protein kinase C-mediated phosphorylation signals activated by increases in diacylglycerol concentrations. In the EMC schema, a PMC event is shown, I(1,4,5)-P$_3$-mediated Ca^{2+} release from the sarcoplasmic reticulum. This is shown because increases in cytosolic $[Ca^{2+}]$ may regulate plasma membrane Ca^{2+} channels. Multiple muscarinic receptors are shown; we assume that different receptors are coupled to different G proteins that direct activation signals to different channels or to PLC. We assume that PMC is entirely due to second messengers generated by inositol phospholipid metabolism, as suggested by the study of Baba *et al.* (1988). The PMC figure emphasizes multiple controls exerted by cGMP and cAMP on inositol phospholipid metabolism.

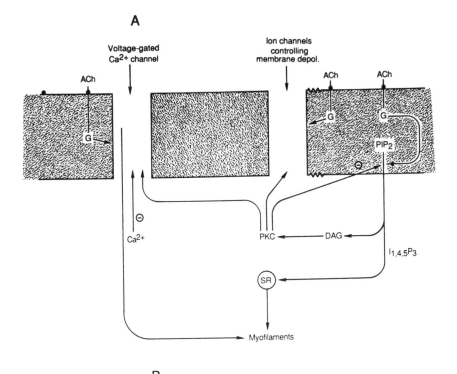

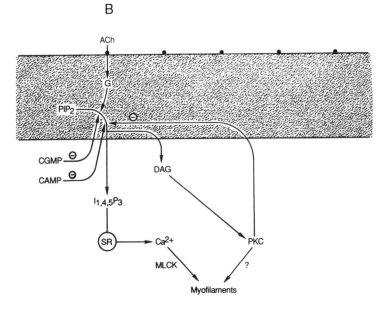

ple G proteins, since it is unknown whether single G proteins or receptors are specific for one effector or for many.

ACKNOWLEDGMENTS. This work was supported by grant 1 R37 HL 37498 from the National Blood, Heart and Lung Institute, National Institutes of Health, Bethesda, Maryland.

VIII. REFERENCES

Ashkenazi, A., Winslow, J. W., Peralta, E. G., Peterson, G. L., Schimerlim, M. I., and Capan, J., 1987, An M2 muscarinic receptor subtype coupled to both adenylcyclase and phosphoinositide turnover, *Science* **238**:672–675.

Baba, K., and Coburn, R. F., 1988, Effects of phorbol ester on the membrane potential and tension in swine tracheal smooth muscle, *FASEB J.* **2**:A331.

Baba, K., Baron, C. B., and Coburn, R. F., 1988, Phorbol ester-induced conversion of pharmacomechanical coupling to electromechanical coupling in the response of swine tracheal smooth muscle to carbachol, *FASEB J.* **2**:A331.

Baron, C. B., and Coburn, R. F., 1987a, Inositol phospholipid turnover during contraction of canine trachealis muscle, *Ann. NY Acad. Sci.* **494**:80–83.

Baron, C. B., and Coburn, R. F., 1987b, Phorbol ester modulates inositol phospholipid metabolism in carbamylcholine-stimulated tracheal smooth muscle, *Fed. Proc.* **46**:704.

Baron, C. B., Cunningham, M., Strauss, J. F. III, and Coburn, R. F., 1984, Pharmacomechanical coupling in smooth muscle may involve phosphatidylinositol metabolism, *Proc. Natl. Acad. Sci. USA* **81**:6899–6903.

Baron, C. B., Pring, M., and Coburn, R. F., 1989, Inositol lipid turnover and compartmentation in canine trachealis muscle, *Am. J. Physiol.* **256**:(in press).

Benham, C. D., Bolton, T. B., and Lang, R. T., 1985, Acetylcholine activates an inward current in single mammalian smooth muscle cells, *Nature (Lond.)* **316**:345–347.

Berridge, M. J., Dawson, R. M. C., Downes, C. P., Heslop, J. P., and Irvine, R. F., 1983, Changes in levels of inositol phosphates after agonist-dependent hydrolysis of membrane phosphoinositides, *Biochem. J.* **212**:473–482.

Bolton, T. B., 1972, The depolarizing action of acetylcholine or carbachol in intestinal smooth muscle, *J. Physiol. (Lond.)* **220**:647–671.

Bolton, T. B., 1979, Mechanism of action of transmitters and other substances on smooth muscle, *Physiol. Rev.* **59**:606–718.

Bruschi, G., Bruschi, M. E., Regolisti, G., and Broghetti, A., 1988, Myoplasmic Ca^{2+}–force relationship studied with fura-2 during stimulation of rat aortic smooth muscle, *Am. J. Physiol.* **254**:H849–854.

Byrne, N. G., and Large, W. A., 1988, Action of noradenaline on single smooth muscle cells freshly dispersed from rat anococcygeus muscle, *J. Physiol. (Lond.)* **389**:513–520.

Chang, J., Musser, J. H., and McGregor, H., 1987, Phospholipase A2, *Biochem. Pharmacol.* **36**:2429–2436.

Chatterjee, M., and Tejada, M., 1986, Phorbol ester-induced contraction in chemically skinned vascular smooth muscle, *Am. J. Physiol.* **251**:C356–361.

Coburn, R. F., 1977, The airway smooth muscle cell, *Fed. Proc.* **36**:2692–2696.

Coburn, R. F., 1979, Electromechanical coupling in canine trachealis muscle: Acetylcholine contractions, *Am. J. Physiol.* **236**:C177–184.

Danthuluri, N. R., Berk, B. C., Brock, T. A., Crago, E. J., Jr., and Deth, R. C., 1987, Protein kinase C-mediated intracellular alkalinization in rat and rabbit aortic smooth muscle cells, *Eur. J. Pharmacol.* **141**:503–506.

Eberlin, L. B., Cherniack, A. D., and Deal, E. C., Jr., 1982, *Am. Rev. Respir. Dis.* **125** (*Suppl.*): 225–233.

Evans, D. H. L., Schild, H. O., and Thesleff, S., 1958, Effects of drugs on depolarized plain muscle, *J. Physiol. (Lond.)* **143**:474–485.

Farese, R. B., Konda, T. S., Davis, J. S., Standaert, M. L., Pollet, R. J., and Cooper, D. R., 1987, Insulin rapidly increases diacylglycerol by activating denovo phosphatidic acid synthesis, *Science* **236**:586–589.

Farley, J. M., and Miles, P. R., 1977, Role of depolarization in acetylcholine-induced contractions of dog tracheal muscle, *J. Pharmacol. Exp. Ther.* **201**:199–205.

Friedman, M. E., Suarez-Kurtz, G., Kaczorowski, G. J., Katz, G. M., and Reuben, J. P., 1986, Two calcium currents in a smooth muscle cell line, *Am. J. Physiol.* **250**:H699–703.

Galizzi, J., Oar, J., Fosset, M., Van Renterghem, C., and Lazdunski, M., 1987, Regulation of calcium channels in aortic muscle cells by protein kinase C activators, *J. Biol. Chem.* **262**: 6947–6950.

Hashimoto, T., Hirata, M., and Ito, Y., 1985, A role for inositol 1,4,5-trisphosphate in the initiation of agonist-induced contractions of dog tracheal smooth muscle, *Br. J. Pharmacol.* **86**:191–199.

Hashimoto, T., Hirata, M., Itoh, T., Kanmura, Y., and Kuriyama, H., 1986, Inositol 1,4,5-trisphosphate activates pharmacomechanical coupling in smooth muscle of the rabbit mesenteric artery, *J. Physiol. (Lond.)* **307**:605–618.

Hescheler, J., Rosenthal, W., Trautwein, W., and Schultz, G., 1987, The GTP-binding protein, Go, regulates neuronal calcium channels, *Nature (Lond.)* **325**:445–447.

Himpens, B., and Somlyo, A. P., 1988, Free calcium and force transients during depolarization and pharmacomechanical coupling in guinea-pig smooth muscle, *J. Physiol. (Lond.)* **395**:507–530.

Irvine, R. F., and Moor, R. M., 1986, Micro-injection of inositol 1,3,4,5-tetrakisphosphate activates sea urchin eggs by a mechanism dependent on external Ca^{2+}, *Biochem. J.* **240**:917–920.

Itoh, T., Kubota, Y., and Kuriyama, H., 1988, Effects of phorbol ester on acetylcholine-induced Ca^{2+} mobilization and contraction in the porcine coronary artery, *J. Physiol. (Lond.)* **397**:401–419.

Ito, Y., Kitamura, K., and Kuriyama, H. 1979, Effects of acetylcholine and catecholamines on the smooth muscle cells of the porcine coronary artery, *J. Physiol. (Lond.)* **294**:594–611.

Jiang, M. J., and Morgan, K. G., 1987, Intracellular calcium levels in phorbol ester-induced contractions of vascular smooth muscle, *Am. J. Physiol.* **253**:H1365–1371.

Kikkawa, U., Takai, Y., Tanaka, Y., Miyake, R., and Nishizuka, Y., 1983, Protein kinase C as a possible receptor protein of tumor-promoting phorbol esters, *J. Biol. Chem.* **258**:11442–11445.

Kotlikoff, M., 1987, Potassium currents in isolated airway smooth muscle cells, *Am. Rev. Respir. Dis.* **135**:A273.

Kotlikoff, M., 1988, Calcium currents in isolated canine airway smooth muscle cells, *Am. J. Physiol.* **254**:C793–801.

Kotlikoff, M. I., Murray, R. K., and Reynolds, E. E., 1987, Histamine-induced calcium release and phorbol antagonists in cultured airway smooth muscle cells, *Am. J. Physiol.* **253**:C561–565.

Lefkowitz, R. J., Stadel, J. M., and Caron, M. C., 1983, Adenylate cyclase-coupled beta adrenergic receptors, *Annu. Rev. Biochem.* **52**:159–186.

Levitan, B. C., 1985, Phosphorylation of ion channels, *J. Membr. Biol.* **87**:177–190.

Litosch, I., and Fain, J. N., 1985, GTP-dependent PLC-mediated hydrolysis, *J. Biol. Chem.* **260**: 16052–16055.

Litten, R. Z., Suba, E. A., and Roth, B. L., 1987, Effects of a phorbol ester on rat aortic contraction and calcium influx in the presence and absence of BAY K 8644, *Eur. J. Pharmacol.* **144**:185–191.

Lundberg, L. M., Cotecchia, F. S., Lomarney, J. F., DeBernadis, J. F., Lefkowitz, F. J., and

Caron, M. G., 1985, Phorbol esters promote alpha$_1$-adrenergic receptor phosphorylation and receptor uncoupling from inositol phospholipid metabolism, *Proc. Natl. Acad. Sci. USA* **82:** 5651–5655.

McCann, J. D., and Welsh, M. J., 1986, Calcium-activated potassium channels in canine airway smooth muscle, *J. Physiol. (Lond.)* **372:**113–126.

Menkes, H., Baraban, J. M., and Snyder, S. H., 1986, Protein kinase C regulates smooth muscle tension in guinea-pig trachea and ileum, *Eur. J. Pharmacol.* **122:**19–28.

Morris, A. P., Gallacher, D. V., Irvine, R. I., and Peterson, O. H., 1987, Synergism of inositol trisphosphate and tetrakisphosphate in activating Ca-dependent K-channel, *Nature (Lond.)* **330:** 653–655.

Nakazawa, K., Saito, H., and Matsuka, N., 1988, Fast and slowly inactivating components of Ca-channel current and their sensitivities to nicardipine in isolated smooth muscle cells from rat vas deferens, *Pflugers Arch.* **411:**289–295.

Nowycky, M. D., Fox, A. P., and Tsien, R. W., 1985, Three types of neuronal calcium channels with different calcium agonist sensitivity, *Nature (Lond.)* **316:**440–443.

Ohya, Y., Kitamura, K., and Kuriyama, H., 1988, Regulation of calcium current by intracellular calcium in smooth muscle cells of rabbit portal vein, *Circ. Res.* **62:**375–383.

Osugi, T., Imaizumi, T., Mizushima, A., Uchida, S., and Yoshida, E., 1986, 1-oleoyl-2-acetyl-glycerol and phorbol diester stimulated Ca^{2+} influx through Ca^{2+} channels in neuroblastoma × glioma hybrid NG108-15 cells, *Eur. J. Pharmacol.* **126:**53–60.

Rapoport, R. M., 1986, Cyclic guanosine monophosphate inhibition of contraction may be mediated through inhibition of phosphatidylinositol hydrolysis in rat aorta, *Circ. Res.* **58:**407–410.

Rasmussen, H., Takuwa, Y., and Park, S., 1987, Protein kinase C in the regulation of smooth muscle contraction, *FASEB J.* **1:**177–185.

Rebecchi, M. J., and Posen, O. M., 1987, Stimulation of polyphosphoinositide hydrolysis by thrombin in membranes from human fibroblasts, *Biochem. J.* **245:**49–57.

Reuter, H., 1974, Localization of beta adrenergic receptors and effects of noradrenaline and cyclic nucleotides on action potential ionic currents and tension in mammalian cardiac muscle, *J. Physiol. (Lond.)* **242:**429–451.

Richards, I. S., and Murlas, C. G., 1987, 8-bromo-cyclic GMP abolishes TEA-induced slow action potentials in canine trachealis muscle, *Eur. J. Pharmacol.* **128:**199–302.

Richards, I. S., Ousterholt, N., Sperelakis, N., and Murlas, C. G., 1987, CAMP suppresses Ca^{2+}-dependent electrical activity of airway smooth muscle induced by TEA, *J. Appl. Physiol.* **62:** 175–179.

Sasakawa, N., Nakaki, T., Yamamoto, S., and Kato, R., 1987, Inositol trisphosphate accumulation by high K^+ stimulation in cultured adrenal chromaffin cells, *FEBS Lett.* **223:**413–416.

Silver, P. J., and Stull, J. T., 1985, Phosphorylation of myosin light chain and phosphorylase in tracheal smooth muscle in response to KCl and carbachol, *Mol. Pharmacol.* **25:**267–274.

Sims, S. M., Singer, J. J., and Walsh, J. V., Jr., 1988, Antagonistic adrenergic-muscarinic regulation of M current in smooth muscle cells, *Science* **239:**190–193.

Smith, C. W. J., Pritchard, K., and Marston, S. B., 1987, The mechanisms of Ca regulation of vascular smooth muscle thin filaments by caldesmon and calmodulin, *J. Biol. Chem.* **262:**116–122.

Somlyo, A. V., and Somlyo, A. P., 1968, Electromechanical and pharmacomechanical coupling in vascular smooth muscle, *J. Pharmacol. Exp. Ther.* **159:**129–145.

Streb, H., Irvine, R. F., Berridge, M. J., and Schulz, I., 1983, Release of Ca^{2+} from a non-mitochondrial intracellular store in pancreatic acinar cells by inositol 1,4,5-trisphosphate, *Nature (Lond.)* **306:**67–68.

Sturek, M., and Hermsmeyer, K., 1986, Calcium and sodium channels in spontaneously contracting vascular muscle cells, *Science* **233:**475–478.

Takuwa, Y., Takuwa, N., and Rasmussen, H., 1987, Measurement of cytosolic free Ca^{2+}

concentration in bovine tracheal smooth muscle using aequorin, *Am. J. Physiol.* **253**:C817–827.

Taylor, C. W., and Merritt, J. E., 1986, Receptor coupling to polyphosphoinositide turnover: A parallel with the adenylate cyclase system, *Trends Pharmacol. Sci.* **7**:238–240.

Williamson, J. R., Hansen, C. A., Verhoeven, A., Coll, K. E., Johanson, R., Williamson, M. T., and Filburn, C., 1987, in: *Cellular Calcium and Membrane Transport* (P. C. Eaton and L. J. Mandel, eds.), pp. 93–116, Rockefeller Press, New York.

Wray, S., 1988, Smooth muscle intracellular pH; measurement, regulation and function, *Am. J. Physiol. Cell* **254**:C213–225.

Yamaguchi, T., Hitzig, B., and Coburn, R. F., 1976, Endogenous prostaglandins and mechanical tension in canine trachealis muscle, *Am. J. Physiol.* **230**:1737–1745.

Yatoni, A., Codina, J., Imoto, Y., Reeves, J. P., Birnbaumer, L., and Brown, A. M., 1987, A G protein directly regulates mammalian cardiac calcium channels, *Science* **238**:1288–1292.

Chapter 11

Postulated Mechanisms Underlying Airway Hyperreactivity

Relationship to Supersensitivity

Douglas W. P. Hay

Department of Pharmacology
Research and Development
Smith, Kline & French Laboratories
King of Prussia, Pennsylvania 19406-0939

I. INTRODUCTION

Airway hyperreactivity is defined as a marked nonspecific increase in sensitivity of the respiratory tract to a variety of stimuli that results acutely in an enhanced bronchospastic response (Benson, 1975; Boushey *et al.*, 1980; Simonsson, 1983). The diverse physical, chemical, and pharmacological stimuli include various smooth muscle spasmogens (Benson, 1975; Boushey *et al.*, 1980), cold air (Wells *et al.*, 1960; Simonsson *et al.*, 1967), dust (Simonsson *et al.*, 1967), exercise (Lee and Anderson, 1985), environmental irritants (Simonsson, 1983), and forced breathing (Simonsson *et al.*, 1967). This phenomenon is a hallmark of asthma and is considered the underlying abnormality in the pathophysiology of this disorder. In addition, airway hyperreactivity is commonly manifest in other pulmonary disorders, including chronic bronchitis (Klein and Salvaggio, 1966; Laitinen, 1974), cystic fibrosis (Mellis, 1978), allergic rhinitis (Laitinen, 1974; Cockcroft *et al.*, 1977), and emphysema (Klein and Salvaggio, 1966; Laitinen, 1974), and also occurs during viral infections (Empey *et al.*, 1976). This chapter deals predominantly with the hyperreactivity of airways associated with asthma.

Asthma is a common and serious disorder of children and adults; it has considerable social and economic impact and not insignificant mortality (Iafrate *et al.*, 1986; Sears, 1986). A disturbing fact, which highlights the importance of elucidating the underlying causes and mechanisms involved in its development and maintenance, is that both the occurrence of, and the mortality from, asthma in several countries, including the United States, is increasing (Sly, 1986; Sears, 1986).

Airway hyperreactivity can be considered a form of supersensitivity. Supersensitivity is defined and manifest as a phenomenon whereby the amount of a substance needed to elicit a set response under certain conditions is less than normal (Fleming et al., 1973; Fleming, 1976; Westfall, 1981). Thus, there is a leftward shift in the dose–response curve and, in some instances, concomitant with this increase in sensitivity, there may be an increase in the maximum response produced by the stimulant and/or a change in the slope of the dose–response curve (Fleming et al., 1973; Fleming, 1976; Westfall, 1981). Supersensitivity, which occurs in many, if not all, excitable tissues, including smooth muscle, can be subdivided into two types: type I (or deviation supersensitivity, also referred to originally as presynaptic or prejunctional supersensitivity), and type II (or nondeviation supersensitivity, also known as disuse, nonspecific, or postjunctional supersensitivity). Type I supersensitivity is brought about by an increase in the percentage of a given amount of agonist interacting with its receptors. That is, there is no inherent alteration in the properties and responsiveness of the effector tissue per se. Type I supersensitivity is specific for a few agonists and has been examined and characterized primarily for the adrenergic system (Fleming et al., 1973; Fleming, 1976; Westfall, 1981). In contrast to type I supersensitivity, type II supersensitivity is nonspecific and occurs as a result of an increase in the responsiveness of the effector tissue rather than an alteration in the drug concentration reaching the receptor. In smooth muscles, type II supersensitivity is usually demonstrated experimentally as a consequence of a chronic suppression in the normal communication between the neurotransmitter and its associated effector cell.

Various mechanisms are postulated to account for the increased cellular or tissue responsiveness with type II supersensitivity (Fleming et al., 1973; Fleming, 1976; Westfall, 1981):

1. An alteration in receptor density and/or affinity (in smooth muscle there is little evidence of a change in the properties of the receptors)
2. Change in the membrane electrophysiological properties
3. Alteration in the permeability and transport of ions (a popular and logical mechanism, for which there is some evidence in smooth muscle, is an alteration in Ca^{2+} homeostasis)
4. Biochemical changes (e.g., in the levels and activity of cyclic nucleotides)
5. Morphological changes in the effector tissue.

The evidence suggests that it is likely that multiple rather than a single mechanism underlies type II supersensitivity, and the relative involvement of these processes will depend on such factors as the particular effector tissue and the agonist under study (Fleming *et al.*, 1973; Fleming, 1976; Westfall, 1981).

In summary, supersensitivity is a widespread phenomenon that may be brought about by various procedures and for which the fundamental processes responsible for it remain to be elucidated. The increased sensitivity may be either a result of a change in the amount of stimulant reaching an unaltered effector cell (type I or deviation supersensitivity) or a consequence of a change in the inherent properties of the effector cell (type II or nondeviation supersensitivity).

In studies in asthmatics there is generally a decrease in the threshold and/or an increase in the steepness of the curve concomitant with elevation in the maximum response to bronchoconstrictor agents *in vivo* (Boushey *et al.*, 1980; Moreno *et al.*, 1986a). These findings, in conjunction with the nonspecific nature of airway hyperreactivity, suggest that airway hyperreactivity is an example of type II, or nondeviation, supersensitivity (Fleming *et al.*, 1973; Fleming, 1976; Westfall, 1981; Moreno *et al.*, 1986a). The enhanced bronchoconstrictor response in airway hyperreactivity suggests that there is an alteration in the smooth muscle contractile system. This bronchospasm is the predominant event in the acute stages of asthma. In this review, the focus is on the possible mechanisms responsible for airway hyperreactivity, in general, and specifically on the associated increased bronchospastic response. It remains to be clarified whether there is hyperreactivity in other systems, such as inflammatory cells and nerves. The evidence strongly suggests that mechanisms other than increased bronchoconstriction, notably inflammation and increased mucus secretion, impart a significant contribution to the airway obstruction and hyperreactivity associated with the chronic stages of the disorder (Chung, 1986; Hargreave *et al.*, 1986b; Holgate, 1986; Holgate *et al.*, 1986b; Page *et al.*, 1986; see also Section II.A).

II. POTENTIAL MECHANISMS OF AIRWAY HYPERREACTIVITY

Despite intensive research, the underlying mechanism(s) responsible for airway hyperreactivity has not been elucidated. A plethora of hypotheses have been proposed some of which are interrelated. The observation that there is nonspecific exaggerated responsiveness of the airways would suggest that a common mechanism is responsible for the phenomenon of airway hyperreactivity. However, the general concensus and prevailing clinical and experimental evidence suggest that several processes are involved. A controversial issue that is not addressed in this chapter is the extent to which airway hyperreactivity is genetically predetermined (Simonsson, 1983; Sibbald, 1986).

Postulated mechanisms underlying airway hyperreactivity include the following:

Dysfunction in receptors
Imbalance in regulation of the autonomic nervous system
Change in electrophysiological characteristics of smooth muscle cells
Abnormalities in Ca^{2+} translocation and handling
Change in airway morphology/geometry: mechanical factors
Decreased airway caliber
Increased release of mediators
Epithelial cell damage or dysfunction
Inflammation

Each of these mechanisms is discussed individually.

A. Dysfunction in Receptors

Airway hyperreactivity could be a consequence of an alteration in the properties of one or more of the numerous membrane receptors in the respiratory tract, which, when stimulated, controls the level of activity in the different cellular systems. This could be manifest as a change in receptor affinity and/or number or in receptor/signal transduction coupling. Research has centered almost exclusively on the hypothesis that an imbalance in the adrenergic system is responsible for the enhanced bronchoconstriction associated with airway hyperreactivity. In addition, the involvement of dysfunction of histamine–H2-receptors has been proposed. These two areas will be discussed below. The increased sensitivity of airways to carbachol following exposure to toluene diisocyanate in guinea pig was attributed to an increase in the number or affinity of muscarinic receptors (McKay and Brooks, 1983).

1. Imbalance between α- and β-Adrenoceptors: The Szentivanyi Hypothesis

Szentivanyi (1968) proposed originally that airway hyperreactivity was a consequence of an abnormality in the bronchodilatory β_2-adrenoceptor system. Concomitant with this deficiency, Szentivanyi (1979, 1980) postulated an increased number or activity of α-adrenoceptors, which mediate bronchoconstriction. Since its introduction, this theory has received rather limited clinical and experimental support. For example, there are several reports of a decreased responsiveness of granulocytes and leukocytes; furthermore, metabolic processes to β-adrenoceptor stimulation in asthmatics (Szentivanyi, 1980) and radioligand binding techniques demonstrated a decreased number of β-adrenoceptors in asthmatics compared with control patients (Kariman and Lefkowitz, 1977; Brooks *et al.*, 1979). One study showed good correlation between the extent of the reduction in β-adrenoceptor activity following epinephrine administration in asthmatics and the severity of the disease (Makino *et al.*, 1970). However, this phenomenon has been, in general, attributed to tachyphylaxis produced in response to treatment with β-adrenoceptor agonists, rather than as a consequence of the disease. For example, exposure of nonasthmatics to β-adrenoceptor agonists can lead to decreased sensitivity to subsequent adrenergic stimulation (Galant *et al.*, 1980b; Svedmyr, 1984). The apparent decrease in β-adrenoceptor responsiveness appears to be more striking and prevalent in nonairway tissues and systems (Szentivanyi, 1980; Svedmyr, 1984), suggesting fundamental differences in the β-adrenoceptors in airways compared with extrapulmonary systems of asthmatics [e.g., increased number of spare receptors resulting in longer periods before the development of tolerance (Svedmyr, 1983)]. Moreover, the reduced sensitivity is not restricted to the β-adrenergic system (Busse and Sosman, 1977; Busse *et al.*, 1979a,b). Some reports indicate that there is no change in β_2-adrenoceptor number or adenylate cyclase activity in lymphocytes and neutrophils from asthmatic patients (Galant *et al.*, 1978, 1980a). Despite this criticism, it has been reported that a decreased β-adrenoceptor responsiveness was apparent in asthmatics not receiving β-agonist medication (Kariman and Lefkowitz, 1977; Brooks *et al.*, 1979).

Perhaps the most controversial aspect of the Szentivanyi hypothesis is the proposal that there is a transformation of β-adrenoceptors to α-adrenocepters. Evidence indicates that α- and β-adrenoceptors are structurally different (Wood *et al.*, 1979) and are unlikely to interconvert (Benfey, 1980). Nonetheless, modulation of α- and/or β-adrenoceptor number, rather than interconversion, has been reported. For example, radioligand studies in human lung membrane fractions indicated a slight decrease in β-adrenoceptor number concomitant with a marked increase in the numbers of α-adrenoceptors in asthmatics compared with control patients (Szentivanyi, 1980). Barnes *et al.* (1980) reported an approximately 10-fold elevation in the number of α_1-adrenoceptors in parenchymal tissue from nine patients with chronic obstructive airway disorders compared with tissue obtained from one nonasthmatic patient. However, these studies were conducted using a tissue containing many different cell types, and no attempt was made to correlate the alterations in receptor number with changes in the responsiveness of airway smooth muscle (Goldie *et al.*, 1985).

A prerequisite to support the hypothesis that dysfunction of the β-adrenoceptor system is important in airway hyperreactivity is decreased sensitivity of asthmatic airways to β-adrenoceptor agonists compared with control airways. This is extremely difficult to study *in vivo* for several reasons, including differences in baseline function between subjects and problems in maintaining the degree of bronchospasm induced by the agonist constant in asthmatic and control groups. No difference in sensitivity to β-adrenoceptor agonists was detected in mild asthmatics compared with controls (Harvey and Tattersfield, 1982), although a difference was evident comparing severe asthmatics and control subjects (Barnes and Pride, 1983). There are conflicting reports from *in vitro* studies of a decrease in sensitivity to β-adrenoceptor agonists in airways from asthmatics (Barnes *et al.*, 1984; Barnes, 1986b; Goldie *et al.*, 1986b).

Another fundamental requirement to validate the Szentivanyi hypothesis is the presence of physiologically significant α-adrenoceptors in airways. *In vitro* evidence suggests that this is not the case, at least in normal airways. For example, in human bronchus from bronchitics *in vitro* α-adrenoceptor stimulation only produced contraction in the presence of a β-adrenoceptor antagonist (Simonsson *et al.*, 1972; Goldie *et al.*, 1985). Phenylephrine produced little or no contractile response in bronchial smooth muscle tissues obtained from either asthmatics or normal patients, even in the presence of propranolol (Goldie *et al.*, 1985). Simonsson *et al.* (1972) reported that, in the absence of β-blockade, human bronchi that were unresponsive to phenylephrine became exquisitely sensitive to α-adrenoceptor activation following preincubation with endotoxin from gram-negative bacteria. It is possible that this may be relevant in asthma resulting from infection. Kneussl and Richardson (1978) reported that norepinephrine contracted large airways obtained from autopsies of patients who had severe respiratory disorders but was without effect in tissues obtained from control subjects.

The evidence from *in vivo* studies for a significant contributory role of α-adrenoceptors in the etiology of disorders of airway hyperreactivity is also conflicting and unconvincing. For example, although it has been shown that α-adrenoceptor antagonists inhibit the bronchospasm elicited by histamine (Kerr *et al.*, 1970; Patel and Kerr, 1973), exercise (Biel and DeKock, 1978; Barnes *et al.*, 1981b), and allergen provocation (Patel *et al.*, 1976), the agents employed were relatively nonselective. Furthermore, although a high dose of aerosolized prazosin exhibits some efficacy in exercise-induced asthma (Barnes *et al.*, 1981b), it was ineffective in patients who responded well to β-adrenoceptor agonists (Barnes *et al.*, 1981a). In several species, including humans, α-adrenoceptor agonists have been reported to elicit bronchoconstriction (Castro de la Mata *et al.*, 1962; Anthracite *et al.*, 1971, Bienfield and Seifter, 1978; Barnes *et al.*, 1983). In asthmatics, α-adrenoceptor agonists were reported to produce bronchospasm, whereas no effect was observed in normal patients (Simonsson *et al.*, 1972; Patel and Kerr, 1973; Snashall *et al.*, 1978; Black *et al.*, 1982).

However, there are four clear clinical findings that are at odds with the postulate that airway hyperreactivity is a consequence of a defect in adrenoceptor function: (1) α-adrenoceptor antagonists are generally ineffective in asthma (Barnes *et al.*, 1981a; Britton *et al.*, 1981; Svedmyr, 1983); (2) chronic treatment with β-adrenoceptor antagonists in atopic patients does not normally result in airway hyperreactivity [in some asthmatics, bronchospasm is produced by β-blockers, but this appears to be largely a result of the increased influence of the bronchoconstrictor systems, notably parasympathetic activation (Simonsson *et al.*, 1972; Boushey *et al.*, 1980)]; (3) epinephrine, which exhibits considerable α_1- and α_2-adrenoceptor agonist activity, is effective in acute asthma and continues to be employed in emergencies in severe asthmatic episodes, (Svedmyr, 1984); and (4) the therapeutic effectiveness of β-adrenoceptor agonists in asthma indicates that the β-adrenoceptor system remains functionally important. Overall, the evidence gathered suggests that Szentivanyi's hypothesis of increased α-adrenoceptor activity concomitant with a decrease in β-adrenoceptor function does not warrant serious consideration as the mechanism underlying airway hyperreactivity.

2. H2-Receptor Deficiency

The predominant effect of histamine administration in airways, both *in vitro* and *in vivo,* is bronchoconstriction mediated via histamine H1-receptor stimulation (Eiser, 1983; White and Eiser, 1983). In addition, there is evidence that activation of histamine H2-receptors mediates relaxation of airway smooth muscle and inhibition of mediator release (Lichtenstein and Gillespie, 1973, 1975; Chand, 1980a,b; Eiser, 1983). Chand (1980b) proposed that airway hyperreactivity is a consequence of a deficiency of H2-receptors concomitant with a corresponding increase in the distribution and activation of H1-receptors. However, the support for this abnormality is largely circumstantial and is based on *in vitro* studies of isolated animal airway tissues and human leukocytes. For example, it has been reported that granulocytes from asthmatics (Busse and Sosman, 1977) and leukocytes from asthmatics with viral respiratory infections (Busse *et al.,* 1979a,b), exhibit decreased responsiveness to H2-receptor activation. *In vivo* and *in vitro* studies have failed to provide convincing evidence of an absence, or at least a decrease in the number, of H2-receptors in the airways of asthmatics compared with normal subjects (Eiser, 1983). In addition, H1-receptor antagonists are generally ineffective in asthma (Karlin, 1972). Although there is evidence, albeit conflicting, as to their existence, it is apparent that the physiological contribution of H2-receptors in airway, both normal and asthmatic, is of negligible significance (Eiser, 1983; White and Eiser, 1983). Accordingly, the postulate that airway hyperreactivity is a consequence of a deficiency of H2-receptors and increase in H1-receptors (Chand, 1980b) is without reasonable foundation. A more attractive proposition, suggested in the same report by Chand (1980b), and based on the many physiological and biochemical similarities between B_2-adrenergic receptors and H2-histamine receptors, is that in asthma there is a deficiency in the adenylate cyclase system(s) sensitive to histamine and epinephrine and other bronchoactive agents.

B. Imbalance in Regulation of the Autonomic Nervous System

Mammalian airways receive their nervous input from three systems: an excitatory parasympathetic innervation mediated by the release of acetylcholine (ACh), an inhibitory sympathetic input involving norepinephrine, and an inhibitory nonadrenergic noncholinergic (NANC) system, for which the neurotransmitter(s) has not been identified (Richardson, 1977, 1979, 1981; Richardson and

Beland, 1976; Nadel and Barnes, 1984). In human airways, pharmacological and morphological studies indicate that the innervation consists predominantly of an excitatory cholinergic input and an inhibitory NANC influence, with a negligible contribution from sympathetic adrenergic nerves (Richardson and Béland, 1976; Richardson, 1979; Davis *et al.*, 1982; Partanen *et al.*, 1982; Sheppard *et al.*, 1983).

The autonomic nervous system is directly involved in determining the level of activity in various systems in airways. Accordingly, an alteration in the neuronal input is a plausible explanation for at least some of the pathophysiological manifestations of airway hyperreactive disorders, notably the enhanced bronchomotor tone. This postulate was proposed many decades ago (Alexander and Paddock, 1921). The functional autonomic nervous system imbalance could result from either (1) increased cholinergic activity or (2) decreased influence of the NANC pathway, or both.

1. Increased Cholinergic Activity

In animals and humans, the resting level of bronchomotor tone in airways is influenced by activity in the parasympathetic nervous system via vagal stimulation (Boushey *et al.*, 1980). In normal persons, atropine abolishes the baseline tone (Severinghaus and Stupfel, 1955; Nadel and Widdicombe, 1963). It is possible that the increased level of bronchoconstriction characteristic of airway hyperreactivity is a result of increased activity in this system. This may be a consequence of a direct effect on the vagal nerve supply at one or more locations from the preganglionic fibers emanating from the central nervous system (CNS) to the postganglionic fibers that arise from ganglia situated in the airway walls (Richardson, 1979; Nadel and Barnes, 1984; Boushey, 1985). Alternatively, it may be a result of increased reflex bronchoconstriction produced by excitation of sensory pathways, present in the epithelium and at subepithelial sites, by mechanical, chemical, and/or pharmacological stimuli or as a result of respiratory tract infection and inflammation (Boushey *et al.*, 1980; Nadel and Barnes, 1984). Atropine and ipratropium have been reported to be effective in asthma, chronic bronchitis, and cystic fibrosis; these agents inhibit the enhanced bronchoconstriction produced by numerous stimuli in asthmatics (Cropp, 1975; Klock *et al.*, 1975; Larsen *et al.*, 1979; Boushey *et al.*,1980; Svedmyr, 1983; Schlueter, 1986). Although there is some dispute to this (Boushey *et al.*, 1980), it appears that anticholinergics are less effective antiasthmatic agents than β-adrenoceptor agonists, suggesting that the pathogenesis of asthma involves mediators and mechanisms other than ACh and excess cholinergic stimulation. However, this may be partly related to the administration of insufficient doses (Smedzyr, 1983; Nadel and Barnes, 1984; Barnes, 1986c). Anticholinergics appear to be more effective than β-adrenoceptors in chronic bronchitis and emphysema (Gross and Skodorin, 1984), which may reflect a significant vagal reflex component in the etiology of these disorders or increased influence of vagal tone in airways that are inherently narrow.

In certain patients anticholinergic agents are effective bronchodilators in airway hyperreactive disorders and are likely to have greatest benefit in chronic bronchitis, emphysema, and stable asthma (Boushey *et al.*, 1980; Gross and Skorodin, 1984; Schlueter, 1986). However, it should be reiterated that the use of anticholinergics as a therapeutic strategy intervenes only at one of the many likely pathways involved in the etiology of airway hyperreactive disorders, hence is likely to be of minimal benefit in most patients.

2. Decreased Contribution of NANC Nervous System

In human airways, the predominant inhibitory neuronal input is via a NANC system (Richardson and Béland, 1976; Richardson, 1979, 1981; Nadel and Barnes, 1984; Barnes, 1984). Nerves characterized as NANC have been identified in human lung associated not only with smooth muscle but also with submucosal glands, ganglia, and blood vessels (Richardson and Béland, 1976; Richardson, 1979, 1981; Dey *et al.*, 1981; Sheppard *et al.*, 1983). There is evidence from *in vitro, in vivo,* and *in situ* studies for a considerable NANC influence in airway function in a variety of species (Barnes 1984, 1987; Nadel and Barnes, 1984). For example, electrical field stimulation of human airways *in vitro* in the presence of cholinergic and adrenergic receptor antagonists produces relaxation that is sensitive to tetrodotoxin (Richardson and Béland, 1976; Davis *et al.*, 1982; Nadel and Barnes, 1984). Airway hyperreactivity has been proposed as a consequence of dysfunction of the NANC system in the respiratory tract (Richardson and Béland, 1976; Diamond and Gillespie, 1982; Barnes, 1984). Thus, the NANC pathway may normally exert a braking influence to counterbalance the effects of excitatory stimuli in airways. The mechanism producing dysfunction in the NANC nervous system is unknown, although Barnes (1987) speculated that it may be the result of inflammation of the airways in which enzymes released from the various inflammatory cells rapidly metabolize the NANC neurotransmitter(s). This alteration in the influence of the NANC system was regarded as a local phenomenon that may contribute to airway hyperreactivity caused by inflammation but was unlikely to be the underlying mechanism of asthma (Barnes, 1987). The identity of the NANC transmitter is unknown. A candidate for which relatively convincing evidence exists is vasoactive intestinal peptide (VIP) (Barnes, 1987; Nadel and Barnes, 1984). Several other neuropeptides are present in airways (Barnes, 1985b, 1987), some of which are associated with nerves. These include substance P, peptide histidine isoleucine (coded by the same gene as VIP) galanin, somatostatin, cholecystokinin, neuropeptide Y, and gastrin-releasing peptide. The physiological function and pathophysiological significance of these peptides as well as VIP remains to be clarified. This will be aided by the development of specific receptor antagonists.

There are no concrete data to support the postulate that dysfunction in the NANC nervous system contributes to airway hyperreactivity. However, a recent preliminary report demonstrated an alteration in the balance between the cholinergic and NANC nervous systems in patients with asthma but not in those with chronic obstructive lung disorders (De Jongste *et al.*, 1987).

C. Change in Electrophysiological Characteristics of Smooth Muscle Cells

The nonspecific supersensitivity in the guinea pig vas deferens resulting from chronic interruption of the innervation of the tissue may be attributed, at least in part, to a reduction in the resting membrane potential (Fleming and Westfall, 1975). This partial depolarization, on the order of 8–10 mV, was associated with attenuation in the electrogenic Na^+, K^+ pump (Urquilla *et al.*, 1978) and in the number of Na^+, K^+ pump binding sites (Gerthoffer *et al.*, 1979; Wong *et al.*, 1981). Extrapolating from these findings, an obvious postulate is that airway hyperreactivity may be a consequence of a fundamental alteration in the electrophysiological characteristics of the smooth muscle cell membrane. For example, it has been suggested that the airway smooth muscle of asthmatics functions as a single unit rather than as a multiunit (Richardson, 1977). However, to date there has been only one published report on electrophysiological studies in human airways (Akasaka *et al.*, 1975). It was observed that bronchial smooth muscle from asthmatic patients was hyperreactive compared with control tissue, as reflected by a marked increase in action potential-like activity. This potentially significant finding has apparently not been followed up by these or other researchers.

A limited amount of research has been conducted in animal airways (Kirkpatrick, 1975; Coburn and Yamaguchi, 1977; Farley and Miles, 1977; Coburn, 1979; Small, 1982). Of potential relevance is the research conducted by Souhrada and co-workers investigating the electrophysiological characteristics of tracheal smooth muscle from ovalbumin-sensitized guinea pigs. Active and passive sensitization results in a marked hyperpolarization of the guinea pig tracheal smooth muscle cell membrane (Souhrada and Souhrada, 1981, 1983, 1984). This hyperpolarization was attributed to a direct alteration of the smooth muscle cell membrane and to increased activity and contribution of the Na^+ pump to the resting membrane potential. It was postulated that by modulating the membrane permeability for Na^+ and increasing the Na^+ gradient there will be stimulation of Ca^{2+} influx into the smooth muscle as a result of increased $Na^+–Ca^{2+}$ exchange (Souhrada and Souhrada, 1984). However, it seems improbable that increased responsiveness of smooth muscle (airway or nonairway) will be a consequence of hyperpolarization of the cell membrane, a phenomenon that is generally attributed to smooth muscle relaxants. Smooth muscle from hyperreactive airways would be expected to be partially depolarized, compared with control airways. In fact it was observed that there was a significant reduction in the resting membrane potential in tracheal smooth muscle obtained from guinea pig repeatedly exposed to antigen *in vivo* (McCaig and Souhrada, 1980; Souhrada and Souhrada, 1981). In addition, further research indicated that antigen produces a biphasic electrical response in ovalbumin-sensitized guinea pig trachea, i.e., an initial depolarization, mediated by Na^+ influx, followed by hyperpolarization of the smooth muscle membrane. Inhibition of the depolarization with amiloride, Li^+, or choline antagonized the antigen-induced contraction (Souhrada and Souhrada, 1985). Although ovalbumin-sensitized guinea pigs are used as a model of airway hyperreactivity, a major concern is that *in vitro* trachea from ovalbumin-sensitized guinea pigs do not show increased responsiveness to contractile agents and, in fact, have been reported to exhibit decreased sensitivity (Cheng and Townley, 1983; Hay *et al.*, 1986b) or a reduced maximum response (Souhrada, 1978) to stimulation by cholinergic agonists or histamine. *In vivo* there was an increase in sensitivity to histamine in ovalbumin-treated guinea pigs compared with that exhibited by control animals (Iwayama *et al.*, 1982). Accordingly, it is not clear what relevance the findings from studies using ovalbumin-sensitized guinea pigs have to the basic mechanisms underlying airway hyperreactivity in humans.

To date, a neglected area of research and a major deficit is the lack of information on the electrophysiological characteristics of control and hyperreactive airway smooth muscle, especially human. These studies may provide valuable insight into fundamental cellular differences in diseased airways compared with control tissues which render the former hyperresponsive to bronchoactive agents.

D. Abnormalities in Ca^{2+} Translocation and Handling

The critical role of Ca^{2+} in maintaining normal cellular function throughout the body and in determining the level of activity in these systems has been unequivocally established. In mammalian airways, the numerous Ca^{2+}-dependent processes include smooth muscle contraction, mediator synthesis and release, mucus secretion, nerve impulse activity, inflammatory cell chemotaxis, and electrolyte and water transport (Middleton, 1980, 1985; Tinkelman, 1985). Increased activation in most, if not all, of these processes may be involved in the pathogenesis of airway hyperreactive disorders. There are two mechanisms by which this can be elicited: (1) increased stimulatory input, and/or (2) inherent enhanced responsiveness of the effector cell. The increased activity in the various airway cells may ultimately result from enhanced availability of Ca^{2+} or increased sensitivity of those cells to Ca^{2+}-dependent stimuli (Middleton, 1985). This concept is the basis of what was termed initially by Middleton (1980) as the calcium hypothesis of asthma, which regards an alteration in the Ca^{2+} translocation and handling in the various cellular systems as the key process that results in all the pathophysiological manifestations of this disorder. The possible mechanisms involving Ca^{2+} transport and mobilization that can result in increased responsiveness are considered below only for contraction of smooth muscle, although they are likely to be relevant to the postulated enhanced activity in other respiratory cell types.

The contractile state of airway smooth muscle is ultimately dependent on the level of free intracellular Ca^{2+} that interacts with the contractile proteins (Bolton, 1979; Middleton, 1980; Andersson, 1983; Rodger, 1986). The intracellular Ca^{2+} levels results from the balance of the influence of several extracellular and intracellular Ca^{2+} translocation pathways that function in concert to control cellular Ca^{2+} homeostasis at rest and following membrane-receptor activation (Bolton, 1979; Middleton, 1980, 1985; Andersson, 1983; Rodger, 1986). The bronchospasm associated with airway hyperreactivity may result from an alteration in one or more of several pathways or processes involving Ca^{2+}, including (1) increased extracellular Ca^{2+} influx; (2) decreased efflux of Ca^{2+}; (3) increased release of Ca^{2+} from intracellular organelles; (4) decreased Ca^{2+} sequestration into intracellular locations; (5) alteration in the number, electrophysiological, and/or biophysical characteristics of the membrane Ca^{2+} channels; and (6) increased sensitivity of the contractile machinery to Ca^{2+}.

However, there have been no published reports using human hyperreactive airways of an alteration in the Ca^{2+} translocation mechanisms, membrane Ca^{2+} channel properties, or sensitivity of the contractile machinery. Therefore, although an attractive concept, the Ca^{2+} hypothesis of airway hyperreactivity awaits experimental validation.

There is some, albeit limited, evidence in studies using animal models to support the theory of a defect in cellular handling of Ca^{2+} producing enhanced bronchoconstriction. There is an increased sensitivity to Ca^{2+} in ovalbumin-sensitized guinea pig K^+-depolarized trachea *in vitro* (Weiss and Viswanath, 1979; Dhillon and Rodger, 1981). Furthermore, an alteration in Ca^{2+} mobilization and binding was reported for histamine-induced contractions in the same preparation (Martorana and Rodger, 1980; Raeburn, 1984). Verapamil was significantly less able to inhibit agonist-induced contractions in lung parenchymal strips obtained from ovalbumin-sensitized compared with control guinea pigs (Perpina *et al.*, 1986). However, the potential inadequacies of this model should be noted (see Section II.C). There was no difference in the basal and K^+-stimulated ^{45}Ca uptake in trachea from control and ovalbumin-sensitized guinea pigs (Raeburn *et al.*, 1987). In trachea from *Ascaris suum* antigen-sensitive (asthmatic) dogs, there was a decrease in the basal levels of cAMP compared with that in control, nonresponsive animals concomitant with a reduction in the ability of isoproterenol to induce relaxation (Rinard *et al.*, 1979). The activity of the β-adrenergic receptor–adenylate cyclase complex did not appear to be affected, and it was concluded that the reduced levels of cAMP over a prolonged period rendered the airways hyporeactive to bronchorelaxants (Rinard *et al.*, 1979), perhaps due to a reduced intracellular sequestration of Ca^{2+}. Treatment with reserpine in rats resulted in an alteration in the density but not the affinity of putative membrane Ca^{2+} channels labeled with [^{3}H]nitrendipine in seminal vesicles (Powers and Colucci, 1985). This finding suggests that smooth muscle supersensitivity produced by reserpine may be the result of a fundamental alteration in the properties of the Ca^{2+} channel.

The extracellular and intracellular targets for altering Ca^{2+} handling are numerous, and a greater understanding of the processes involved in Ca^{2+} translocation and mobilization in the various airway cell types in both control and diseased states may lead to the development of selective therapeutic agents to act at these sites.

E. Change in Airway Morphology/Geometry: Mechanical Factors

The level of bronchoconstriction produced by an excitatory stimulus *in vivo*, in addition to being governed by the intrinsic biophysical properties of the airway smooth muscle cell, may also be directly dependent on several geometrical, morphological and/or mechanical considerations, including (1) the proportion of airway smooth muscle present in the airway circumference, (2) the thickness of the airway wall, (3) the interdependence between parenchyma and the airway wall, (4) the smooth muscle force, and (5) the load on airway smooth muscle and its resting length (Macklem, 1985; Hogg, 1986; Moreno *et al.*, 1986a). An alteration or imbalance in the influence of one or more of these parameters may contribute to airway hyperreactivity. (See Chapter 13, this volume, which deals with some of these parameters at length.)

Changes in some of the above parameters have been documented. For example, smooth muscle hypertrophy or hyperplasia, which will increase airway wall thickness, are well-known features of the airway of asthmatics who have died from the disorder (Dunnill, 1960; Takizawa and Thurlbeck, 1971), and also in patients with chronic bronchitis (Hossain and Heard, 1970). This alteration in the smooth muscle mass is likely to occur progressively in diseased airways and may contribute significantly to the chronic symptoms of asthma. However, it is unlikely to be a major cause of hyperreactive disorders, and it will not be a factor in airway hyperreactivity that manifests itself rapidly in normal persons with respiratory tract infections (Empey *et al.*, 1976; Empey, 1982) or after exposure to pollutants (Orehek *et al.*, 1976; Golden *et al.*, 1978) in which there is insufficient time for an alteration in smooth muscle mass or function. Thickening of the airway wall, which is prevalent in asthmatics (Dunnill, 1960; Dunnill *et al.*, 1969; James *et al.*, 1987), may also be caused by edema and/or hyperemia of acute inflammation (Orehek, 1983; Hogg, 1986). Hogg and co-workers recently reported that airways obtained from asthmatics showed thickening of the airway wall but not excess muscle shortening (James *et al.*, 1987). The possibility exists that certain persons may possess inherently thicker airway walls while lacking an inflammatory condition or hypertrophied smooth muscle (Moreno *et al.*, 1986a). Whatever the underlying mechanism, thickening of the airway wall will result in increased airway narrowing by contractile agents and may also be associated with a reduction in baseline caliber (see Section II.F). Mucus hypersecretion and formation of mucus plugs will produce similar effects on the responsiveness of airway smooth muscle as an increase in wall thickness (Moreno *et al.*, 1986a). (The effects of increased secretions and changes in wall thickness on airway responsiveness are discussed in greater depth in Chapter 13, this volume.)

An important influence of cartilage on the baseline caliber and reactivity of airways is suggested by the experimental observation by Moreno *et al.* (1985, 1986b) that papain, which is thought to soften cartilage, produces increased responsiveness to ACh in rabbits *in vivo* in conjunction with elevation in the resting airway function. The cartilage normally exerts a significant influence on baseline airway caliber and airway reactivity, probably as a consequence of the substantial preload it exerts on the smooth muscle, thereby limiting its shortening (Moreno *et al.*, 1986a). An alteration in the rigidity and/or the amount of cartilage will be associated with changes in the preload of the airway smooth muscle, subsequently affecting the response to contractile agents. Chronic inflammation may alter the physical and mechanical properties of cartilage (Moreno *et al.*, 1986b), and it has been reported that patients with chronic obstructive lung disease have reduced quantities of cartilage (Thurlbeck *et al.*, 1974). Smaller bronchial cartilage plates and similar alterations in the geometry of the cartilage may also decrease the preload (Moreno *et al.*, 1986a). In addition to exerting a significant mechanical and biophysical influence in airways, cartilage was recently proposed to be a source of mobilizable Ca^{2+} for contraction of airway smooth muscle (Raeburn *et al.*, 1986d). In view of the central importance of Ca^{2+} in airway function, this postulate introduces another mechanism by which cartilage may affect the airway responsiveness.

Another postulated mechanism for a reduction in the airway smooth muscle preload is a decreased restraining influence of the lung parenchyma. This may be because of a reduction in the elasticity of the lung, a feature of emphysema, fibrosis, and asthma (Hogg, 1986; Moreno *et al.*, 1986a).

Although smooth muscle hypertrophy and hyperplasia, increased wall thickness, mucus hypersecretion, and formation of mucus plugs are known features of airway hyperreactive disorders, these manifestations probably occur at chronic stages of the disease and/or in severe conditions, and are therefore likely to be a consequence, rather than a primary cause, of the airway hyperreactivity. It remains to be assessed whether changes in the other postulated properties of the airways (e.g., increase in proportion of airway smooth muscle, softening of cartilage, decreased cartilage plates) are prevalent in hyperreactive compared with control airways.

F. Decreased Airway Caliber

Airway resistance is inversely proportional to the fourth power of the airway radius (Freedman, 1972). Accordingly, a further decrease in the radius of narrow airway will result in a greater increase in resistance than an identical alteration in the radius of a dilated airway. It has been postulated that airway hyperreactivity may be a consequence of an inherent reduction in baseline airway caliber resulting in an increase in responsiveness and airway resistance upon exposure to bronchoconstrictor stimuli (Benson, 1975). This basal partial obstruction of the airways may be attributable to, and/or exacerbated by, several factors, including an increase in mucosal secretions, mucosal edema (Bouhuys, 1963; Freedman, 1972; Garland, 1984), smooth muscle hypertrophy and hyperplasia (McFadden, 1984), increased cholinergic neural input, and/or attenuation in the influence of the inhibitory nervous pathway. Some studies demonstrated an apparent direct correlation between the baseline level of airway obstruction and the magnitude of the response to bronchoconstrictors (Parker et al., 1965; Simonsson, 1965; Makino, 1966). However, many subjects are hyperresponsive to bronchoconstrictor stimuli, e.g., asthmatics in clinical remission, hay fever sufferers, and nonasthmatics with upper respiratory tract viral infections or after exposure to ozone (see Boushey et al., 1980), even though they possess no apparent baseline bronchial obstruction. Furthermore, there are profound differences in the airway reactivity of patients with similar baseline pulmonary function (Cade and Pain, 1971; Orehek et al., 1977; Rubinfeld and Pain, 1977). Differences in basal lung function may be partly responsible for the inherent inter- and intra-animal variability characteristic of in vivo studies.

It appears that the changes in the geometry of airway lumen have a less marked effect on smooth muscle reactivity than would be anticipated from a theoretical consideration (Boushey et al., 1980). The relationship between airway caliber and smooth muscle reactivity may play a significant role in patients with severe hyperreactive disorders at the chronic stage of the disease, when there is significant airway obstruction at rest. Evidence from animal studies support this contention (DeKock et al., 1966; Benson and Graf, 1977; Hahn et al., 1978).

G. Increased Release of Mediators

Several mediators, both primary and secondary, have been proposed as being involved in the pathophysiology of airway hyperreactive disorders (Austen and Orange, 1975; Michel et al., 1982; Hargreave et al., 1985; Lewis and Robin, 1985; Church and Holgate, 1986; Holgate et al., 1986a,b; Page et al., 1986; Schulman, 1986). Those for which the strongest case has been made are histamine, the peptidoleukotrienes (LTC_4, LTD_4, and LTE_4) and, more recently, platelet-activating factor (PAF). Other bronchoactive agents touted include

LTB$_4$, the prostaglandins, especially prostaglandin D$_2$ (PGD$_2$), thromboxanes, adenosine, bradykinin, and 15-hydroxyeicosatetranoic acid (15-HETE). These mediators produce many of the pathophysiological manifestations of airway hyperreactivity, including enhanced smooth muscle contraction, increased mucus secretion, increased microvascular permeability and edema, epithelial cell damage, and/or inflammatory cell chemotaxis (Austen and Orange, 1975; Michel *et al.*, 1982; Hargreave *et al.*, 1985; Holgate *et al.*, 1986b; Page *et al.*, 1986). There are numerous potential cellular origins for mediator release, although for some mediators this remains to be elucidated. The mast cell has been extensively studied over the years and was considered the most important cell category in the etiology of airway hyperreactive disorders, as well as a viable therapeutic target (Pepys, 1973; Holgate *et al.*,1986a; Schulman, 1986). However, the focus of research on the mast cell was recently criticized (Page *et al.*, 1986), and other cell types such as eosinophils, platelets, basophils, and macrophages warrant consideration as potential sites for therapeutic intervention (Simonsson, 1984; Page *et al.*, 1986; Holgate *et al.*, 1986b).

Considerable evidence in humans and from studies using various animal models suggests that various mediators play a role in both the early and late phases of airway hyperreactive disease (Michel *et al.*, 1982; Barnes, 1985a; Hargreave *et al.*, 1985, 1986a,b; Holgate *et al.*, 1986b; Page *et al.*, 1986; Larsen *et al.*, 1987). Although several individual mediators have been championed as the primary causative agent responsible for airway hyperreactivity, it is likely that several agents rather than a single one contribute to the overall pathogenesis of this phenomenon. There is likely to be a complex interrelationship between the effects of the numerous mediators (Austen and Orange, 1975; Barnes, 1985a; Holgate *et al.*, 1986b). For example, PGD$_2$ increases airway responsiveness to histamine and methacholine in mild asthmatics (Fuller *et al.*, 1986), PAF increases the *in vivo* sensitivity of normal subjects to methacholine (Cuss *et al.*, 1986), and LTB$_4$ enhances the reactivity of guinea pig airways *in vivo* and *in vitro* to bronchoactive agents (Thorpe and Murlas, 1986).

The increased influence of mediators, postulated to contribute significantly to airway hyperreactivity, may be attributed to (1) increased sensitivity of the various effector systems to their actions, and/or (2) increased basal and/or stimulated release of mediators (Neijens *et al.*, 1980; Simonsson, 1984; Hargreave *et al.*, 1986a). The initial mechanism is discussed in Sections II.A–II.F. There is experimental and clinical evidence to support the latter postulate. For example, there are elevated levels of several mediators in the airways and body fluids of patients with asthma and other hyperreactive disorders compared with those of controls. This observation is often used as the basis for ascribing a pathophysiological role of an individual mediator in airway disorders. However, this reasoning is scientifically tenuous and, in the absence of additional evidence, one should be cautious about ascertaining that for a mediator presence necessarily implies cause. Notwithstanding this caveat, increased amounts of peptidoleukotrienes have been detected in the plasma, sputum, and bronchial lavages of patients with asthma, chronic bronchis, cystic fibrosis, and bronchiectasis (Cromwell *et al.*, 1982; Zakrzewski *et al.*, 1984, 1985a; Ishihara *et al.*, 1985; Isono *et al.*, 1985; Schönfeld *et al.*, 1985). Significantly, it was observed that in asthmatic patients there was a positive correlation between the amounts of peptidoleukotrienes detected and the severity of the disorder (Ishihara *et al.*, 1985; Isono *et al.*, 1985; Zakrzewski *et al.*, 1985a). LTB$_4$ has also been detected in the biological fluids of some patients with asthma but not in control subjects (Zakrzewski *et al.*, 1985a,b). A recent study reported the detection of tissue kallikrein in bronchial lavages of asthmatic but generally not in nonasthmatics (Christiansen *et al.*, 1987). PAF is detected in the sputum of patients with respiratory tract disorders, including asthma (Grandel *et al.*, 1985). Increased plasma levels of histamine are detected in asthmatics (Barnes *et al.*, 1982). Allergen challenge in asthmatics results in the release of several mediators, including histamine (Atkins *et al.*, 1980; Brown *et al.*, 1982), peptidoleukotrienes (Lewis and Austen, 1984), PAF (Knauer *et al.*, 1981; Thompson *et al.*, 1984), adenosine (Mann *et al.*, 1983), and cyclo-oxygenase products of arachidonic acid (Shephard *et al.*, 1985).

There is evidence of the involvement of several inflammatory cell types in airway hyperreactivity. Thus, asthmatic subjects have increased numbers of mast cells in the mucosa and submucosa of their airways (Dalquen, 1985). Elevated levels of mast cells are found in bronchioalveolar lavages from asthmatics compared with control subject (Tomioka *et al.*, 1984; Wardlaw *et al.*,1985). There is enhanced histamine release from basophils obtained from asthmatics (Haruhisa *et al.*, 1985). Increased levels of eosinophils and neurophils and their contents in blood, sputum, and/or bronchoalveolar lavage of allergic and nonallergic asthmatics have been reported (Frigas *et al.*, 1981; De Monchy *et al.*, 1985; Durham *et al.*, 1985; Boschetto *et al.*, 1986; Frigas and Gleich, 1986).

Within the airways, both in normal and in diseased conditions, there is likely to be a complex interplay between the various inflammatory cells and their contents, some of which individually produce several of the pathophysiological features of airway hyperreactivity. To date there is limited knowledge on the intercellular communication pathways in airways. A greater insight into this intricate multifunctional system and how it may change in airway hyperreactive disorders may introduce avenues for novel therapeutic approaches in these diseases.

H. Epithelial Cell Damage or Dysfunction

The epithelium is the inner cell layer present throughout the respiratory tract. It serves as a considerable barrier to foreign substances in addition to mechanical stimuli (Hogg, 1983, 1985). It has been proposed that the epithelium plays a central role in the pathogenesis of hyperreactive disorders of the airways.

The main trigger for this postulate is that epithelial damage or loss is a classic pathological hallmark of asthma, even of a mild nature (Dunnill, 1960; Naylor, 1962; Laitinen et al., 1985), and also of other respiratory diseases, including sarcoidosis and allergic alveolitis (Laitinen et al., 1983). This excessive shedding of epithelium may result in an increased turnover of epithelial cells that may produce a thickened basement membrane, another feature of asthma (Callerame et al., 1971; Hogg, 1984). Inflammation of the airways is associated with a loss of epithelium (Hulbert et al., 1981; Hogg, 1984, 1985; Laitinen et al., 1985) and an increase in epithelial cell turnover (Hulbert et al., 1981). The mechanism(s) responsible for the damage and/or shedding of the epithelium is unknown. Theories advocated include spasm of airway smooth muscle (Houston et al., 1953), edema of the submucosal region, and resultant exudation of plasma into the lumen of the airways (Dunnill, 1960; Barnes, 1986a). Another postulate for which there is more convincing support is that epithelium damage is a result of the cytotoxic effects of constituents of eosinophils, notably major basic protein (MBP) (Frigas et al., 1980; Frigas and Gleich, 1986). In asthma, there is infiltration of eosinophils into subepithelial sites (Huber and Koessler, 1922; Hogg et al., 1977; Gleich, 1986; Frigas and Gleich, 1986). Furthermore, the levels of MBP located in the airways of asthmatics were elevated compared with nonasthmatic patients, and there was a correlation between the amount of MBP and the extent of the reactivity of the airways in vivo (Horn et al., 1975; Frigas et al., 1981; Gleich, 1986). In vitro, MBP produces almost identical alterations in the integrity of the epithelium and subepithelial regions as occurs in asthma and the MBP concentration in the plasma of asthmatics is in the same range that is cytotoxic to epithelial cells in vitro (Frigas et al., 1981; Dor et al., 1984; Frigas and Gleich, 1986; Gleich, 1986). These findings form part of the increasing evidence that eosinophils play a significant role in the pathophysiology of airway hyperreactivity.

Various postulates, some of which are interrelated, have been proposed to explain how epithelial damage of dysfunction could result in airway hyperreactivity:

Increased mucosal and epithelial permeability
Sensitization of vagal sensory nerve endings (axon reflex); release of neuropeptides
Loss of influence of epithelium-derived inhibitory factor(s)
Promotion of inflammatory cell chemotaxis and release of bronchoactive mediators
Inability to adapt to osmotic stress

1. Increased Mucosal and Epithelial Permeability

It has been speculated that increased permeability of the epithelial cell layer and also of the mucosal region may produce airway hyperreactivity by facilitating the access of bronchoactive stimuli to the subepithelial sites, such as mucosal mast cells, irritant sensory receptors, afferent C fibers, and the smooth muscle layer (Simonsson, 1984; Hogg, 1983). One result of this increased permeability may be a reduction in the threshold, in conjunction with increased stimulation, of the efferent limb of the reflex arc, which produces bronchoconstriction (Empey et al., 1976; Orehek et al., 1976; Golden et al., 1978; Hogg, 1981). This increased permeability may be the result of an inflammatory response (Empey et al., 1976; Hogg, 1981, 1983, 1984) and can be produced by irritant agents, such as smoke, sulfur dioxide, and ozone (Orehek et al., 1976; Golden et al., 1978; Simonsson, 1984). Exposure to allergen in monkeys (Boucher et al., 1979) and humans (Cockcroft et al., 1977) results in increased mucosal permeability and airway hyperreactivity. This may be produced by an interaction of the antigen with mast cells present in the epithelium (Hogg et al., 1977; Jeffrey and Corrin, 1984) and by the release of mediators that open up the junctions between the epithelial cells to permit penetration of larger quantities of antigen to subepithelial sites (Hogg et al., 1977; Hogg, 1983). However, it has been demonstrated that some smokers do not exhibit enhanced airway responsiveness despite having increased airway permeability (Kennedy et al., 1984). Furthermore, stable asthmatics with airway hyperreactivity have been shown not to exhibit increased epithelial cell permeability (Elwood et al., 1983). In addition, an effect on epithelium permeability cannot explain the hyperreactivity of airways to intravenously administered agonist (Murlas and Roum, 1985a). These findings suggest that increased epithelial or mucosal permeability is not a causative factor but may merely be a consequence of airway hyperreactivity. After a comprehensive study in guinea pigs exposed to smoke, Hogg and co-workers concluded that there was no correlation between airway permeability and responsiveness to histamine unless the autonomic input to the airways was blocked (Hulbert et al., 1981, 1982; Hogg, 1983). This suggests that the nervous supply to the airways is normally able to counteract any influence that changes in permeability may have on airway reactivity.

2. Sensitization of Vagal Sensory Afferent Nerve Endings (Axon Reflex);
   Release of Neuropeptides

   In several species, including humans, the epithelium contains an appreciable nerve supply. These predominantly sensory afferent nerve endings generally lie between and circumvent the epithelial cells and are close to the airway lumen (Richardson, 1979; Sant'Ambrogio, 1982; Hogg, 1986). Barnes (1986a) postulated that a local noncholinergic axon reflex—perhaps produced by inflammatory mediators—that stimulates the release of neuropeptides may be the crucial link between epithelial injury and airway hyperreactivity. The identity of the neuropeptide(s) is unknown; candidates include substance P, neurokinin A, and calcitonin gene-related peptide, which produce many of the symptoms of asthma (Barnes, 1986a).

3. Loss of Influence of Epithelium-Derived Inhibitory Factor(s)

Increasing research during the past 3 years or so has examined the possible role of the epithelium in controlling airway smooth muscle reactivity (Fedan et al., 1988; Farmer, 1987). Mechanical removal of the epithelium increases the sensitivity to a variety of bronchoconstrictor agents in guinea pig (Goldie et al., 1986a; Hay et al., 1986a; Holroyde, 1986), dog (Flavahan et al., 1985), bovine (Barnes et al., 1985), rabbit (Raeburn et al., 1986c), and, importantly, human airways (Raeburn et al., 1986a). An increased responsiveness to antigen in guinea pig trachea denuded of epithelium was reported by Hay et al. (1986b). This alteration in response following epithelial removal is generally reflected by a decrease in the agonist EC_{50}, i.e., leftward shift in the concentration response curve; in some cases an increase in the maximum response is produced. This influence of the epithelium on airway smooth muscle reactivity has been attributed to the presence of an epithelium-derived inhibitory factor. As a consequence of damage to or loss of epithelium, reduction in the influence of this inhibitory system may contribute to airway hyperreactivity. There are regional differences in the magnitude of the influence of the epithelium on the responsiveness of dog (Hay et al., 1987a) and rabbit airways (Raeburn et al., 1986c). Two recent reports provided more direct evidence from sandwich-type experiments (Furchgott and Zawadzki, 1980) in guinea pig trachea for the release of an inhibitory substance from the epithelium that modulates responses produced by antigen (Hay et al., 1987a) and substance P (Tschirhart and Landry, 1986). The identity of the postulated inhibitory factor is unknown. The involvement of a prostanoid in the effects of epithelial removal on responsiveness of guinea pig trachea (Hay et al., 1986a) or dog bronchus (Flavahan et al., 1985) has been proposed. There is some evidence that more than one inhibitory factor may modulate the reactivity of guinea pig tracheal smooth muscle (Hay et al., 1986a; Raeburn et al., 1986b).

It has been shown that exposure to ozone, which causes damage to epithelial cells, produces airway hyperreactivity in vivo in guinea pigs (Lee and Murlas, 1985; Murlas and Roum, 1985b) and dogs (Holtzman et al., 1983a,b; O'Byrne et al., 1984). In guinea pigs there was a good correlation between the damage to the epithelium and the extent of airway hyperreactivity (Murlas and Roum, 1985a). Accordingly, it was proposed that epithelial cell damage was closely linked to the associated airway hyperreactivity induced by ozone (Lee and Murlas, 1985; Murlas and Roum, 1985a). However, it has been advocated by Holtzman and co-workers that, at least in dogs, ozone-induced airway hyperreactivity is not a consequence of epithelium damage per se but is related to an inflammatory reaction and neutrophil infiltration (Holtzman et al., 1983a,b; Fabbri et al., 1984; O'Byrne et al., 1984).

4. Promotion of Inflammatory Cell Chemotaxis and Release of Bronchoactive
   Mediators

Damage to airway epithelial cells will result in the release of potent chemotactic factors, such as LTB_4, which subsequently attract inflammatory cells, including neutrophils, eosinophils, and polymorphonuclear leukocytes, in the airway lumen (Holtzman et al., 1983a,b; Nadel, 1984; Barnes, 1986a). These inflammatory cells may then release their array of potent bronchoactive contents, such as cyclo-oxygenase and lipoxygenase products of arachidonic acid metabolism and PAF. The ability of these chemical mediators to produce the effects on subepithelial sites, such as mucosal mast cells, smooth muscle, and nerves, may be assisted by the opening of the epithelium cell tight junctions as a result of an inflammatory reaction (Hogg, 1983; Simonsson, 1984; Barnes, 1986a). This scenario has been advocated for airway hyperreactivity produced by ozone in animals (Holtzman et al., 1983a,b; Fabbri et al., 1984; Lee and Murlas, 1985; Murlas and Roum, 1985a,b) and humans (Seltzer et al., 1986), in which there is inflammatory cell infiltration in the airways and release of products of arachidonic acid metabolism. Thus, the epithelium appears to be a major stimulus and site for the infiltration of various inflammatory cell types as a result of both immunological and nonimmunological provocation.

5. Inability to Adapt to Osmotic Stress

Challenge with aerosolized hypotonic or hypertonic solutions in asthmatics, but not normal patients, produces bronchoconstriction (Cade and Pain, 1972; Allegra and Bianco, 1980; Schoeffel et al., 1981; Elwood et al., 1982). Hogg and Eggleston (1984) speculated that asthma is a result of a fundamental defect in the epithelium related to its inability to control the ionic concentration and osmolarity of the inner surface of the airways. The associated bronchoconstriction may result from disruption of the tight junctions in the epithelium and stimulation of the irritant reflex (Hogg and Eggleston, 1984) or release of bronchoactive mediators from subepithelial sites (Hogg and Eggleston, 1984). Exercise-induced asthma may be a consequence of water rather than of temperature loss (Lee and Anderson 1985) in conjunction with inherent susceptibility of the epithelium to osmotic stress (Hogg and Eggleston, 1984).

In summary, the available evidence suggests that the epithelium does not act merely as a physical barrier but possesses a multifunctional role in airways and is intimately involved in airway hyperreactivity. The precise mechanism(s) whereby the epithelium influences the responsiveness of the airways is unknown, although it is likely that more than one mechanism is involved.

I. Inflammation

To date, experimental studies and the therapeutic control of asthma have focused predominantly on the acute bronchospasm. However, it has become increasingly apparent and accepted that a crucial aspect is the inflammatory response that appears to be associated with the development and long-term pathophysiological manifestations of asthma (Hogg, 1984, 1986; Chung, 1986; Hargreave *et al.*, 1986b; Holgate, 1986; Holgate *et al.*, 1986b; Page *et al.*, 1986). The etiology of airway inflammation is largely unclear, but it may be the final common pathway in respiratory hyperreactive disorders. There is considerable evidence from morphological and histopathological studies that many of the classic features of inflammation are prevalent in the airways of patients with airway hyperreactive disorders (Houston *et al.*, 1953; Dunnill, 1960; Glynn and Michaels, 1960; Hogg, 1984; Laitinen *et al.*, 1985). For example, inflammatory bronchial edema is another hallmark of asthmatics (Hulbert and Koessler, 1922; Vaughan *et al.*, 1973; Dunnill, 1975). The edema may result from increased permeability to macromolecules in both the mucosal region and the microvasculature (Persson, 1986; Persson and Svensjo, 1983) that will result in extravasation of plasma proteins and the formation of an exudate that can enter the airway lumen (Persson and Svensjo, 1983; Hogg, 1984; Persson, 1986). This may contribute to airway hyperreactivity by a geometrical influence via narrowing of airway lumen and reduction in baseline caliber (Benson, 1975; Persson and Svensjo, 1983; Hargreave *et al.*, 1986a; see Section II.F) or by sensitizing and increasing the activity in exposed afferent nerves (Sellick and Widdecombe, 1969; Empey *et al.*, 1976; Coleridge and Coleridge, 1977). In addition, the exudate may affect the function of the ciliated epithelium (Dunnill, 1960; Dulfano *et al.*, 1982). It has been suggested that bronchial edema plays a significant role in the pathogenesis of airway hyperreactivity (Dunnill, 1960; Dunnill *et al.*, 1969). Mucus hypersecretion and shedding of the epithelial cell layer are pathological manifestations of airway hyperreactive disorders that are specific features of inflammation of mucosal surfaces (Florey *et al.*, 1932; Hogg, 1984, 1986). The enhanced production of mucus in combination with exudate accumulation and decreased ability of the epithelium and airways to clear the mucus will result in the formation of mucus plugs, which play a considerable role in the pathophysiology of the chronic stages of airway hyperreactive disorders (Hogg, 1984, 1986; Persson, 1986).

In a significant number of asthmatics, especially patients with chronic severe asthma, a delayed late asthmatic response (LAR) is prevalent a few hours following exposure to an allergen (Pepys and Hutchcroft, 1975; Boulet *et al.*, 1983; Cockcroft, 1985). The pathogenesis of LAR remains to be clarified, although clinical and laboratory evidence suggests that it is associated with the release of bronchoactive mediators and the cellular phase of inflammatory response (Kaliner, 1984; Hargreave *et al.*, 1985, 1986a; Holgate *et al.*, 1986b; Page *et al.*, 1986; Larsen *et al.*, 1987). In conjunction with increased airway reactivity, elevation in the numbers of inflammatory cells, notably eosinophil (De Monchy *et al.*, 1985), has been detected in the body fluids of patients with LAR (Abraham *et al.*, 1983; Hargreave *et al.*, 1985, 1986a,b; Larsen *et al.*,

1987). The influence of inflammation on airway responsiveness is further substantiated by the finding that several stimuli that produce inflammation, such as infection, industrial chemicals, and ozone, result in persistent airway hyperreactivity in humans and animals (Holtzman *et al.*, 1983b; Skoogh, 1984; Hargreave *et al.*, 1986a,b). The mechanism whereby airway inflammation affects pulmonary function remains to be clarified. Postulates suggested by Boushey *et al.* (1980) and Hargreave *et al.* (1986a) include (1) enhanced spontaneous and/or stimulated release of mediators from the various inflammatory cells (see Section II.G), (2) increased epithelial and mucosal permeability (see Section II.H), (3) increased sensitivity of bronchial irritant receptors and altered neuronal reflexes and neuronal control, and (4) effects of inflammatory edema (i.e., decreased airway caliber, sensitization of afferent nerves, decreased epithelium function). Similarly, the specific mediators producing inflammation and their cellular sources remain uncertain (Chung, 1986; Holgate *et al.*, 1986b; Page *et al.*, 1986).

In summary, inflammation appears to play a critical role in the development and maintenance of airway hyperreactivity and in most associated clinical manifestations and pathophysiological changes in the respiratory tract. An inflammatory response is characteristic of mild as well as severe chronic asthma (Dunnill *et al.*, 1969; Hogg *et al.*, 1977; Woolcock, 1986); the prevalence and magnitude of the numerous features of airway inflammation appear to correlate with the severity of the disease state. At the extreme—patients who have died of asthma—airway inflammation is severe, and histopathological and anatomical examination of the airways indicates that all the classic features of an inflammatory response are evident (Houston *et al.*, 1953; Dunnill, 1960; Hogg, 1984, 1986). Mucus plugs are prevalent and, by causing significant airway narrowing throughout the respiratory tree, may be a major cause of death (Hogg, 1984, 1986). Therefore, it is likely that it is not the enhanced bronchoconstriction *per se,* but the consequences of the combined and culminating effects of the cellular and exudative phases of the inflammatory response, notably the formation of mucus plugs, that are responsible for the chronic aspects of airway hyperreactive disorders that may lead to eventual death (Hogg, 1984, 1986).

III. SUMMARY

Airway hyperreactivity, a hallmark of several common respiratory disorders including asthma, is a complex multifaceted phenomenon. The etiology of airway hyperreactivity is associated with numerous pathological and physiological abnormalities. Airway hyperreactivity is defined and characterized by enhanced responsiveness to allergic and nonallergic stimuli. The diverse pathological manifestations of airway hyperreactivity include increased epithelial and mucosal permeability, edema, epithelial cell damage and shedding, smooth muscle hypertrophy and hyperplasia, mucus hypersecretion, increased vascular permeability, and/or mucus plugs. In asthma, the airway narrowing and resistance to airflow manifest in the acute response and early stages of the disease are probably exclusively the result of an increased level of bronchoconstriction, whereas in the chronic stages of the disorder and in the common late asthmatic response (LAR), additional influences such as inflammation and edema contribute significantly to obstruction of the airways. Despite intensive and increasing multidisciplinary research during the past several years, the underlying mechanism(s) responsible for the development and maintenance of airway hyperreactivity remains largely a mystery. This is probably a reflection of the multiple mechanisms and abnormalities likely to be involved in producing the clinical disease. The pathogenesis of airway hyperreactivity will be the result of the complex interaction and integration of the influence of neuronal and hormonal input and the effects of released primary and secondary mediators in various airway cells, the properties of which may be inherently altered. Numerous individual mechanisms or precipitating factors have been advocated, and in some cases vigorously championed, as being responsible for, or contributing significantly to, airway hyperreactivity. Some of these postulated mechanisms are not mutually exclusive, and, in view of the diversity and complexity of the pathophysiological manifestations of airway hyperreactivity, it is likely that several mechanisms or functional abnormalities, rather than a single one, are responsible.

Overall, there is minimal or no experimental and clinical evidence that airway hyperreactivity is the result of dysfunction in one or all of the various receptor populations present in the respiratory tract. Similarly, it is unlikely that increased influence of the effects of a single mediator or neurotransmitter would be sufficient to elicit airway hyperreactivity. Accordingly, a therapeutic strategy based on antagonizing the effects of one bronchoactive chemical is likely to be of minimal benefit. The inflammatory response in the airways is increasingly being recognized as playing a central, critical role in airway hyperreactivity, especially that of a chronic nature. Major research efforts are focusing on the inflammatory aspect of airway hyperreactive disorders. The unraveling of the relative involvement of the various inflammatory cells and of their contents in airway hyperreactivity, as well as the complex cell–cell interactions, may lead to the development of novel therapeutic agents for airway obstruction associated with LAR and the chronic stages of airway hyperreactive disorders.

An important area of research will be analysis of the morphological, physiological, and pharmacological properties of human airway smooth muscle from controls and from patients with airway hyperreactive disorders to explore whether there are any inherent abnormalities in diseased airway smooth muscle. It remains to be clarified *in vitro* whether the airway hyperreactivity is an example of deviation or nondeviation supersensitivity; furthermore, denervation-induced supersensitivity has not been examined in airways. Information is also lacking on the electrophysiological properties and the excitation–contraction coupling mechanisms (including the role of cyclic nucleotides) that occur in hyperreactive and nonhyperreactive airways. In addition to examining potential abnormalities of the smooth muscle cell, possible differences in the characteristics of other cell types (e.g., inflammatory) should be explored. Other areas of research that should aid our understanding of the mechanisms underlying airway hyperreactivity include (1) the role of the NANC transmitter, (2) the physiological role of the many neuropeptides present in airways, (3) identification of epithelium-derived factors and characterization of their effects on airway smooth muscle reactivity, and (4) the role of the numerous endogenous mediators on neurotransmission.

The complexity and diversity of the functional abnormalities and the pathogenesis are likely to contribute to the difficulty in the clinical management of airway hyperreactivity. Airway hyperreactivity should be considered as possessing two major components—enhanced bronchoconstriction and airway inflammation—and should be treated accordingly. The hope is that present and future research will provide a clear insight into the basis of airway hyperreactivity and will lead to the definitive identification of critical chemical and cellular targets and the development of novel strategies and therapeutic agents. This is of particular concern, as deaths from the classic airway hyperreactive disorder, asthma, are on the increase.

ACKNOWLEDGMENTS. The author wishes to thank Dr. David Westfall for helpful discussion, as well as Dotti Lavan, Robin Brown, and Christine Stefankiewicz for expert typing of the manuscript. The assistance of Roseanna M. Muccitelli in preparing this chapter is gratefully appreciated.

IV. REFERENCES

Abraham, W. M., Delehunt, J. C., Yerger, L., and Marchette, B., 1983, Characterization of a late-phase pulmonary response after antigen challenge in allergic sheep, *Am. Rev. Respir. Dis.* **128:** 839–844.

Akasaka, K., Konno, K., Ono, Y., Mue, S., Abe, C., Kumagai, M., and Ise, I., 1975, Electromyography study of bronchial smooth muscle in bronchial asthma, *Tohoku J. Exp. Med.* **117:** 55.

Alexander, H. S., and Paddock, R., 1921, Bronchial asthma; response to pilocarpine and epinephrine, *Arch. Intern. Med.* **27:**184–191.

Allegra, L., and Bianco, S., 1980, Non-specific bronchoreactivity obtained with an ultrasonic aerosol of bottled water, *Eur. J. Respir. Dis* **61**(Suppl. 106):41–49.

Andersson, K. E., 1983, Airway hyperreactivity, smooth muscle and calcium, *Eur. J. Respir. Dis.* **64**(Suppl. 131):49–70.

Anthracite, R. F., Vachon, L., and Knapp, P. H., 1971, Alpha-adrenergic receptors in the human lung, *Psychosomat. Med.* **33:**481–489.

Atkins, P. C., Rosenblum, F., and Dunskey, E. H., 1980. Comparison of plasma histamine and cyclic nucleotides after antigen and methacholine inhalation in man, *J. Allergy Clin. Immunol.* **66:**478–485.

Austen, K. F., and Orange, R. P., 1975, State of the Art. Bronchial asthma: the possible role of the chemical mediators of immediate hypersensitivity in the pathogenesis of subacute chronic disease, *Am. Rev. Respir. Dis.* **112**:423–436.

Barnes, P. J., 1984, The third nervous system in the lung: Physiology and clinical perspectives, *Thorax* **39**:561–567.

Barnes, P. J., 1985a, Mediators and asthma, *Br. J. Hosp. Med.* **34**:337–344.

Barnes, P. J., 1985b, Peptides in asthma, *Br. J. Clin. Pharmacol.* **20**:519P–520P. (abst.).

Barnes, P. J., 1986a, Asthma as an axon reflex, *Lancet* **1**:242–245.

Barnes, P. J., 1986b, Airway receptors and asthma, *N. Engl. Regul. Allergy Proc.* **7**:219–227.

Barnes, P. J., 1986c, Asthma therapy: Basic mechanisms, *Eur. J. Respir. Dis.* **68**(Suppl. 144):217–265.

Barnes, P. J., 1987, Airway neuropeptides and asthma, *Trends Pharmacol. Sci.* **8**:24–27.

Barnes, P. J., and Pride, N. N., 1983, Dose–response curves to inhaled β-adrenoceptor agonists in normal and asthmatic subject, *Br. J. Clin. Pharmacol.* **15**:677–682.

Barnes, P. J., Kaliner, J. S., and Dollery, C. T., 1980, Human lung adrenoceptors studied by radioligand binding, *Clin. Sci.* **58**:457–461.

Barnes, P. J., Ind, P. W., and Dollery, C. T., 1981a, Inhaled prazosin in asthma, *Thorax* **36**:378–381.

Barnes, P. J., Wilson, N. M., and Vickers, H., 1981b, Prazosin, an alpha₁-adrenoceptor antagonist, partially inhibits exercise-induced asthma, *J. Allergy Clin. Immunol.* **68**:411–415.

Barnes, P. J., Ind, P. W., and Brown, M. J., 1982, Plasma histamine and catecholamines in stable asthmatic subjects, *Clin. Sci.* **62**:661–665.

Barnes, P. J., Skoogh, B. E., Nadel, J. A., and Roberts, J. M., 1983, Prostsynaptic alpha₂-adrenoceptors predominate over alpha₁-adrenoceptors in canine tracheal smooth muscle and mediate neuronal and hormonal alpha-adrenergic contraction, *Mol. Pharmacol.* **23**:570–575.

Barnes, P. J., Ind, P. W., and Dollery, C. T., 1984, Beta-adrenoceptors, in: *Asthma* (A. B. Kay, K. F. Austen, and L. M. Lichtenstein, eds.), pp. 339–358, Academic, London, New York.

Barnes, P. J., Cuss, F. M., and Palmer, J. B., 1985, The effect of airway epithelium on smooth muscle contractility in bovine trachea, *Br. J. Pharmacol.* **86**:685–691.

Benfey, B. G., 1980, The evidence against interconversion of α- and β- adrenoceptors, *Trends Pharmacol. Sci.* **1**:193–194.

Benson, M. K., 1975, Bronchial hyperreactivity, *Br. J. Dis. Chest* **69**:227–239.

Benson, M. K., and Graf, P. D., 1977, Bronchial reactivity: Interaction between vagal stimulation and inhaled histamine, *J. Appl. Physiol.* **43**:643–647.

Biel, M., and DeKock, M. A., 1978, Role of alpha-adrenergic receptors in exercise-induced bronchoconstriction, *Respiration* **35**:78–86.

Bienfield, W. W., and Seifter, J., 1978, Contraction of dog trachealis muscle *in vivo:* Role of alpha-adrenergic receptors, *J. Appl. Physiol.* **48**:723–733.

Black, J. L., Salome, C. M., Yan, K., and Shaw, J., 1982, Comparison between airways response to an α-adrenoceptor agonist and histamine in asthmatic and non-asthmatic subjects, *Br. J. Clin. Pharmacol.* **14**:464–466.

Bolton, T. B., 1979, Mechanisms of action of transmitters and other substances on smooth muscle, *Physiol. Rev.* **59**:606–718.

Boschetto, P., Zocca, E., Milani, G. F., Licata, B., Pivorotti, F., Mapp, C. E., and Fabbri, L. M. 1986, Bronchoalveolar neutrophilia during late, but not early, asthmatic reactions induced by toluene diisocyanate (TDI), *J. Allergy Clin Immunol.* **77**:244 (abst.).

Boulet, L. P., Cartier, A., Thomson, N. C., Roberts, R. S., Dolovich, J., and Hargreave, F. E., 1983, Asthma and increases in nonallergic bronchial responsiveness from seasonal pollen exposure, *J. Allergy Clin. Immunol.* **71**:399–406.

Boushey, H. A., 1985, Role of the vagus nerves in bronchoconstriction in humans, *Chest* **87**:1975–2015.

Boushey, H. A., Holtzman, M. J., Sheller, J. R., and Nadel, J. A., 1980, State of the art, Bronchial hyperreactivity, *Am. Rev. Respir. Dis.* **121**:389–413.

Boucher, R. C., Pare, P. D., and Hogg, J. C., 1979, Relationship between airway hyperreactivity and hyperpermeability in *Ascaris*-sensitive monkeys, *J. Allergy Clin. Immunol.* **64**:197–201.

Bouhuys, A., 1963, Effect of posture in experimental asthma in man, *Am. J. Med.* **34**:470–476.

Britton, J., Ayres, J., and Cochrane, G. M., 1981, Effect of inhaled α-blocker on airflow obstruction in asthma, *J. R. Soc. Med.* **74**:646–648.

Brooks, S. M., McGowan, K., Bernstein, I. L., Altenau, P., and Peagler, J., 1979, Relationship between numbers of beta adrenergic receptors in lymphocytes and disease severity in asthma, *J. Allergy Clin. Immunol.* **63**:401–406.

Brown, M. J., Ind, P. W., Canson, R., and Lee, T. H., 1982, A novel double-isotope technique for enzymatic assay of plasma histamine: Application to estimation of mast cell activation as assessed by antigen challenge in asthmatics, *J. Allergy Clin. Immunol.* **69**:20–24.

Busse, W. W., and Sosman, J., 1977, Decreased H_2-histamine response of granulocytes of asthmatic patients, *J. Clin. Invest.* **59**:1080–1087.

Busse, W. W., Bush, R. K., and Cooper, W., 1979a, Granulocyte response *in vitro* to isoproterenol, histamine, and prostaglandin E_1 during treatment with beta-adrenergic aerosols in asthma, *Am. Rev. Respir. Dis.* **120**:337–384.

Busse, W. W., Cooper, W., Warshauer, D. M., Dick, E. C., Wallow, I. H. L., and Albrecht, R., 1979b, Impairment of isoproterenol, H_2-histamine, and prostaglandin E_1 response of human granulocytes after incubation *in vitro* with live influenza vaccines, *Am. Rev. Respir. Dis.* **119**:561–569.

Cade, J. F., and Pain, M. C. F., 1971, Role of bronchial reactivity in aetiology of asthma, *Lancet* **2**:186–188.

Cade, J. F., and Pain, M. C. F., 1972, Lung function in provoked asthma: Response to inhaled urea, methacholine and isoprenaline, *Clin. Sci.* **43**:759–769.

Callerame, M. L., Condemi, J. J., Bohrod, M. G., and Vaughan, J. H., 1971, Immunological reactions of bronchial tissues in asthma, *N. Engl. J. Med.* **284**:459–464.

Castro de la Mata, R., Penna, M., and Aviada, D. M., 1962, Reversal of sympathomimetic bronchodilatation by dichloroisoproterenol, *J. Pharmacol. Exp. Ther.* **135**:197–203.

Chand, N., 1980a, Distribution and classification of airway histamine receptors. The physiological significance of histamine H_2-receptors, *Adv. Pharmacol. Chemother.* **17**:103–131.

Chand, N., 1980b, Is airway hyperreactivity in asthma due to histamine H_2-receptor deficiency? *Med. Hypoth.* **6**:1105–1112.

Cheng, J. B., and Townley, R. G., 1983, Decreased sensitivity and response of isolated tracheal smooth muscle to methacholine and histamine without changing the activity of pulmonary muscarinic receptors in the egg albumin sensitized guinea pig, *Int. Arch. Allergy Appl. Immunol.* **72**:303–309.

Christiansen, S. C., Proud, D., and Cochrane, C. G., 1987, Detection of tissue kallikrein in the bronchoalveolar lavage fluid of asthmatic subjects, *J. Clin. Invest.* **79**:188–197.

Chung, K. F., 1986, Role of inflammation in hyperreactivity of the airways in asthma, *Thorax* **41**:657–662.

Church, M. K., and Holgate, S. T., 1986, Adenosine and asthma, *Trends Pharmacol. Sci.* **7**:49–50.

Coburn, R. F., and Yamaguchi, T., 1977, Membrane potential-dependent and -independent tension in the canine tracheal muscle, *J. Pharmacol. Exp. Ther.* **201**:276–284.

Coburn, R. F., 1979, Electromechanical coupling in canine trachealis muscle: Acetylcholine contractions, *Am. J. Physiol.* **236**:177–184.

Cockcroft, D. W., 1985, The bronchial late response in the pathogenesis of asthma and its modulation by therapy, *Ann. Allergy* **55**:857–862.

Cockcroft, D. W., Killian, D. N., Mellon, J. J. A., and Hargreave, F. E., 1977, Bronchial reactivity to inhaled histamine: A clinical survey, *Clin. Allergy* **7**:235–243.

Coleridge, H. M., and Coleridge, J. C. G., 1977, Apparent vagal C-fibers in dog lung. Their
discharge during spontaneous breathing and their stimulation by alloxan and pulmonary congestion, in: *Krogh Centenary Symposium of Respiratory Adaptions, Capillary Exchange and Reflex
Mechanisms* (A. S. Paintal and P. Gill-Kumar, eds.), pp. 369–406, University of Delhi, Delhi,
India.

Cromwell, O., Walport, M. J., Taylor, G. W., Morris, H. R., O'Driscoll, B. R. C., and Kay, A. B.,
1982, Identification of leukotrienes in the sputum of patients with cystic fibrosis, *Adv. Prostaglandin Thromboxane Leuk. Res.* **9:**251–259.

Cropp, G. J. A., 1975, The role of the parasympathetic nervous system in the maintenance of chronic
airway destruction in asthmatic children, *Am. Rev. Respir. Dis.* **112:**599–605.

Cuss, F. M., Dixon, C. M., and Barnes, P. J., 1986, Inhaled platelet-activating factor causes
bronchoconstriction and increased bronchial reactivity in man, *Am. Rev. Respir. Dis.* **133:**A212
(abst.).

Dalquen, P., 1985, Morphological findings in fatal anaphylactic and anaphylactoid reactions in:
Asthma and Bronchial Hyperreactivity (H. Herzog and A. P. Perruchond, eds.), *Prog. Respir.
Dis.* **19:**189–193.

Davis, C., Kannan, M. S., Jones, T. R., and Daniel, E. E., 1982, Control of human airway smooth
muscle:*In vitro* studies, *J. Appl. Physiol.* **53:**1080–1087.

De Jongste, J. C., Mons, H., Bonta, I. L., and Kerrebijn, K. F., 1987, Cholinergic- and nonadrenergic inhibitory nerve-mediated responses of isolated human airways in COPD and asthma, *Am. Rev. Respir. Dis.* **135:**A181 (abst.).

DeKock, M. A., Nadel, J. A., Zwi, S., Colebatch, H. J. H., and Olsen, C. R., 1966, New method
for perfusing bronchial arteries: Histamine bronchoconstriction and apnea, *J. Appl. Physiol.* **21:**
185–194.

De Monchy, J. G. R., Kauffman, H. F., Venge, P., Koeter, G. H., Jansen, H. M., Sluiter, H. J.,
and DeVries, K., 1985, Broncho-alveolar eosinophilia during allergen-induced late asthmatic
reactions, *Am. Rev. Respir. Dis.* **131:**373–376.

Dey, R. D., Shannon, W. A., Jr., and Said, S. I. 1981, Localization of VIP-immunoreactive nerves
in airways and pulmonary vessels of dogs, cats and human subjects, *Cell Tissue Res.* **220:**231–
238.

Dhillon, D. S., and Rodger, I. W., 1981, Hyperreactivity of guinea-pig isolated airway smooth
muscle, *Br. J. Pharmacol.* **74:**180P (abst.).

Diamond, L., and Gillespie, M., 1982, The lung nonadrenergic inhibitory nervous system, *Trends
Pharmacol. Sci.* **3:**237–239.

Dor, P. J., Ackerman, S. J., and Gleich, G. J., 1984, Charcot–Leyden crystal protein and eosinophil
granule major basic protein in sputum of patients with respiratory disease, *Am. Rev. Respir. Dis.*
130:1072–1077.

Dulfano, M. J., Luk, C. K., Beckage, M., and Wooten, O., 1982, Ciliary inhibitory effect of asthma
patients' sputum, *Clin. Sci.* **62:**393–396.

Dunnill, M. S., 1960, The pathology of asthma, with special reference to changes in the bronchial
mucosa, *J. Clin. Pathol.* **13:**27–33.

Dunnill, M. S., 1975, The morphology of airways in bronchial asthma, in: *New Directions in Asthma*
(M. Stein, ed.), pp. 213–222, American College of Chest Physicians, Park Ridge, Illinois.

Dunnill, M. S., Massarella, G. R., and Anderson, J. A., 1969, A comparison of the quantitative
anatomy of the bronchi in normal subjects, in status asthmaticus in chronic bronchitis and in
emphysema, *Thorax* **24:**176–179.

Durham, S. R., Loegering, D. A., Gleich, G. J., and Kay, A. B., 1985, Eosinophils, non-specific
bronchial responsiveness and allergen-induced late phase asthmatic reactions, *J. Allergy Clin.
Immunol.* **75:**148 (abst.).

Eiser, N. M., 1983, Hyperreactivity—Its relationship to histamine receptors, *Eur. J. Respir. Dis.* **64**
(Suppl. 131):99–114.

Elwood, R. K., Hogg, J. C., and Paré, P. D., 1982, Airway response to osmolar challenge in asthma, *Am. Rev. Respir. Dis.* **125**:61 (abst.).

Elwood, R. K., Kennedy, S., Belzberg, A., Hogg, J. C., and Paré, P. D., 1983, Respiratory mucosal permeability in asthma, *Am. Rev. Respir. Dis.* **128**:523–527.

Empey, D., 1982, Mechanisms of bronchial hyperreactivity, *Eur. J. Respir. Dis.* **63**(Suppl. 117): 34–40.

Empey, D. W., Laitinen, L. A., Jacobs, L., Gold, W. M., and Nadel, J. A., 1976, Mechanisms of bronchial hyperreactivity in normal subjects after upper respiratory tract infection, *Am. Rev. Respir. Dis.* **113**:131–139.

Fabbri, L. M., Aizawa, H., Alpert, S. E., Walters, E. H., O'Byrne, P. M., Gold, B. D., Nadel, J. A., and Holtzman, M. J., 1984, Airway hyperresponsiveness and changes in cell counts in bronchoalveolar lavage after ozone exposure in dogs, *Am. Rev. Respir. Dis.* **129**:288–291.

Farley, J. M., and Miles, P. R., 1977, Role of depolarization in acetylcholine-induced contractions of dog trachealis muscle, *J. Pharmacol. Exp. Ther.* **201**:199–205.

Farmer, S. G., 1987, Airway smooth muscle responsiveness: Modulation by the epithelium, *Trends Pharmacol. Sci.* **8**:8–10.

Fedan, J. S., Hay, D. W. P., Farmer, S. G., and Raeburn, D., 1988, Modulation of airway smooth muscle reactivity by epithelial cells, in: *Asthma—Basic Mechanisms and Clinical Management* (I. W., Rodger, P. T. Barnes, and N. C. Thompson, eds.), pp. 143–162, Croom Helm, Amsterdam.

Flavahan, N. A., Aarhus, L. L., Rimele, T. J. and Vanhoutte, P. M., 1985, Respiratory epithelium inhibits bronchial smooth muscle tone, *J. Appl. Physiol.* **58**:834–838.

Fleming, W. W., 1976, Variable sensitivity of excitable cells: Possible mechanisms and biological significance in: *Reviews of Neuroscience,* Vol. 2 (S. Ehrenpreis and I. J. Kopin, eds.), pp. 43–90, Raven, New York.

Fleming, W. W., and Westfall, D. P., 1975, Altered resting membrane potential in the supersensitive vas deferens of the guinea-pig, *J. Pharmacol. Exp. Ther.* **192**:381–389.

Fleming, W. W., McPhillips, J. J., and Westfall, D. P., 1973, Postjunctional supersensitivity and subsensitivity of excitable tissues to drugs, *Ergeb. Physiol.* **68**:55–119.

Florey, H., Carlston, H. M., and Wells, A. Q., 1932, Mucous secretion in the trachea, *Br. J. Exp. Pathol.* **13**:269–284.

Freedman, B. J., 1972, The functional geometry of the bronchi, *Bull. Physiopathol. Respir.* **8**:545–551.

Frigas, E., and Gleich, F. J., 1986, The eosinophil and the pathophysiology of asthma, *J. Allergy Clin. Immunol.* **77**:527–537.

Frigas, E., Loegering, D. A., and Gleich, G. J. 1980, Cyctotoxic effects of the guinea pig eosinophil major basic protein on tracheal epithelium, *Lab. Invest.* **42**:35–43.

Frigas, E., Loegering, D. A., Solley, G. O., Farrow, G. M., and Gleich, G. J., 1981, Elevated levels of the eosinophil granule major basic protein in the sputum of the patients with bronchial asthma, *Mayo Clin. Proc.* **56**:345–353.

Fuller, R. W., Dixon, C. M. S., Dollery, C. T., and Barnes, P. J., 1986, Prostaglandin D_2 potentiates airway responsiveness to histamine and methacholine, *Am. Rev. Respir. Dis.* **133**: 252–254.

Furchgott, R. F., and Zawadzki, J. V., 1980, The obligatory role of endothelial cells in the relaxation of arterial smooth muscle by acetylcholine, *Nature (Lond.)* **288**:373–376.

Galant, S. P., Duriseti, L., Underwood, S., and Insel, P. A., 1978, Decreased beta adrenergic receptors on PMN leukocytes after adrenergic therapy, *N. Engl. J. Med.* **299**:933–936.

Galant, S. P., Alfred, S., and Griffiths, R., 1980a, Characterization of adenylate cyclase activity in asthmatic neutrophil sonicates, *Am. Rev. Respir. Dis.* **122**:231–238.

Galant, S. P., Duriseti, L., Underwood, S., and Insel, P. A., 1980b, Beta-adrenergic receptors of polymorphonuclear particulates in bronchial asthma, *J. Clin. Invest.* **65**:577–585.

Garland, L. G., 1984, The pharmacology of airway hyper-reactivity, *Trends Pharmacol. Sci.* **5**:338–340.

Gerthoffer, W. T., Fedan, J. S., Westfall, D. P., Goto, K., and Fleming, W. W., 1979, Involvement of the sodium–potassium pump in the mechanism of postjunctional supersensitivity of the vas deferens of the guinea pig, *J. Pharmacol. Exp. Ther.* **210**:27–36.

Gleich, G. J., 1986, The pathology of asthma: With emphasis on the role of the eosinophil, *N. Engl. Reg. Allergy* **1**:421–424.

Glynn, A. A., and Michaels, L., 1960, Bronchial biopsy in chronic bronchitis and asthma, *Thorax* **15**:142–153.

Golden, J. A., Nadel, J. A., and Boushey, H. A., 1978, Bronchial hyperirritability in healthy subjects after exposure to ozone, *Am. Rev. Respir. Dis.* **118**:287–294.

Goldie, R. G., Lulich, K. M., and Paterson, J. W., 1985, Bronchial α-adrenoceptor function in asthma, *Trends Pharmacol. Sci.* **6**:469–472.

Goldie, R. G., Papadimitriou, J. M., Paterson, J. W., Rigby, P. J., Self, H. M., and Spina, D., 1986a, Influence of the epithelium on responsiveness of guinea-pig isolated trachea to contractile and relaxant agonists, *Br. J. Pharmacol.* **87**:5–14.

Goldie, R. G., Spina, D., Henry, P. J., Lulich, K. M., and Paterson, J. W., 1986b, *In vitro* responsiveness of human asthmatic bronchus to carbachol, histamine, β-adrenoceptor agonists and theophylline, *Br. J. Clin. Pharmacol.* **22**:669–676.

Grandel, K. E., Schwartz, E., Greene, D., Wardlow, M. L., Pinckard, R. N., and Farr, R. S., 1983, A platelet activating factor (PAF) extracted from acidified human plasma, *Fed. Proc.* **42**:1026 (abst.).

Gross, N. J., and Skorodin, M. S., 1984, State of the art. Anticholinergic, antimuscarinic bronchodilators, *Am. Rev. Respir. Dis.* **129**:856–870.

Hahn, H. L., Wilson, A. G., Graf, P. D., Fischer, S. P., and Nadel, J. A., 1978, Interaction between serotonin and efferent vagus nerves in dog lungs, *J. Appl. Physiol.* **44**:144–149.

Hargreave, F. E., O'Byrne, P. M., and Ramsdale, E. H., 1985, Mediators, airway responsiveness, and asthma, *J. Allergy Clin. Immunol.* **76**:272–276.

Hargreave, F. E., Dolovich, J., O'Byrne, P. M., Ramsdale, E. H., and Daniel, E. E., 1986a, The origin of airway hyperresponsiveness, *J. Allergy Clin. Immunol.* **78**:825–832.

Hargreave, F. E., Ramsdale, E. H., Kirby, J. G., and O'Byrne, P. M., 1986b, Asthma and the role of inflammation, *Eur. J. Respir. Dis.* **69**(Suppl. 147):16–21.

Haruhisa, M., Yui, Y., Taniguchi, N., Yasueda, H., and Shida, T., 1985, Increased activity of 5-lipoxygenase in polymorphonuclear leucocytes from asthmatic patients, *Life Sci.* **37**:907–914.

Harvey, J. W., and Tattersfield, A. E., 1982, Airway response to salbutamol: Effect of regular salbutamol inhalations in normal, atopic and asthmatic subjects, *Thorax* **37**:280–287.

Hay, D. W. P., Farmer, S. G., Raeburn, D., Robinson, V. A., Fleming, W. W., and Fedan, J. S., 1986a, Airway epithelium modulates the reactivity of guinea-pig respiratory smooth muscle, *Eur. J. Pharmacol.* **129**:11–18.

Hay, D. W. P., Raeburn, D., Farmer, S. G., Fleming, W. W., and Fedan, J. S., 1986b, Epithelium modulates the reactivity of ovalbumin-sensitized guinea-pig airway smooth muscle, *Life Sci.* **38**:2461–2468.

Hay, D. W. P., Muccitelli, R. M., Horstemeyer, D. L., Wilson, K. A., and Raeburn, D., 1987, Demonstration of the release of an epithelium-derived inhibitory factor from a novel preparation of guinea-pig trachea, *Eur. J. Pharmacol.* **136**:247–250.

Hogg, J. C., 1981, Bronchial mucosal permeability and its relationship to airways hyperreactivity, *J. Allergy Clin. Immunol.* **67**:421–425.

Hogg, J. C., 1983, Bronchial mucosal permeability in airways hyperreactivity, *Eur. J. Respir. Dis.* **64**(Suppl. 131):171–181.

Hogg, J. C., 1984, The pathology of asthma, *Clin. Chest Med.* **5**:567–571.

Hogg, J. C., 1986, Airway structure and function in asthma, *N. Engl. Reg. Allergy Proc.* **7**:228–235.

Hogg, J. C., and Eggleston, P. A. 1984, Is asthma an epithelial disease?, *Am. Rev. Respir. Dis.* **129**:207–208.

Hogg, J. C., Paré, P. D., Boucher, R., Michoud, M. C., Guerzon, G., and Moroz, L., 1977, Pathologic abnormalities in asthma, in: *Asthma: Physiology, Immunopharmacology and Treatment*, Vol. 23 (L. M. Lichtenstein and K. A. Austen, eds.), pp. 1–4, Academic, New York.

Holgate, S. T., 1986, The pathophysiology of bronchial asthma and targets for its drug treatment, *Agents Actions* **18**:281–287.

Holgate, S. T., Hardy, C. C., Robinson, C., Agius, R. M., and Howarth, P. H., 1986a, The mast cell as a primary effector cell in the pathogenesis of asthma, *J. Allergy. Clin. Immunol.* **77**:274–282.

Holgate, S. T., Howarth, P. H., Beasly, R., Agius, R. and Church, M. K., 1986b, Cellular and biochemical events in the pathogenesis of asthma, *Eur. J. Respir. Dis.* **68**(Suppl. 144):37–76.

Holroyde, M. C., 1986, The influence of epithelium on the responsiveness of guinea-pig isolated trachea, *Br. J. Pharmacol.* **87**: 501–507.

Holtzman, M. J., Aizawa, H., Nadel, J. A., and Goetzl, E. J. 1983a, Selective generation of leukotriene B_4 by tracheal epithelial cells from dogs, *Biochem. Biophys. Res. Commun.* **114**: 1071–1076.

Holtzman, M. J., Fabbri, L. M., O'Byrne, P. M., Gold, B. D., Aizawa, H., Walters, E. H., Albert, S. E., and Nadel, J. A., 1983b, Importance of airway inflammation for hyperresponsiveness induced by ozone, *Am. Rev. Respir. Dis.* **127**:686–690.

Horn, B. R., Robin, E. D., Theodore, J., and Van Kessel, A., 1975, Total eosinophil counts in the management of bronchial asthma, *N. Engl. J. Med.* **292**:1152–1155.

Hossain, S., and Heard, B. E., 1970, Hyperplasia of bronchial muscle in chronic bronchitis, *J. Pathol.* **101**:171–184.

Houston, J. C., DeNavasquez, S., and Trounce, J. R., 1953, A clinical and pathologic study of fatal cases of status asthmaticus, *Thorax* **8**:207–213.

Huber, H. L., and Koessler, K. K., 1922, The pathology of bronchial asthma, *Arch. Intern. Med.* **30**:689–760.

Hulbert, W. C., Walker, D. C., Jackson, A., and Hogg, J. C., 1981, Airway permeability to horseradish peroxidase in guinea pigs: The repair phase after injury by cigarette smoke, *Am. Rev. Respir. Dis.* **123**:320–326.

Hulbert, W. C., Paré, P. D., Amour, C., Wiggs, B., and Hogg, J. C., 1982, The relationship between airway permeability and hyperreactivity, *Am. Rev. Respir. Dis.* **125**:226 (abst.).

Iafrate, R. P., Massey, K. L., and Hendeles, L., 1986, Current concepts in clinical therapeutics: Asthma, *Clin. Pharm.* **5**:206–227.

Ishihara, Y., Uchida, Y., and Kitamura, S., 1985, Measurement of leukotrienes in peripheral venous blood from patients with bronchial asthma, *Clin. Res.* **33**:466A (abst.).

Isono, T., Koshihara, Y., Murota, S.-I., Fukuda, Y., and Furukawa, S., 1985, Measurement of immunoreactive leukotrine C_4 in blood of asthmatic children, *Biochem. Biophys. Res. Commun.* **130**:486–492.

Iwayama, Y., Chung, C. Z., and Takayanagi, I., 1982, Effects of anti-asthmatic drugs on airway resistance and plasma level of cyclic AMP in guinea pig, *Jpn. J. Pharmacol.* **32**:329–334.

James, A. L., Paré, P. D., and Hogg, J. C., 1987, Airway wall thickness and smooth muscle shortening in normal and asthmatic subjects, *Am. Rev. Respir. Dis.* **135**:A181 (abst.).

Jeffrey, P., and Corrin, B., 1984, Structural analysis of the respiratory tract, in: *Immunology of the Lung and Upper Respiratory Tract* (J. Bienenstock, eds.), pp. 1–27, McGraw-Hill, New York.

Kaliner, M., 1984, Hypothesis on the contribution of late-phase allergic responses to the understanding and treatment of allergic diseases, *J. Allergy Clin. Immunol.* **73**:311–315.

Kariman, K., and Lefkowitz, R. J., 1977, Decreased beta adrenergic receptor binding in lymphocytes of patients with bronchial asthma, *Clin. Res.* **25**:503A (abst.).

Karlin, J. M., 1972, The use of antihistamines in asthma, *Ann. Allergy* **30**:342–345.

Kennedy, S. M., Elwood, R. K., Wiggs, B. J. R., Paré, P. D., and Hogg, J. C., 1984, Increased airway mucosal permeability: Relationship to airway reactivity, *Am. Rev. Respir. Dis.* **129**: 143–148.

Kerr, J. W., Govindoraj, M., and Patel, K. R., 1970, Effect of alpha-receptor blocking drugs and disodium cromoglycate on histamine sensitivity in bronchial asthma, *Br. Med. J.* **2**:139–141.

Kirkpatrick, C. T., 1975, Excitation and contraction in bovine tracheal smooth muscle, *J. Physiol. (Lond.)* **244**:263–281.

Klein, R. C., and Salvaggio, J. E., 1966, Nonspecificity of the bronchoconstricting effect of histamine and acetyl-beta-methylcholine in patients with obstructive airway disease, *J. Allergy* **37**:158–168.

Klock, L. E., Miller, T. D., Morris, A. H., Watanabe, S., and Dickman, M., 1975, A comparative study of atropine sulfate and isoproterenol hydrochloride in chronic bronchitis, *Am. Rev. Respir. Dis.* **112**:371–376.

Knauer, K. A., Lichtenstein, L. M., Adkinson, N. F., Jr., and Fish, J. E., 1981, Platelet activation during antigen-induced airway reactions in asthmatic subjects, *N. Engl. J. Med.* **304**:1404–1407.

Kneussl, M. P., and Richardson, J. B., 1978, Alpha-adrenergic receptors in human and canine tracheal and bronchial smooth muscle, *J. Appl. Physiol.* **45**:307–311.

Laitinen, L. A., 1974, Histamine and methacholine challenge in the testing of bronchial reactivity, *Scand. J. Respir. Dis.* **86**:1–47.

Laitinen, L. A., Haahtela, R., Kava, T., and Laitinen, A., 1983, Non-specific bronchial reactivity and ultrastructure of the airway epithelium in patients with sarcoidosis and allergic alveolitis, *Eur. J. Respir. Dis.* **64**(Suppl. 137):267–284.

Laitinen, A. S., Heino, M., Laitinen, A., Kava, T., and Haahtela, T., 1985, Damage of airway epithelium and bronchial reactivity in patients with asthma, *Am. Rev. Respir. Dis.* **131**:599–606.

Larsen, G. L., Barron, R. J., Cotton, E. K., and Brooks, J. G., 1979, A comparative study of inhaled atropine sulfate and isoproterenol hydrochloride in cystic fibrosis, *Am. Rev. Respir. Dis.* **119**:399–407.

Larsen, G. L., Wilson, M. C., Clark, R. A. F., and Behrens, B. L., 1987, The inflammatory reaction in the airways in an animal model of the late asthmatic response, *Fed. Proc.* **46**:105–112.

Lee, H. Y., and Murlas, C., 1985, Ozone-induced bronchial hyperreactivity in guinea pigs is abolished by BW755C or FPL 55712 but not by indomethacin, *Am. Rev. Respir. Dis.* **132**: 1005–1009.

Lee, T. H., and Anderson, S. D., 1985, Heterogeneity of mechanisms in exercise induced asthma, *Thorax* **40**:481–487.

Lewis, R. A., and Austen, K. F., 1984, The biologically active leukotrienes: Biosynthesis, metabolism, receptors, functions and pharmacology, *J. Clin. Invest.* **73**:889–897.

Lewis, R. A., and Robin, J.-L., 1985, Arachidonic acid derivatives as mediators of asthma, *J. Allergy Clin. Immunol.* **76**:259–264.

Lichtenstein, L. M., and Gillespie, E., 1973, Inhibition of histamine release by histamine controlled by H_2-receptors, *Nature (Lond.)* **244**:287–288.

Macklem, P. T., 1985, Bronchial hyperresponsiveness, *Chest* **87**:158S–159S.

Makino, S., 1966, Clinical significance of bronchial sensitivity to acetylcholine and histamine in bronchial asthma, *J. Allergy* **38**:127–142.

Makino, S., Oulette, J. J., Reed, C. E., and Fishel, C. W., 1970, Correlation between increased bronchial responses to acetylcholine and diminished metabolic and eosinopenic responses to epinephrine in asthma, *J. Allergy* **46**:178–189.

Mann, J. S., Renwick, A. G., and Holgate, S. T., 1983, Antigen bronchial provocation causes an increase in plasma adenosine levels in asthma, *Clin. Sci.* **65**:22P–23P (abst.).

Martorana, M. G., and Rodger, I. W., 1980, Effects of Ca^{2+} withdrawal and verapamil on excitation-contraction coupling in sensitized guinea-pig airway smooth muscle, *Br. J. Pharmacol.* **72**:175P (abst.).

McCaig, D. J., and Souhrada, J. F., 1980, Alteration of electrophysiological properties of airway smooth muscle from sensitized guinea pigs, *Respir. Physiol.* **41**:49–60.

McFadden, E. R., Jr., 1984, Pathogenesis of asthma, *J. Allergy Clin. Immunol.* **4**:413–424.

McKay, R. T., and Brooks, S. M., 1983, Induction of tracheal smooth muscle hyperresponsiveness to carbachol following exposure to toluene diisocyanate (TDI) vapours, *Am. Rev. Respir. Dis.* **127**:174 (abst.).

Mellis, C. M., 1978, Bronchial reactivity in cystic fibrosis, *Pediatrics* **61**:446–450.

Michel, F. B., Godard, P., and Bousquet, J., 1982, How important are mediators? in: *Perspectives in Asthma. Bronchial Hyperreactivity* (J. Morley, ed.), pp. 53–68, Academic, London.

Middleton, E., Jr., 1980, Antiasthmatic drug therapy and calcium ions: Review of pathogenesis and role of calcium, *J. Pharmaceut. Sci.* **69**:243–251.

Middleton, E., Jr., 1985, Calcium antagonists and asthma, *J. Allergy Clin. Immunol.* **76**:341–346.

Moreno, R. H., Dahlby, R., Hogg, J. C., and Paré, P. D., 1985, Increased airway responsiveness caused by airway cartilage softening in rabbits, *Am. Rev. Respir. Dis.* **131**:A288 (abst.).

Moreno, R. H., Hogg, J. C., and Paré, P. D., 1986a, Mechanics of airway narrowing, *Am. Rev. Respir. Dis.* **133**:1171–1180.

Moreno, R. H., McCormack, G. S., Mullen, J. B. M., Hogg, J. C., Bert, J., and Paré, P. D., 1986b, Effect of intravenous papain on the tracheal pressure–volume curves in rabbits, *J. Appl. Physiol.* **60**:247–252.

Murlas, C., and Roum, J. H., 1985a, Sequence of pathological changes in the airway mucosa of guinea pig during ozone-induced bronchial hyper-reactivity, *Am. Rev. Respir. Dis.* **131**:314–320.

Murlas, C., and Roum, J. H., 1985b, Bronchial hyperreactivity occurs in steroid-treated guinea pigs depleted of leukocytes by cyclophosphamide, *J. Appl. Physiol.* **58**:1630–1637.

Nadel, J. A., 1984, Inflammation and asthma, *J. Allergy Clin. Immunol.* **73**:651–653.

Nadel, J. A., and Barnes, P. J., 1984, Autonomic regulation of the airways, *Annu. Rev. Med.* **35**:451–467.

Nadel, J. A., and Widdicombe, J. G., 1963, Reflex control of airway size, *Ann. NY Acad. Sci.* **109**:712–722.

Nadel, J. A., Davis, B., and Phipps, R. J., 1979, Control of mucus secretion and ion transport in airways, *Am. Rev. Physiol.* **41**:369–381.

Naylor, B., 1962, The shedding of the mucosa of the bronchial tree in asthma, *Thorax* **17**:69–72.

Neijens, H. J., Degenhart, H. J., Raatgeep, R., and Kerrebijin, K. F., 1980, The correlation between increased reactivity of the bronchi and of mediator releasing cells in asthma, *Clin. Allergy* **10**:535–539.

O'Byrne, P. M., Walters, E. H., Aizawa, H., Fabbri, L. M., Holtzman, M. J., and Nadel, J. A., 1984, Indomethacin inhibits the airway hyperresponsiveness but not the neutrophil influx by ozone in dogs, *Am. Rev. Respir. Dis.* **130**:220–224.

Orehek, J., 1983, The concept of airway "sensitivity" and "reactivity," *Eur. J. Respir. Dis.* **64** (Suppl. 131):27–48.

Orehek, J., Massari, J. P., Gayrard, P., Grimaud, C., and Charpin, C., 1976, Effect of short-term, low level nitrogen dioxide exposure on bronchial sensitivity of asthmatic patients, *J. Clin. Invest.* **57**:301–307.

Orehek, J., Gayrard, P., Smith, A. P., Grimaud, C., and Charpin, J. 1977, Airway response to carbachol in normal and asthmatic subjects, *Am. Rev. Respir. Dis.* **115**:937–943.

Page, C. P., Sanjar, S., Alvemini, D., and Morley, J., 1986, Inflammatory mediators of asthma, *Eur. J. Respir. Dis.* **68**(Suppl. 144):163–189.

Parker, C. D., Bilbo, R. E., and Reed C. E., 1965, Methacholine aerosol as test for bronchial asthma, *Arch. Intern. Med.* **115**:452–458.

Partanen, M., Laitinen, A., Hervonen, A., Toivanen, M., and Laitinen, L. A., 1982, Catcholamine- and acetylcholinesterase-containing nerves in human lower respiratory tract, *Histochem.* **76**: 175–188.

Patel, K. R., and Kerr, J. W., 1973, The airways response to phenylephrine after blockade of alpha and beta receptors in extrinsic bronchial asthma, *Clin. Allergy* **3**:439–448.

Patel, K. R., Kerr, J. W., and MacDonald, J. D., 1976, The effect of thymoxamine and cromolyn sodium on postexercise bronchoconstriction in asthma, *J. Allergy Clin. Immunol.* **57**:285–292.

Pepys, J., 1973, Immunopharmacology of allergic lung disease, *Clin. Allergy* **3**:1–22.

Pepys, J., and Hutchcroft, B. J., 1975, Bronchial provocation tests in the etiologic diagnosis and analysis of asthma, *Am. Rev. Respir. Dis.* **112**:829–859.

Perpina, M., Cortifo, J., Esplugues, J., and Morillo, E. J., 1986, Different ability of verapamil to inhibit agonist-induced contraction of lung parenchymal strips from control and sensitized guinea pig, *Respiration* **50**:174–184.

Persson, C. G. A., 1986, The role of microvascular permeability in the pathogenesis of asthma, *Eur. J. Respir. Dis.* **68**(Suppl. 144):190–216.

Persson, C. G. A., and Svensjo, E., 1983, Airway hyperreactivity and microvascular permeability to large molecules, *Eur. J. Respir. Dis.* **64**(Suppl. 131):183–214.

Powers, R. E., and Colucci, W. S., 1985, An increase in putative voltage dependent calcium channel number following reserpine treatments, *Biochem. Biophys. Res. Commun.* **132**:844–849.

Raeburn, D., 1984, Studies on excitation-coupling in airway smooth muscle, Ph.D. thesis, University of Strathclyde, Glasgow, Scotland.

Raeburn, D., Hay, D. W. P., Farmer, S. G., and Fedan, J. S., 1986a, Epithelium removal increases the reactivity of human isolated tracheal muscle to methacholine and reduces the effect of verapamil, *Eur. J. Pharmacol.* **123**:451–453.

Raeburn, D., Hay, D. W. P., Mawhinney, M. G., and Fedan, J. S., 1986b, Comparison of the effects of epithelium (EPI) removal on reactivity to histamine (H) and methacholine (Mch) of guinea-pig trachea, from normal and vitamin A-deficient animals, *Physiologist* **29**:173 (abst.).

Raeburn, D., Hay, D. W. P., Robinson, V. A., Farmer, S. G., Fleming, W. W., and Fedan, J. S., 1986c, The effect of verapamil is reduced in isolated airway smooth muscle preparations lacking the epithelium, *Life Sci.* **38**:809–816.

Raeburn, D., Rodger, I. W., Hay, D. W. P., and Fedan, J. S., 1986d, The dependence of airway smooth muscle on extracellular Ca^{2+} for contraction is influenced by the presence of cartilage, *Life Sci.* **38**:1499–1505.

Raeburn, D., Hay, D. W. P., and Fedan, J. S., 1987, Calcium uptake into guinea-pig tracheals: The effects of epithelium removal, *Cell Calcium* **8**:429–436.

Richardson, J. B., 1977, The neural control of human tracheobronchial smooth muscle, in: *Asthma: Physiology, Immunopharmacology and Treatment* (L. M. Lichtenstein and K. F. Austen, eds.), pp. 232–237, Academic, New York.

Richardson, J. B., 1979, State of the art. Nerve supply to the lungs, *Am. Rev. Respir. Dis.* **119**:785–802.

Richardson, J. B., 1981, Nonadrenergic inhibitory innervation of the lung, *Lung* **159**:315–322.

Richardson, J., and Béland, J., 1976, Nonadrenergic inhibitory nervous system in human airways, *J. Appl. Physiol.* **41**:764–771.

Rinard, G. A., Rubinfeld, A. R., Brunton, L. L., and Mayer, S. E., 1979, Depressed cyclic AMP levels in airway smooth muscle from asthmatic dogs, *Proc. Natl. Acad. Sci. USA* **76**:1472–1476.

Rodger, I. W., 1986, Calcium ions and contraction of airway, in: *Asthma Clinical Pharmacology and Therapeutic Progress* (A. B. Kay, ed.), pp. 114–127, Blackwell Scientific Publications, Oxford.

Rubinfeld, A. R., and Pain, M. C. F., 1977, Relationship between bronchial reactivity, airway caliber, and sensitivity of asthma, *Am. Rev. Respir. Dis.* **115**:381–387.

Sant' Ambrogio, G., 1982, Inflammation arising from the tracheobronchial tree of mammals, *Physiol. Rev.* **62**:531–539.

Schlueter, D. P., 1986, Ipratropium bromide in asthma. A review of the literature, *Am. J. Med.* **81** (Suppl. 5A):55–60.

Schönfeld, W., Koller, M., Knoller, J., Konig, W., Muller, W., and Van Der Hardtt, H., 1985, Leukotrienes and their metabolism in plasma and bronchial fluids, *J. Allergy Clin. Immunol.* **75**: 126 (abst.).

Schulman, E. S., 1986, The role of mast cell derived mediators in airway hyperresponsiveness, *Chest* **90**:578–583.

Sears, M. R., 1986, Why are deaths from asthma increasing? *Eur. J. Respir. Dis.* **69**(Suppl. 147): 175–181.

Sellick, H., and Widdicombe, J. G., 1969, The activity of lung irritant receptors during pneumothorax, hyperpnoea and pulmonary vascular congestion, *J. Physiol. (Lond.)* **203**:359–381.

Seltzer, J., Bigby, B. G., Stulbarg, M., Holtzman, M. J., Nadel, J. A., Ueki, I. F., Leikauf, G. D., Goetzl, E. J., and Boushey, H. A., 1986, O_3-induced change in bronchial reactivity to methacholine and airway inflammation in humans, *J. Appl. Physiol.* **60**:1321–1326.

Severinghaus, J. W., and Stupfel, M., 1955, Respiratory dead space increase following atropine in man, and atropine, vagal, or ganglionic blockade and hypothermia in dogs, *J. Appl. Physiol.* **8**: 81–87.

Shephard, E. G., Malan, L., Macfarlane, C. M., Mouton, W., and Joubert, J. R., 1985, Lung function and plasma levels of thromboxane B_2, 6-keto prostaglandin $F_{1\alpha}$ and β-thromboglobulin in antigen-induced asthma before and after indomethacin pretreatment, *Br. J. Clin. Pharmacol.* **19**:459–470.

Sheppard, M. N., Kurian, S. S., Henzen Logmans, S. C., Michetti, F., Cocchia, D., Cole, P., Rush, R. A., Marangos, P. J., Bloom, S. R., and Polak, J. M. 1983, Neuron-specific enolase and S-100. New markers for delineating the innervation of the respiratory tract in man and other animals, *Thorax* **38**:333–340.

Sibbald, B., 1986, Genetic basis of asthma, *Semin. Respir. Med.* **7**:307–315.

Simonsson, B. G., 1965, Clinical and physiological studies on bronchitis. III. Bronchial reactivity to inhaled acetylcholine, *Acta Allergol.* **20**:325–348.

Simonsson, B. G., 1983, Airway hyperreactivity. Definition and short review, *Eur. J. Respir. Dis.* **64**(Suppl. 131):9–25.

Simonsson, B. G., 1984, Non-specific bronchial hyperreactivity: Correlation to asthma and modifying factors, *Eur. J. Respir. Dis.* **65**(Suppl. 136):17–24.

Simonsson, B. G., Jacobs, F. M., and Nadel, J. A., 1967, Role of the autonomic nervous system and the cough reflex in the increased responsiveness of airways in patients with obstructive airway disease, *J. Clin. Invest.* **46**:1812–1818.

Simonsson, B. G., Svedmyr, N., Skoogh, B.-E., Andersson, R., and Bergh, N. P., 1972, *In vivo* and *in vitro* studies on α-receptors in human airways. Potentiation with bacterial endotoxin, *Scand. J. Respir. Dis.* **53**:227–236.

Skoogh, B. E., 1984, Control of airway smooth muscle and mechanisms of bronchial hyperreactivity, *Eur. J. People Dis.* **65**(Suppl. 136):9–15.

Sly, R. M., 1986, Effects of treatment on mortality from asthma, *Ann. Allergy* **56**:207–212.

Small, R. C., 1982, Electrical slow waves and tone of guinea-pig isolated trachealis muscle: Effects of drugs and temperature changes, *Br. J. Pharmacol.* **77**:45–54.

Snashall, R., Boother, F. A., and Sterling, G. M., 1978, The effect of alpha adrenergic stimulation on the airways of normal and asthmatic man, *Clin. Sci. Mol. Med.* **54**:283–289.

Souhrada, J. F., 1978, Changes in airway smooth muscle in experimental asthma, *Respir. Physiol.* **32:**79–90.

Souhrada, J. F., and Souhrada, M., 1983, Significance of the sodium pump for airway smooth muscle, *Eur. J. Respir. Dis.* **64**(Suppl. 128):196–205.

Souhrada, M., and Souhrada, J. F., 1981, Reassessment of electrophysiological and contractile characteristics of sensitized airway smooth muscle, *Respir. Physiol.* **46:**17–27.

Souhrada, M., and Souhrada, J. F., 1984, Immunologically induced alterations of airway smooth muscle cell membrane, *Science* **225:**723–725.

Souhrada, M., and Souhrada, J. F., 1985, Sensitization-induced sodium influx in airway smooth muscle cells of guinea pigs, *Respir. Physiol.* **60:**157–168.

Svedmyr, N., 1983, Airway hyperreactivity. Cholinergic and adrenergic receptors, *Eur. J. Respir. Dis.* **64**(Suppl. 131):71–98.

Svedmyr, N., 1984, Szentivanyi's hypothesis of asthma, *Eur. J. Respir. Dis.* **65**(Suppl. 36):59–65.

Szentivanyi, A., 1968, The beta-adrenergic theory of the atopic abnormality in bronchial asthma, *J. Allergy* **42:**203–232.

Szentivanyi, A., 1979, The conformational flexibility of adrenoceptors and the constitutional basis of atopy, *Triangle* **18:**109–115.

Szentivanyi, A., 1980, The radioligand binding approach in the study of lymphocytic adrenoceptors and the constitutional basis of atopy, *J. Allergy Clin. Immunol.* **65:**5–11.

Takizawa, T., and Thurlbeck, W. M., 1971, Muscle and mucous gland size in the major bronchi of patients with chronic bronchitis, asthma and asthmatic bronchitis, *Am. Rev. Respir. Dis.* **104:**331–336.

Thompson, P. J., Hanson, J. M., Bilani, H., Turner-Warwick, M., and Morley, J., 1984, Platelets, platelet activating factor, and asthma, *Am. Rev. Respir. Dis.* **129:**A3 (abst.).

Thorpe, J. E., and Murlas, C. G., 1986, Leukotriene B_4 potentiates airway muscle responsiveness *in vivo* and *in vitro, Prostaglandins* **31:**899–908.

Thurlbeck, W. M., Pun, R., Toth, J., and Frazer, R. G., 1974, Bronchial cartilage in chronic obstructive lung disease, *Am. Rev. Respir. Dis.* **109:**73–80.

Tinkelman, D. G., 1985, Calcium channel blocking agents in the prophylaxis of asthma, *Am. J. Med.* **78:**35–38.

Tomioka, M., Ida, S., Yriko, S., Ishihara, T., and Takishima, T., 1984, Mast cells in bronchoalveolar lavage of patients with bronchial asthma, *Am. Rev. Respir. Dis.* **129:**1000–1005.

Tschirhart, E., and Landry, Y., 1986, Airway epithelium releases a relaxant factor: Demonstration with substance P, *Eur. J. Pharmacol.* **132:**103–104.

Urquilla, R. R., Westfall, D. P., Goto, K., and Fleming, W. W., 1978, The effects of ovalbumin and alterations in potassium concentration on the sensitivity to drugs and the membrane potential of the smooth muscle of the guinea pig and rat vas deferens, *J. Pharmacol. Exp. Ther.* **207:**347–355.

Vaughan, J. H., Tan, E. M., Mathison, D. A., Stevenson, D. D., Bergman, S. Z., and Brande, A. J., 1973, Bronchial asthma: Pathophysiology, in: *Asthma: Physiology, Immunopharmacology and Treatment* (K. F. Austen and L. M. Lichtenstein, eds.), pp. 1–13, Academic, New York.

Wardlaw, A. J., Fitzharris, P., Cromwell, O., Collins, J. V., and Kay, A. B., 1985, Histamine release from mucosal-type human lung mast cells, *J. Allergy Clin. Immunol.* **75:**193 (abst.).

Weiss, E. B., and Viswanath, S. G., 1979, Calcium hypersensitivity in airways smooth muscle: Isometric tension responses following anaphylaxis, *Respiration* **38:**266–272.

Wells, R. E., Walker, J. E. C., and Hickler, R. B., 1960, Effects of cold air on respiratory airflow resistance in patients with respiratory tract disease, *N. Engl. J. Med.* **263:**268–273.

Westfall, D. P., 1981, Supersensitivity of smooth muscle, in: *Smooth Muscle: An Assessment of Current Knowledge* (E. Bulbring, A. F. Brading, A. W. Jones, and T. Tomita, eds.), pp. 285–310, University of Texas Press, Austin.

White, J., and Eiser, N. M., 1983, The role of histamine and its receptors in the pathogenesis of asthma, *Br. J. Clin. Chest* **77:**215–226.

Wong, S. K., Westfall, D. P., Fedan, J. S., and Fleming, W. W., 1981, The involvement of the sodium–potassium pump in postjunctional supersensitivity of the guinea-pig vas deferens as assessed by [^{3}H]ouabain binding, *J. Pharmacol. Exp. Ther.* **219**:163–169.

Wood, C. L., Caron, M. G., and Lefkowitz, R. J., 1979, Separation of solubilized α-adrenergic and β-adrenergic receptors by affinity chromatography, *Biochem. Biophys. Res. Commun.* **88**:1–8.

Woolcock, A. J., 1986, Therapies to control the airway inflammation of asthma, *Eur. J. Respir. Dis.* **69**(Suppl. 147):166–174.

Zakrzewski, J. T., Barnes, N. C., Piper, P. J., and Costello, J. F., 1984, Sputum leukotrienes and prostanoids: Possible synergistic mediators in cystic fibrosis, bronchiectasis or chronic bronchitis, *Prostaglandins* **28**:641 (abst.).

Zakrzewski, J. T., Barnes, N. C., Piper, P. J., and Costello, J. F., 1985a, Quantitation of leukotrienes in asthmatic sputum, *Br. J. Pharmacol.* **19**:549P (abst.).

Zakrzewski, J. T., Barnes, W. C., Piper, P. J., and Costello, J. F., 1985b, Measurement of leukotrienes in arterial and venous blood from normal and asthmatic subjects by radioimmunoassay, *Br. J. Clin. Pharmacol.* **19**:574P (abst.).

Chapter 12

Airway Epithelial Metabolism and Airway Smooth Muscle Hyperresponsiveness

David B. Jacoby

Pulmonary Division
Department of Medicine
University of Maryland
Baltimore, Maryland 21201

and

Jay A. Nadel

Cardiovascular Research Institute
Moffitt Hospital
University of California, San Francisco
San Francisco, California 94143-0130

I. INTRODUCTION

Asthma is a disease of the airways characterized by reversible airway narrowing. Airway smooth muscle hyperresponsiveness to a variety of stimuli plays an important role in asthma. The clinical response in asthma depends on both the degree of the stimulus and the degree of responsiveness of the smooth muscle. Thus, stimuli that normally produce no clinical response can produce symptoms in the presence of smooth muscle hyperresponsiveness.

The involvement of the airway epithelium in the production of smooth muscle hyperresponsiveness is suggested by the consistent finding of abnormal epithelium in asthma. During acute asthma attacks, many epithelial cells are desquamated and appear in the sputum in clumps known as Creola bodies (Naylor, 1962) or Curschmann's spirals (Curschmann, 1883). Airway tissues of patients who have died in status asthmaticus show extensive epithelial cell shedding associated with prominent epithelial inflammation (Dunnill, 1960). Even stable asthmatics have abnormal airway epithelium (Laitinen *et al.*, 1985a), with a marked increase in the number of goblet cells, desquamation of the normal ciliated columnar epithelium, and exposure of underlying basal cells and nerve endings.

Further support for an association between epithelial damage and airway smooth muscle hyperresponsiveness comes from the observation that a variety of clinical conditions causing airway hyperresponsiveness are associated with airway inflammation and epithelial damage. Among these conditions are viral infections (Empey *et al.*, 1976), allergen inhalation (Behrens *et al.*, 1985), and exposure to irritants such as ozone (Golden *et al.*, 1978), toluene diisocyanate (Gordon *et al.*, 1985), and cigarette smoke (Gerrard *et al.*, 1980).

The functions of the airway epithelial cell can be divided into three broad categories: (1) transport of fluid and solute; (2) formation of a barrier between the external milieu and other cells of the airway, including nerves, muscles, and inflammatory cells; and (3) elaboration of mediators which may modulate both the inflammatory response via an effect on leukocytes and smooth muscle responsiveness via direct effects on nerves and muscles. Although fluid and solute transport are markedly altered in the face of epithelial inflammation, there is no evidence connecting such perturbations with smooth muscle responsiveness. This discussion focuses on alterations of the epithelial permeability and on epithelial mediator production in the setting of airway inflammation. The role of inflammatory cells, recruited and activated following epithelial damage are also considered, as well as the specific case of viral infection of the airway epithelium. Finally, the possible role of products of intraepithelial sensory nerves and the role of the epithelium in modulating the effects of these products are discussed.

II. EPITHELIAL PERMEABILITY

The observation that conditions leading to airway hyperresponsiveness are frequently associated with increased airway permeability led to speculation that loss of the epithelial barrier might increase access of agonists to subepithelial tissues including nerves, muscles, and mast cells (Empey *et al.*, 1976; Hogg, 1981). The finding that virus-induced airway hyperresponsiveness to histamine could be blocked by atropine initially suggested that exposure of subepithelial and intraepithelial nerves following epithelial damage might account for bronchoconstriction via a vagal reflex (Empey *et al.*, 1976). Similar findings in humans exposed to ozone (Golden *et al.*, 1978) and nitrogen dioxide (Orehek *et al.*, 1976), as well as the finding that exposure of *Ascaris suum*-sensitive rhesus monkeys to specific antigen increased both epithelial permeability and airway smooth muscle responsiveness (Boucher *et al.*, 1977; Boucher *et al.*, 1979) supported this contention.

The fact that cigarette smokers have both increased epithelial permeability and airway smooth muscle hyperresponsiveness led Boucher *et al.* (1980) to study the effects of cigarette smoke on guinea pig airways. Electron microscopic examination of the airways of smoke-exposed guinea pigs showed disruption of the epithelial tight junctions, allowing penetration of the marker horseradish peroxidase (HRP) to the vicinity of the intraepithelial nerve endings. Thus, it appeared that, even in the absence of gross epithelial desquamation, the access of agonists to target tissues might be increased. Indeed, in cigarette smoke-exposed guinea pigs, the maximal increase in epithelial permeability occurred before gross epithelial damage was visible, while the later severe epithelial disruption occurred at the same time that permeability was returning to normal (Hulbert *et al.*, 1981). Furthermore, the increase in airway responsiveness corresponded temporally with the increase in epithelial permeability (Hulbert *et al.*, 1985).

While the well-documented relationship of loss of the epithelial barrier with airway smooth muscle hyperresponsiveness in the guinea pig provided an attractive model explaining the involvement of the epithelium in asthma, current evidence in other species, including humans, does not support this hypothesis. Thompson *et al.* (1986) pretreated guinea pigs with hydroxyurea before exposing them to toluene diisocyanate. Although histologic sections of the airways showed prominent epithelial damage, the development of airway hyperresponsiveness was inhibited. The airway hyperresponsiveness seen in dogs exposed to ozone is dependent on the development of airway inflammation (Holtzman *et al.*, 1983a). When dogs are pretreated with indomethacin (blocking cyclo-oxygenase activity) (O'Byrne *et al.*, 1984a) or hydroxyurea (to produce leukocyte depletion) (O'Byrne *et al.*, 1984b), epithelial damage still occurs in response to ozone, but the development of airway hyperresponsiveness is blocked.

Studies have been undertaken in humans attempting to relate epithelial permeability to airway responsiveness. Elwood *et al.* (1983) instilled technetium-99m-(^{99m}Tc)-labeled diethylenetriaminepentaacetate (DPTA) into the airways of stable asthmatics and normal controls. As clearance of DPTA from the airway is dependent on absorption through the epithelium, measurement of the disappearance of radioactivity from the lung and its appearance in the blood provide an *in vivo* measurement of epithelial permeability. Despite marked differences in responsiveness to methacholine, the epithelial permeability was similar in the two groups. A subsequent study by the same group (Kennedy *et al.*, 1984) examined epithelial permeability and airway responsiveness to histamine in smokers and nonsmokers. Although epithelial permeability was markedly increased in smokers, there was no difference in responsiveness. O'Byrne *et al.* (1984c), using the same technique, found no difference in epithelial permeability between asthmatics and normal controls. Despite the finding of increased epithelial permeability in smokers, these investigators found no correlation between responsiveness to histamine and epithelial permeability among asthmatics, smokers, and nonsmoking nonasthmatic controls.

It should be noted that patients with airway hyperresponsiveness are hyperresponsive to histamine administered via either inhaled or intravenous routes

(Weiss *et al.*, 1929). Epithelial permeability would not affect the access of parenterally administered agonists to the effector tissues; thus, it appears that increased epithelial permeability is neither necessary nor sufficient to produce airway smooth muscle hyperresponsiveness.

III. EPITHELIAL MEDIATOR RELEASE

Epithelium-derived mediators may affect the degree of airway responsiveness both by direct effects on smooth muscle and nerves and by modulating airway inflammation. Because epithelial damage is followed by an influx of inflammatory cells, and because these cells may affect smooth muscle tone and responsiveness, the effects of both epithelial cells products and inflammatory cell products are discussed.

An early suggestion that epithelial prostaglandin production might inhibit airway smooth muscle contraction was the finding by Orehek *et al.* (1975) that indomethacin treatment of guinea pig tracheas increased the contractile response to acetylcholine (ACh), histamine, and 5-hydroxytryptamine (5-HT). Gently scratching the epithelial surface of the tracheas produced large amounts of prostaglandin E_2 (PGE$_2$). Dog and human (Leikauf *et al.*, 1985a) airway epithelial cells also produce PGE$_2$; release of this mediator is markedly augmented by stimulation with several inflammatory mediators, including bradykinin (Leikauf *et al.*, 1985b), leukotrienes C4 and D4 (Leikauf *et al.*, 1986), platelet-activating factor (PAF) (I. F. Ueki, personal communication), and eosinophil major basic protein (Jacoby *et al.*, 1987, 1988a). The effects of increased production of PGE$_2$ in inflamed airways are predominantly inhibitory, both on the inflammatory response, on decreasing mast cell mediator release (Hubscher, 1975), and on smooth muscle responsiveness, by decreasing cholinergic neurotransmission (Walters *et al.*, 1984) as well as by postjunctional effects (Shore *et al.*, 1987).

Barnett *et al.*, (1987) studied the effect of supernatants from cultured dog airway epithelial cells on dog trachealis muscle *in vitro*. Supernatants from unstimulated cultures did not significantly affect the contractile response to ACh, histamine, or electrical field stimulation, although there was a trend toward reduction of the response to electrical field stimulation. When the epithelial cultures were stimulated with bradykinin, supernatants strongly inhibited the response to electrical field stimulation, with no effect on the response to ACh or histamine. Measurement of PGE$_2$ concentration in the epithelial cell supernatants revealed a 10-fold increase in the bradykinin-stimulated cultures, yielding PGE$_2$ concentrations similar to those used by Walters *et al.* (1984) to achieve a similar degree of inhibition. Pretreatment of epithelial cell cultures with indomethacin abolished the effect, also blocking the PGE$_2$ production in both the baseline and bradykinin-stimulated cultures, and eliminating even the trend toward inhibition of the response to electrical field stimulation seen with supernatants from unstimulated cultures. On the basis of this series of experiments, it was concluded that the principal inhibitory effect of the airway epithelium on smooth muscle contractility is the result of epithelial cell PGE_2 inhibiting cholinergic neurotransmission.

Butler *et al.* (1987) found that rabbit tracheal rings, precontracted with ACh, relaxed when exposed to arachidonic acid. This effect was accompanied by an increase in PGE_2 and PGI_2. Mechanical removal of the epithelium prevented both the relaxation and the prostaglandin production, once again suggesting that epithelial prostaglandin production decreased airway smooth muscle responsiveness. Whether, in fact, rabbit airway epithelium produces PGI_2 as well as PGE_2 remains to be determined.

It has been suggested recently that the normal airway epithelium inhibits airway smooth muscle contractility. Flavahan *et al.* (1985) removed the epithelium from dog bronchial rings and found this increased the contractile response to ACh, histamine, 5-HT *in vitro*. Likewise, Barnes *et al.* (1985) found that removal of the epithelium from strips of bovine trachealis muscle enhanced the response to the same agonists. Both studies found a decrease in isoproterenol-induced relaxation without epithelium, and neither study found any effect on cholinergic neurotransmission. Although the mediator responsible for this effect is unknown, the effect was not inhibited by indomethacin, suggesting a mechanism other than PGE_2 release. A similar effect of vascular endothelium on vascular smooth muscle was recently attributed to release of nitric oxide by the endothelium (R. M. J. Palmer *et al.*, 1987).

Airway epithelial cells also produce smaller amounts of $PGF_{2\alpha}$ (Leikauf *et al.*, 1985a), which stimulates smooth muscle contraction, and, at lower concentrations, can cause smooth muscle hyperresponsiveness in dogs (O'Byrne *et al.*, 1984d). A recent study (Eling *et al.*, 1986) showed production of PGD_2 by acutely disaggregated dog airway epithelial cells stimulated with arachidonic acid. Confirmation of this finding is confirmed would provide another mechanism by which the epithelium could cause smooth muscle hyperresponsiveness, as PGD_2 causes both airway hyperresponsiveness *in vivo* (Fuller *et al.*, 1986) and increased cholinergic neurotransmission *in vitro* (Tamaoki *et al.*, 1987c) (see discussion of PGD_2 in Section IV.C).

Among the lipoxygenase products, the chemotactic leukotriene B4 produced by dog airway epithelial cells (Holtzman *et al.*, 1983b) is not produced by human airway epithelial cells. In humans, the 15-lipoxygenase pathway predominates (Hunter *et al.*, 1985), leading to the generation of 15-hydroxyeicosatetraenoic acid (15-HETE), 8,15-diHETE, and 14,15-diHETE. Goetzl *et al.* (1983) found that 15-HETE increased leukotriene C4 production by dog mastocytoma cells. If this finding is applicable to the mast cells of the human airway, it would suggest that epithelial damage leading to 15-HETE production could lead to airway inflammation and smooth muscle hyperresponsiveness via effects on the mast cell. Furthermore, 8,15-diHETE is chemotactic for neutrophils

(Shak *et al.*, 1983) and may therefore be important in initiating an inflammatory cascade after an insult to the epithelium. Once recruited to the airway, neutrophils may release leukotriene B4 (Ford-Hutchinson *et al.*, 1980) and PAF (Lotner *et al.*, 1980), attracting eosinophils (Wardlaw *et al.*, 1986; Sigal *et al.*, 1987) and monocytes, as well as platelets (Lellouch-Tubiana *et al.*, 1985) and more neutrophils, to the airway. Thus, epithelial products are capable, directly or indirectly, of recruiting and activating a wide variety of inflammatory cells.

IV. AIRWAY INFLAMMATION

During airway inflammation, resident inflammatory cells (e.g., monocytes, mast cells) as well as cells recruited into the airways from the bloodstream (e.g., neutrophils, eosinophils, platelets) are activated and play a role in smooth muscle hyperresponsiveness, as well as in other tissue responses. Each of these cells is considered separately.

A. Neutrophils

Many conditions leading to airway hyperresponsiveness are also associated with neutrophil influx into the airways. Thus, bronchoalveolar lavage fluid after ozone inhalation in both humans (Seltzer *et al.*, 1986) and dogs (Fabbri *et al.*, 1984) contains large numbers of neutrophils. The same is true in allergic asthmatics (Metzger *et al.*, 1987), allergic dogs (Chung *et al.*, 1985a), and passively sensitized rabbits (Murphy *et al.*, 1986) after specific allergen challenge. In dogs, the time course of the development of airway hyperresponsiveness after ozone inhalation corresponds to the time of neutrophil influx into the airways (Holtzman *et al.*, 1983c), and the number of neutrophils seen in airway mucosal biopsies correlates well with the degree of airway hyperresponsiveness to inhaled ACh (Holtzman *et al.*, 1983a). Likewise, inhalation of leukotriene B4 causes neutrophil influx into the airways of dogs, with the accompanying development of airway hyperresponsiveness (O'Byrne *et al.*, 1985). Neutrophil depletion in dogs (O'Byrne *et al.*, 1984b) and rabbits (Murphy *et al.*, 1986) blocks the development of airway hyperresponsiveness in response to inhalation of ozone and specific allergen, respectively. In the rabbit studies, transfusion of neutrophils to the neutrophil-depleted animals before allergen challenge permitted the return of the development of airway hyperresponsiveness.

While the above studies support the role of the neutrophil in the genesis of airway hyperresponsiveness, the precise mediators involved are uncertain. In dogs, inhibition of cyclo-oxygenase (using indomethacin) (O'Byrne *et al.*, 1984a), thromboxane synthetase (using OKY-046) (Aizawa *et al.*, 1985), or both cyclo-oxygenase and lipoxygenase (using BW755C) (Fabbri *et al.*, 1985) prevents the development of airway hyperresponsiveness after ozone exposure.

Pretreatment with OKY-046 also prevents allergen-induced hyperresponsiveness in dogs (Chung *et al.*, 1986a). The neutrophil influx and airway hyperresponsiveness caused by inhalation of leukotriene B4 in dogs is accompanied by a marked increase in thromboxane B2, the stable metabolite of thromboxane A_2 (TXA_2), in bronchoalveolar lavage fluid (O'Byrne *et al.*, 1985). The same was found in humans after ozone exposure (Seltzer *et al.*, 1986). Aizawa *et al.* (1985) showed that inhalation of U-46619, a stable analogue of TXA_2, produced a fourfold increase in airway responsiveness to inhaled ACh in dogs. *In vitro* response of canine airway smooth muscle to electrical field stimulation was also increased by the thromboxane analogue, indicating that cholinergic neurotransmission was facilitated (Chung *et al.*, 1985b).

Studies in humans also suggest a role for TXA_2 in the genesis and maintenance of airway hyperresponsiveness. Fujimura *et al.* (1986) found that the thromboxane synthetase inhibitor OKY-046 significantly decreased the airway response to inhaled ACh in a group of 25 asthmatics. Iwamoto *et al.* (1986) found that pretreatment of allergic asthmatics with OKY-046 before allergen challenge partially attenuated the early-phase fall in forced expiratory volume in 1 sec (FEV_1) and completely blocked the late-phase fall in FEV_1.

Although all the above studies suggest that TXA_2 is capable of producing airway hyperresponsiveness, none defined the cell responsible for the production of TXA_2 *in vivo*. Thus, the neutrophil (Goldstein *et al.*, 1978), the platelet (Hamberg *et al.*,1975), or the macrophage (MacDermot *et al.*, 1984), all of which can release TXA_2, can be implicated in this effect.

Other neutrophil products may also be important in the production of airway hyperresponsiveness. PAF (see Section IV.E) may cause bronchoconstriction and airway hyperresponsiveness (Chung *et al.*, 1986b; Patterson *et al.*, 1984; Denjean *et al.*, 1983; Cuss *et al.*, 1986)—effects probably mediated by platelets. PAF is also chemotactic for a variety of cells, with a particularly strong effect on eosinophils (Wardlaw *et al.*, 1986; Sigal *et al.*, 1987) and, along with leukotriene B4 and interleukin-1 (IL-1) (Tiku *et al.*, 1986), may perpetuate and amplify the inflammatory response in areas of neutrophil influx. Activated neutrophils also release a variety of proteases and toxic oxygen radicals that might damage airway epithelium.

A recent addition to the list of arachidonic acid metabolites with the potential to act as mediators in asthma is lipoxin A. Lipoxins are products of human leukocytes resulting from the action of 5-lipoxygenase on 15-HPETE, another product of 15-lipoxygenase (Serhan *et al.*, 1986). Under conditions of airway inflammation, 15-HPETE released by damaged airway epithelial cells may be metabolized into lipoxin A by inflammatory cell in the airway (Kim *et al.*, 1987). Lipoxin A causes contraction of guinea pig lung strips (Serhan *et al.*, 1986). Lipoxin A is also a potent stimulant to neutrophil superoxide generation and may thereby serve to amplify the inflammatory response.

B. Monocytes/Macrophages

The principal product of arachidonic acid metabolism in the macrophage is TXA_2 (MacDermot et al., 1984). As noted above, several lines of evidence suggest that TXA_2 is important in the production of airway hyperresponsiveness. Evidence that TXA_2 produced by alveolar macrophages can produce smooth muscle hyperresponsiveness was provided by Tamaoki et al. (1987a), who found that alveolar macrophages obtained from dogs by bronchoalveolar lavage and activated with the calcium ionophore A23187 increased the contractile response of canine bronchial smooth muscle to electrical field stimulation in vitro. This effect was blocked by pretreating the macrophages with indomethacin or by pretreating the smooth muscle with SQ29548, a TXA_2 receptor antagonist.

Among the other arachidonic acid metabolites of the macrophage are the bronchoconstricting leukotriene C4 and $PGF_{2\alpha}$, which also causes airway hyperresponsiveness in dogs (O'Byrne et al., 1984d). Macrophages also produce PGE_2 (MacDermot et al., 1984), which inhibits smooth muscle responsiveness, leukotriene B4 (Martin et al., 1984; MacDermot et al., 1984), a potent chemotactic factor for neutrophils (Malmsten et al., 1980) and eosinophils, and PAF (Arnoux et al., 1980), whose ability to promote both inflammation and smooth muscle hyperresponsiveness is discussed in Section IV.E.

Human lung macrophages in culture spontaneously secrete a substance of 18,000 M_r that stimulates human basophils and human lung mast cells to secrete histamine (Schulman et al., 1985; Liu et al., 1986). This effect is mediated via an interaction of the substance with immunoglobin E (IgE) bound to the target cell.

Activated monocytes and macrophages also produce toxic oxygen radicals (Nakagawara et al., 1981), as do neutrophils, eosinophils, and platelets. Engels et al. (1985) demonstrated that stimulated guinea pig alveolar macrophages decreased isoproterenol-induced relaxation of guinea pig tracheal smooth muscle. The decrease in β-adrenergic effect was blocked by thioureum (a hydroxyl radical scavenger). Superoxide dismutase, which catalyzes the conversion of superoxide to hydrogen peroxide, did not affect the decrease in isoproterenol-induced relaxation. They concluded that the hydroxyl radical was responsible for this effect. Such toxic substances may also damage tissues and amplify the inflammatory response.

C. Mast Cells

Apart from histamine, the bronchoconstricting effects of which are well known, mast cells release a variety of other mediators. Some of these mediators have direct effects on airway smooth muscle tone. Others may participate in modulating the inflammatory response.

In general, mast cell-derived mediators are divided into two categories:

preformed mediators (those stored in cytoplasmic granules, awaiting an appropriate stimulus for their release); and newly formed mediators (those membrane-derived lipid mediators that are not stored but that are produced at the time of stimulation for immediate release). Among the preformed mediators in human pulmonary mast cells, only histamine is clearly involved in the early-phase bronchospasm seen after antigen challenge in allergic asthmatics. Among the newly formed mediators, leukotriene C4 and PGD_2 act directly to produce bronchospasm. In addition, PAF may produce bronchospasm by stimulating platelets, and perhaps neutrophils and monocytes, to produce TXA_2.

Prostaglandin D_2 may actually have more than one effect on smooth muscle. Inhaled PGD_2 has a direct effect on the production of bronchospasm. This effect is relatively mild in normal subjects but is much more pronounced in asthmatics (Hardy *et al.*, 1984). Aside from this direct effect, however, PGD_2 also causes airway smooth muscle hyperresponsiveness. Fuller *et al.* (1986) found that stable asthmatics who inhaled PGD_2 became hyperresponsive to inhaled histamine and methacholine. Tamaoki *et al.* (1987c) found that the response of canine airway smooth muscle to electrical field stimulation *in vitro* was markedly increased after exposure to a dose of PGD_2 too small to produce muscle contraction in itself. This finding suggests that PGD_2 facilitates cholinergic neurotransmission, an effect that has also been demonstrated in canine mesenteric arteries (Nakajima and Toda, 1984) and in cat nictitating membrane (Hemker and Aiken, 1981), and *in vitro* using co-cultured neuroblastoma and muscle cells (Higashida *et al.*, 1984).

Likewise, although PAF probably has no direct effect on airway smooth muscle, the TXA_2 generated by platelets or leukocytes in the presence on PAF can cause both bronchospasm and hyperresponsiveness (Tamaoki *et al.*, 1987b) (see Section IV.E). It should be noted that both Schleimer *et al.* (1986) and D. Elias (personal communication) found that the PAF generated by human lung mast cells and by canine mastocytoma cells, respectively, remained cell associated and was not released into the fluid phase. The significance of mast cell PAF production is therefore unsettled.

Thus, although the early-phase asthmatic response is not generally associated with airway hyperresponsiveness, several of the mediators thought to be active during this phase have the potential to produce hyperresponsiveness. Other mast cell mediators may induce an influx of other inflammatory cells producing mediators which may themselves promote the airway hyperresponsiveness characteristic of the late-phase asthmatic response.

D. Eosinophils

Eosinophil migration and activation can be elicited by substances released from a variety of cells. The mast cell tetrapeptide eosinophil chemotactic factor of anaphylaxis (ECF-A) (Goetzl and Austen, 1975), as well as a higher-molecular-weight mast cell peptide (Boswell *et al.*, 1978), are both weakly chemotactic for eosinophils, while leukotrinene B4 (Ford-Hutchinson *et al.*, 1980), a product of mast cells, neutrophils, and macrophages, is a potent eosinophil chemotaxin. PAF, a product of many cells, including the eosinophil itself (Lee *et al.*, 1984), is strongly chemotactic for eosinophils, the chemotactic effect for eosinophils being considerably stronger than that for neutrophils (Wardlaw *et al.*, 1986; Sigal *et al.*, 1987). Activated fifth component of complement (Kay, 1970) and the lymphokine eosinophil stimulation promoter (Colley, 1973) are also eosinophil chemotaxins, as are a variety of other lymphokines (Rand and Colley, 1983). All these chemotaxins may be important in attracting eosinophils to the airways during the late-phase asthmatic response as well as in nonallergic asthma.

The eosinophil produces an array of substances with the ability to both limit and promote inflammatory reactions. Among the eosinophil's enzymes are histaminase (Zeiger *et al.*, 1976), which degrades histamine, lysophospholipase and phospholipase D (Weller and Goetzl, 1980), which inactivate leukotrienes and PAF, and eosinophil peroxidase (Henderson *et al.*, 1982), which, with hydrogen peroxide and a halide, inactivates leukotrienes B4, C4, and D4. 15-HETE, the major lipoxygenase product of the eosinophil (Henderson *et al.*, 1984), inhibits neutrophil 5-lipoxygenase activity (Vanderhoek *et al.*, 1980) and may therefore be important in decreasing production of leukotrienes. PGE_2, the major cyclo-oxygenase product, inhibits mast cell degranulation (Hubscher, 1975). The eosinophil may also be able to ingest whole mast cell granules (Mann, 1969).

By contrast, an assortment of lipid and protein products of the eosinophil can serve to damage tissues and amplify the inflammatory reaction. These cytotoxic and pro-inflammatory functions of the eosinophil can be stimulated by several stimuli. Eosinophil chemotactic factor of anaphylaxis, while only weakly chemotactic, increases eosinophil-mediated schistosomula killing significantly (Capron *et al.*, 1981a). Eosinophils bear receptors for IgG-Fc and IgE-Fc, and stimulation with specific antigen increases cytotoxicity (Capron *et al.*, 1981b; Khalife *et al.*, 1985). Leukotriene B4 and f-Met-Leu-Phe both stimulate the eosinophil to generate oxygen radicals (Palmblad *et al.*, 1984). Eosinophils also express the receptor for the activated third component of complement (Weller and Goetzl, 1980).

The principal protein of the eosinophil granule is major basic protein (Gleich *et al.*, 1973), a cytotoxic protein found in the sputum of asthmatic patients during acute exacerbations (Frigas *et al.*, 1981), as well as in areas of epithelial denudation in the airways of patients who had died in status asthmaticus (Filley *et al.*, 1982). An extension of the mast cell–eosinophil antagonism is the inactivation of major basic protein by heparin (Frigas *et al.*, 1980), a product of mast cells and basophils. Major basic protein, as well as two other eosinophil granule proteins, eosinophil cationic protein, and eosinophil peroxidase (along with hydrogen peroxide and a halide), can all induce degranulation of rat peritoneal mast cells (O'Donnell *et al.*, 1983; Zheutlin *et al.*, 1984; Henderson *et al.*, 1980), although the effect on airway mast cells is unknown. Major basic protein can damage airway epithelium (Frigas *et al.*, 1980). Subtoxic concentrations of major basic protein stimulate production of PGE_2 by airway epithelium (Jacoby *et al.*, 1987, 1988a). Cationic protein and the peroxidase–hydrogen peroxide–halide systems are also toxic (McLaren *et al.*, 1981; Jong and Klebanoff, 1980) and might damage airway epithelium in asthma. Eosinophil peroxidase can also enhance macrophage cytotoxicity (Nathan and Klebanoff, 1982).

Among the lipid mediators that might promote bronchospasm and hyperresponsiveness are PAF (see Section IV.E) (Lee *et al.*, 1984), leukotriene C4 (Weller *et al.*, 1983; Verhagen *et al.*, 1984), and 15-HETE (Henderson *et al.*, 1984), which may stimulate leukotriene production by mast cells (Goetzl *et al.*, 1983). The 15-lipoxygenase activity of eosinophils might also allow them to participate in the generation of lipoxins. Furthermore, stimulated eosinophils show a prolonged increase in generation of toxic oxygen metabolites (Palmblad *et al.*, 1984), which might also contribute to tissue injury.

E. Platelets

In addition to their well-known functions in hemostasis, increasing attention has been paid to the role of platelets as inflammatory cells (Page *et al.*, 1984; Capron *et al.*, 1985). Platelet chemotaxis has been demonstrated *in vitro* (Lowenhaupt *et al.*, 1982). Intravenous injection of PAF causes extravascular accumulation of platelets in guinea pig lungs (Lellouch-Tubiana *et al.*, 1985). Platelet activation in both antigen-induced (Knauer *et al.*, 1981; Gresele *et al.*, 1982) and exercise-induced bronchospasm (Johnson *et al.*, 1986) has prompted investigation of both the mechanism of activation of the platelet and the role of platelet-derived mediators in asthma and airway hyperresponsiveness. Although the precise cause of the platelet activation in asthma has not been determined, work in this area has principally addressed two possibilities: release of PAF by other inflammatory cells and direct IgE-mediated stimulation of platelets by antigen.

Platelet-activating factor is a phospholipid (1-*O*-alkyl-2-acetyl-*sn*-glyceryl-3-phosphorylcholine) produced by a variety of cells, including mast cells (Schleimer *et al.*, 1986), basophils (Hanahan *et al.*, 1980), neutrophils (Lotner *et al.*, 1980), eosinophils (Lee *et al.*, 1984), macrophages (Arnoux *et al.*, 1980), and platelets themselves (Chignard *et al.*, 1980). It has numerous biological effects, including eosinophil and neutrophil chemotaxis (Wardlaw *et al.*, 1986; Sigal *et al.*, 1987), increasing vascular permeability (Bjork and Smedegard, 1983), and stimulation of airway epithelial PGE_2 production (I. F. Ueki, personal communication), as well as its originally described function of causing platelet aggregation.

Denjean *et al.* (1983) found that intratracheal administration of PAF caused bronchospasm in baboons. This was accompanied by a fall in the platelet count, suggesting platelet activation and, perhaps, sequestration in the lungs (McManus *et al.*, 1980). Inhaled PAF also causes bronchospasm and airway hyperresponsiveness in dogs (Chung *et al.*, 1986b), monkeys (Patterson *et al.*, 1984), and humans (Cuss *et al.*, 1986). In rabbits, platelet depletion before intravenous PAF blocks the development of bronchospasm (McManus *et al.*, 1980), while in dogs pretreatment with the thromboxane synthetase inhibitor OKY-046 blocks the development of both bronchospasm and airway hyperresponsiveness (Chung *et al.*, 1986b). Thus, the bronchospasm produced by PAF appears to be platelet dependent. Although it is not clear that the same can be said for the hyperresponsiveness, as the source of the thromboxane that appears to be critical is not known and could be another cell such as the neutrophil or macrophage, platelets stimulated with PAF do produce large amounts of thromboxane (Macconi *et al.*, 1985).

Alternatively, platelets may be activated, especially in allergic asthma, via an IgE-dependent mechanism. Human platelets contain low-affinity IgE-Fc receptors that appear to be antigenically and functionally identical to the IgE-Fc receptors of lymphocytes, macrophages, and eosinophils, but different from the high-affinity IgE-Fc receptors of basophils and mast cells (Joseph *et al.*, 1986). Crosslinking these receptors with anti-IgE initiates a variety of platelet functions, including aggregation, secretion, and production of oxygen metabolites (Cines *et al.*, 1986; Joseph *et al.*, 1986). Platelets can also be stimulated with specific antigen. Interestingly, although only about 10% of platelets from normal donors bind IgE, about twice as many of the platelets obtained from allergic asthmatic patients bind IgE. In a few allergic asthmatic patients, as many as 50% of the platelets were found to bind IgE (Joseph *et al.*, 1986).

Among the mediators released from platelets, TXA_2 has received the most attention as a possible cause of airway hyperresponsiveness. TXA_2 is produced by a variety of cells that may be involved in asthma. That the platelet is at least capable of causing hyperresponsiveness by releasing TXA_2 was demonstrated by Tamaoki *et al.* (1987b). Again, using an *in vitro* system, aggregating platelets increased the response of canine airway smooth muscle to electrical field stimulation. This facilitation of cholinergic neurotransmission was prevented both by preincubating the platelets with indomethacin to inhibit cyclo-oxygenase activity (which effectively eliminated TXA_2 production) and by pretreating the muscle with the TXA_2 receptor antagonist SQ 29548, strongly suggesting that TXA_2 was the active mediator.

Alternatively, other platelet-derived mediators may be important. PGD_2, a product of mast cells (Lewis *et al.*, 1982; Peters *et al.*, 1983), macrophages (MacDermot *et al.*, 1984), and perhaps epithelial cells (Eling *et al.*, 1986), as well as platelets (Oelz *et al.*, 1977), facilitates cholinergic neurotransmission in canine airway smooth muscle (Tamaoki *et al.*, 1987b). Other platelet products might be important in their effects on other inflammatory cells. Platelet factor 4, released during platelet activation in asthma (Knauer *et al.*, 1981; Johnson *et al.*, 1986), is chemotactic for neutrophils and monocytes (Deuel *et al.*, 1981), while PAF, another product of stimulated platelets, is chemotactic for both eosinophils and neutrophils as noted above (Wardlaw *et al.*, 1986; Sigal *et al.*, 1987). Furthermore, a cationic protease released from platelets directly activates the fifth component of complement, leading both to chemotactic activity and to the formation of the terminal attack complex of complement (Weksler and Coupal, 1973). Platelets also release a soluble factor that stimulates the release of histamine from human basophils (Orchard *et al.*, 1986). Finally, stimulated platelets elaborate toxic oxygen metabolites that might contribute to airway hyperresponsiveness by damaging the airway epithelium.

Thus, a variety of inflammatory cells are recruited and activated after epithelial damage. Products released by these cells may affect airway smooth muscle responsiveness.

V. VIRAL INFECTION OF AIRWAY EPITHELIUM

Viral respiratory infections frequently exacerbate asthma, both in children (Henderson *et al.*, 1979; Horn *et al.*, 1979; Welliver *et al.*, 1979; Welliver, 1983) and in adults (Welliver, 1983; Migueres *et al.*, 1985). A role for viral infection in the pathogenesis of asthma is suggested both by the clinical association of viral infection in early childhood with asthma and airway hyperresponsiveness later in life (Weiss *et al.*, 1985; Pullan and Hey, 1982) and by the finding of temporary airway hyperresponsiveness following acute viral infections in previously normal subjects (Empey *et al.*, 1976; Little *et al.*, 1978).

On a cellular basis, the effects of viral infections on airway responsiveness fall into seven categories:

Induction of airway inflammation
Damage to airway epithelium
Impaired β-adrenoceptor function
Induction of IgE response
Increased release of histamine by mast cells
Induction of T-lymphocyte response
Increased sensitivity to the bronchoconstricting effects of substance P

A. Induction of Airway Inflammation

Walsh *et al.* (1960) found infiltration of the airway epithelium with mononuclear cells and neutrophils during acute influenza infection in humans. The effects of the products of these cells on airway responsiveness has been discussed in the sections on these cells.

While the precise mediators responsible for recruitment of inflammatory cells into the virus-infected epithelium are not known, Ward *et al.* (1972) found that African green monkey kidney cells, infected with mumps virus, released substances that were chemotactic for neutrophils and macrophages independent of complement activation. On the basis of experiments using herpes simplex-infected rabbit kidney cells, Snyderman *et al.* (1972) suggested four possible mechanisms for the production of chemotactic activity: (1) viruses might destroy cells, releasing intracellular proteolytic enzymes that activate complement; (2) virus-induced antigens on the cell surface bind antibodies, possibly leading to complement activation; (3) viruses bound to the cell surface, even in the absence of cell infection, might interact with antibodies and complement; and (4) lymphocytes might release chemotactic lymphokines in response to stimulation with specific viral antigens.

B. Damage to Airway Epithelium

Walsh *et al.* (1960) found that the pathological change found in the airways on the first day of symptoms of influenza infection was desquamation of the airway epithelium. Damage to the epithelium by viruses might be expected to have similar results to those found after exposure to ozone, which also causes extensive epithelial desquamation. Both the release of inflammatory mediators by the epithelium and loss of the inhibitory influence of the epithelial cell on smooth muscle contraction may be important, and are discussed in Section III.

C. Impaired β-Adrenoceptor Function

Much of the work on this subject has used the neutrophil β-adrenergic response as a model for what was presumed to be occurring in the airway smooth muscle. Stimulation of β-adrenoceptor in neutrophils with isoproterenol before exposure to opsonized zymosan inhibits the release of lysosomal enzymes. Busse (1977) found that in neutrophils from asthmatic patients the isoproterenol-induced inhibition of lysozyme release was reduced compared with the inhibition seen in neutrophils from nonasthmatic donors. During viral respiratory infections associated with increased asthmatic symptoms, the isoproterenol response of the neutrophils was further impaired. The same group (Bush *et al.*, 1978) found a similar effect following experimental rhinovirus infections in normal subjects and subsequently demonstrated that impairment of the neutrophil isoproterenol response could be produced by incubating the neutrophils with live influenza vaccine *in vitro* (Busse *et al.*, 1979).

A direct effect of viral infection on airway smooth muscle β-adrenoceptor function has still to be demonstrated, but Buckner *et al.* (1981) found *in vivo* loss

loss of β-agonist sensitivity in the airways of guinea pigs infected with parainfluenza virus. This was manifested as a loss of the ability of β-adrenoceptor stimulation to block antigen-induced contraction on isolated airway smooth muscle; whether this represents an effect of the virus on the muscle, of the virus on mast cells, or of airway inflammation on the muscle remains open to question. Nonetheless, the possibility of a direct effect of respiratory viral infection on airway smooth muscle is suggested by the recent demonstration of the ability of influenza virus to infect cultured human striated muscle cells (Klavinskis *et al.*, 1985). This has not yet been demonstrated using smooth muscle.

D. Induction of IgE Response

Frick *et al.* (1979) followed a group of children considered at high risk of the development of allergies on the basis of biparental histories of allergies. These investigators found not only a high incidence of allergies in the children but also that the development of allergic symptoms, as well as the development of immunological evidence for allergic sensitization, occurred during the 1–2 months following viral upper respiratory infection. Welliver *et al.* (1980, 1981) demonstrated an increase in respiratory syncytial virus-specific IgE in the nasopharyngeal secretions of patients with acute respiratory syncytial virus infections. This was accompanied by the release of histamine into the nasopharyngeal secretions. In patients with wheezing associated with the infection, the presence of virus-specific IgE persisted longer after the resolution of the infection than in patients without wheezing. Histamine was found more frequently, and in higher concentrations, in patients with wheezing. Measurement of arterial blood gases in these patients demonstrated a correlation of peak titer of virus-specific IgE and histamine with the degree of hypoxia.

E. Increased Release of Histamine by Mast Cells

Neuraminidase, an enzyme that cleaves sialic acid from oligosaccharides, is a component of the influenza virus. During influenza replication in epithelial cells, viral neuraminidase, as well as hemagglutinin, is expressed on the epithelial cell surface (Maeno and Kilbourne, 1970), and is ultimately released as part of the progeny virus during viral budding. Rat peritoneal mast cells treated with neuraminidase release more histamine than untreated mast cells in response to several stimuli (Jensen *et al.*, 1986). Furthermore, Pecoud *et al.* (1981) found that treatment of rat basophilic leukemia cells and rat peritoneal mast cells with neuraminidase led to increased affinity of the IgE receptor for IgE. By contrast, release of 5-HT by rat peritoneal mast cells in response to polycationic peptides (Coleman, 1982) and in response to substance P (Coleman *et al.*, 1986) was significantly reduced by pretreatment with neuraminidase.

F. Induction of T-Lymphocyte Response

Welliver *et al.* (1979) examined the cell-mediated immune response to respiratory syncytial virus infection in 39 infants. A significantly stronger cell-mediated immune response, as assessed by *in vitro* lymphocyte transformation in the presence of respiratory syncytial virus antigen, was found in those infants who developed wheezing than in those who did not. The significance of this correlation is not known, but an increased T-lymphocyte response would be expected to be associated with a multiplicity of effects on the other inflammatory cells of the airway.

G. Increased Sensitivity to the Bronchoconstricting Effects of Substance P

Buckner *et al.* (1986) found an increased contractile response of guinea pig bronchial rings to substance P after parainfluenza infection. It has recently been shown that influenza A infection increases the contractile response to substance P in ferret tracheas by decreasing enkephalinase production by the airway tissues (Jacoby *et al.*, 1988) (see discussion in Section VI). Thus, viral respiratory infections, which are commonly associated with wheezing and airway hyperresponsiveness, may produce these effects by acting on the epithelium, smooth muscle, or inflammatory cells.

VI. SENSORY NERVE PRODUCTS

In addition to the cholinergic and adrenergic influences on airway smooth muscle tone, nonadrenergic noncholinergic neural effects influence the airways. These effects may be either excitatory or inhibitory. The precise transmitters and their roles in airway hyperresponsiveness have not yet been worked out in detail. However, a number of candidates have been identified in airway nerves, including substance P (Lundberg *et al.*, 1983a, 1984) and other tachykinins, and calcitonin gene-related peptide (excitatory) (J. B. D. Palmer *et al.*, 1987) and vasoactive intestinal peptide (VIP) and peptide histidine methionine (inhibitory) (Laitinen *et al.*, 1985b; Palmer *et al.*, 1986).

The tachykinins are found in sensory nerve endings found in smooth muscle, in and around blood vessels, in glands, and in the lateral intercellular spaces of the surface epithelium (Fig. 1). The intraepithelial sensory nerve endings are exposed after epithelial damage; this has been demonstrated in asthmatics (Laitinen *et al.*, 1985a). Substance P is widely distributed in the sensory nerves of the lower respiratory tract, including the sensory nerves of the airway epithelium, smooth muscle, and blood vessels (Lundberg *et al.*, 1984a). The following discussion briefly reviews the role of substance P as a mediator of the

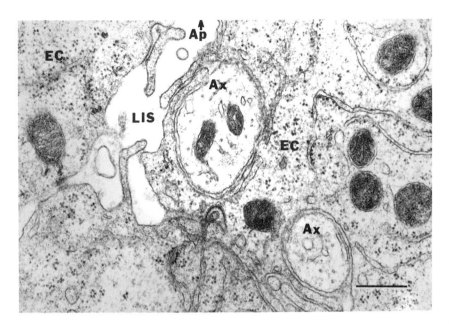

FIGURE 1. Electron micrograph of guinea pig tracheal epithelium, showing intraepithelial sensory nerve axons (Ax), one communicating with the lateral intercellular space (LIS) between two epithelial cells (EC). Arrow points toward the apical (luminal) surface of the epithelium. Bar: 0.27 μm. (From Basbaum, 1984.) (×75,000)

inflammatory response and outlines the current understanding of the effects of substance P on smooth muscle tone and responsiveness.

In general, substance P increases inflammation via effects on a variety of leukocytes. Substance P is chemotactic for neutrophils (Marasco *et al.*, 1981), and monocytes (Ruff *et al.*, 1985) although the effect on neutrophils is seen at relatively high concentrations and is therefore of questionable significance. Neutrophil and macrophage phagocytosis and lysosomal enzyme release are potentiated by substance P (Bar-Shavit *et al.*, 1980) as is macrophage thromboxane release (Hartung and Toyka, 1983). Substance P also stimulates histamine release and leukotriene C4 production by mast cells derived from mouse bone marrow progenitors (Sydbom, 1982; Fewtrell *et al.*, 1982). The effects of substance P of human lung mast cells have not been studied. However, human skin mast cells release histamine on stimulation by substance P (Foreman and Jordan, 1983). Substance P is also a specific T-lymphocyte mitogen (Payan *et al.*, 1983, 1984). Thus, by all these stimulatory effects, as well as by increasing vascular permeability (Lundberg *et al.*, 1984b), substance P tends to amplify the inflammatory response.

Release of substance P in the airways can be induced by capsaicin or by antidromic vagal stimulation. The stimuli responsible for substance P release in the airways under physiological conditions are unknown. While it has been suggested that cigarette smoke and other chemical irritants may release substance P (Lundberg and Saria, 1983b), further work is needed in this area. It has been suggested that bradykinin causes release of substance P in the eye (Wahlestedt *et al.*, 1985), and if this applies to the airways, it may explain bradykinin-induced bronchoconstriction in humans (Fuller *et al.*, 1987). The finding that bradykinin stimulates C fibers in dog airways (Coleridge and Coleridge, 1984) also supports a possible role for bradykinin in releasing substance P. Whether epithelial damage itself releases substance P by exposing C-fiber nerve endings remains to be determined, as does the possible role of epithelial archidonic acid metabolites such as PGE_2 and 8R,15S-diHETE (Levine *et al.*, 1986), which induce hyperalgesia, perhaps via a substance P-dependent mechanism.

Both capsaicin and antidromic stimulation of the vagus cause bronchospasm (Lundberg *et al.*, 1983c). Depletion of substance P (with capsaicin) from C-fiber nerve endings blocks the atropine-resistant part of the bronchospasm caused by vagal stimulation (that part not attributable to ACh release) (Martling *et al.*, 1984). Furthermore, pretreatment of guinea pigs with a substance P antagonist inhibits the airway smooth muscle response to both capsaicin and vagal stimulation (Lundberg *et al.*, 1983d), supporting the contention that substance P or another tachykinin is the transmitter involved. Exogenous substance P causes bronchospasm in the guinea pigs and contraction of ferret and human airway smooth muscle *in vitro*, although the latter effect is not seen at concentrations less than $1\mu M$ (Sekizawa *et al.*, 1987; Lundberg *et al.*, 1983c). At lower doses, substance P facilitates cholinergic neurotransmission in rabbit (Tanaka and Grunstein, 1986) and ferret (Sekizawa *et al.*, 1987) airways, leading to an increased response to electrical field stimulation.

Enkephalinase (EC 3.4.24.11) is an enzyme that degrades substance P and other tachykinins into inactive metabolites (Skidgel *et al.*, 1984). The discovery of enkephalinase in human airway tissues (Johnson *et al.*, 1985) provides a mechanism for limiting the effects of released substance P. When the substance P-degrading activity of enkephalinase in ferret airways is inhibited using leucine-thiorphan, a contractile effect is seen with concentrations of substance P as low as 5×10^{-7} M (as compared with $>10^{-6}$ M in the absence of leucine-thiorphan) (Sekizawa *et al.*, 1987). This contraction is partially attenuated by atropine, suggesting that release of acetylcholine is responsible for part of the effect of substance P. Furthermore, while leucine-thiorphan itself did not cause contraction, treatment with leucine-thiorphan does lead to an increased response to electrical field stimulation. This effect can be blocked by pretreatment with D-Pro[2], D-Trp[7,9]-SP, a tachykinin receptor antagonist, suggesting that endogenous tachykinins (possibly substance P), when freed from the controlling influence of airway enkephalinase, augment cholinergic neurotransmission. While tachykinins are also degraded by other enzymes, including angiotensin-converting enzyme and serine proteases, inhibition of these enzymes does not potentiate the effects on the response to electrical field stimulation, suggesting that enkephalinase is the enzyme responsible for regulation of tachykinin activity.

A decrease in airway enkephalinase activity may be one mechanism leading to airway hyperresponsiveness associated with viral infections. Infection of ferret tracheal rings with human influenza A virus *in vitro* leads to a markedly increased response to substance P. Treatment of tissues with thiorphan increases the substance P-contractile response of infected and control tissues to the same level. The enkephalinase activity of the infected tissues is markedly decreased. It is likely that removal of the airway epithelium accounts for this decrease in enkephalinase activity, as histological examination reveals extensive epithelial denudation of infected tracheal rings. Despite this loss of epithelium, however, the response to ACh is not enhanced, excluding an increase in epithelial permeability or loss of a nonspecific epithelial derived relaxant factor as the mechanism of the increase substance P response (Jacoby *et al.*, 1988b).

Other possible excitatory nonadrenergic noncholinergic mediators in the airways are calcitonin gene-related peptide and neurokinin A. Calcitonin gene-related peptide is a product of alternative processing of messenger RNA (mRNA) transcribed from the calcitonin gene (Rosenfeld *et al.*, 1983). It has been identified in the airways of guinea pigs (Lundberg *et al.*, 1985), where it coexists with substance P in C-fiber nerve endings, as well as in human airways (Palmer *et al.*, 1987), and is a potent bronchoconstrictor in both species. Neurokinin A is a tachykinin that is also a bronchoconstrictor (Lundberg *et al.*, 1985). The physiological regulation of the effects of these substances has not yet been worked out, but neurokinin A, like substance P, can be degraded by enkephalinase.

In addition to these excitatory mediators, there are inhibitory nonadrenergic noncholinergic mediators. VIP and the related peptide histidine methionine (pre-pro-VIP) both relax human bronchi that were precontracted by histamine (J. B. D. Palmer *et al.*, 1986). If production of these peptides were decreased after epithelial damage, or if they were degraded by inflammatory cell proteases, this might contribute to the smooth muscle hyperresponsiveness associated with epithelial inflammation.

VII. SUMMARY

The airway epithelium serves many roles in the control of airway smooth muscle responsiveness. While the barrier function of the epithelium is often abnormal in asthma, epithelial permeability changes do not appear to be important in smooth muscle hyperresponsiveness. The normal airway epithelium decreases smooth muscle responsiveness by releasing PGE_2 and possibly by releasing a nonspecific relaxant factor and inhibiting the actions of tachykinins by producing enkephalinase. The damaged epithelium seen in asthma exposes airway sensory nerve endings and releases 15-lipoxygenase products that produce airway smooth muscle hyperresponsiveness by promoting inflammation. Inflammatory cells recruited and activated by products of the damaged epithelium release a variety of products, increasing smooth muscle responsiveness. The balance of the positive and negative influences of the airway epithelium on smooth muscle responsiveness and airway inflammation determines the net effect of the epithelium on airway responsiveness in health and disease.

VIII. REFERENCES

Aizawa, H., Chung, K. F., Leikauf, G. D., Ueki, I., Bethel, R. A., O'Byrne, P. M., Hirose, T., and Nadel, J. A. 1985, Significance of thromboxane generation in ozone-induced airway hyperresponsiveness in dogs, *J. Appl. Physiol.* **59**:1918–1923.

Arnoux, B., Duval, D., and Benveniste, J., 1980, Release of platelet-activating factor (PAF-acether) from alveolar macrophages by the calcium ionophore A23187 and phagocytosis, *Eur. J. Clin. Invest.* **10**:437–441.

Bar-Shavit, Z., Goldman, R., Stubinsky, Y., Gottlieb, P., Fridkin, M., Teichberg, V. I., and Blumberg, S., 1980, Enhancement of phagocytosis—A newly found activity of substance P residing in its N-terminal tetrapeptide sequence, *Biochem. Biophys. Res. Commun.* **4**:1445–1451.

Barnes, P. J., Cuss, F. M., and Palmer, J. B., 1985, The effect of airway epithelium on smooth muscle contractility in bovine trachea, *Br. J. Pharmacol.* **86**:685–691.

Barnett, K., Jacoby, D. B., Lazarus, S. C., and Nadel, J. A., 1987, Bradykinin stimulates release of an epithelial cell product that inhibits smooth muscle contraction, *Am. Rev. Respir. Dis.* **135**:A274.

Basbaum, C. B., 1984, Innervation of the airway mucosa and submucosa, *Semin. Respir. Med.* **5**:308–313.

Behrens, B. L., Clark, R. A. F., Feldsein, D. L., Presley, D. M., Glezen, L. S., Graves, J. P., and Larsen, G. L., 1985, Comparison of the histopathology of the immediate and late asthmatic and cutaneous responses in a rabbit model, *Chest* **87**:153S–153S.

Bjork, J., and Smedegard, G., 1983, Acute microvascular effects of Paf-acether, as studied by intravital microscopy, *Eur. J. Pharmacol.* **96**:87–94.

Boswell, R. N., Austen, K. F., and Goetzl, E. J., 1978, Intermediate molecular weight eosinophil chemotactic factions in rat peritoneal mast cells: Immunologic release, granule association, and demonstration of structural heterogeneity, *J. Immunol.* **120**:15–20.

Boucher, R. C., Pare, P. D., Gilmore, N. J., Moroz, L. A., and Hogg, J. C., 1977, Airway mucosal permeability in the Ascaris-suum sensitive rhesus monkey, *J. Allergy Clin. Immunol.* **60**:134–140.

Boucher, R. C., Pare, P. D., and Hogg, J. C., 1979, Relationship between airway hyperreactivity and hyperpermeability in *Ascaris*-sensitive monkeys, *J. Allergy Clin. Immunol.* **64**:197–201.

Boucher, R. C., Johnson, J., Inoue, S., Hulbert, W., and Hogg, J. C., 1980, The effect of cigarette smoke on the permeability of guinea pig airways, *Lab. Invest.* **43**:94–100.

Buckner, C. K., Clayton, D. E., Ain-Shoka, A. A., Busse, W. W., Dick, E. C., and Shult, P., 1981, Parainfluenza 3 infection blocks the ability of a beta-adrenergic receptor agonist to inhibit antigen-induced contraction of guinea pig isolated airway smooth muscle, *J. Clin. Invest.* **67**:376–384.

Buckner, C. K., Saban, R., and Dick, E. C., 1986, Increased airway contraction to substance P (SP) after parainfluenza 3 (P3) infection of the guinea pig, *Fed. Proc.* **45**:324.

Bush, R. K., Busse, W., Flaherty, D., Warshauer, D., Dick, E. C., and Reed, C. E., 1978, Effects of experimental rhinovirus 16 infection on airways and leukocyte function in normal subjects, *J. Allergy Clin. Immunol.* **61**:80–87.

Busse, W. W., 1977, Decreased granulocyte response to isoproterenol in asthma during upper respiratory infections, *Am. Rev. Respir. Dis.* **115**:783–791.

Busse, W. W., Cooper, W., Warshauer, D. M., Dick, E. C., Wallow, I. H. L., and Albrecht, R., 1979, Impairment of isoproterenol, H2 histamine, and prostaglandin E_1 response of human granulocytes after incubation *in vitro* with live influenza vaccines, *Am. Rev. Respir. Dis.* **119**: 561–569.

Butler, G. B., Adler, K. B., Evans, J. N., Morgan, D. W., and Szarek, J. L., 1987, Modulation of rabbit airway smooth muscle responsiveness by respiratory epithelium, *Am. Rev. Respir. Dis.* **135**:1099–1104.

Capron, M. A., Capron, A., Goetzl, E. J., and Austen, K. F., 1981a, Eosinophil Fc receptor: Enhancement by the tetrapeptides of the eosinophil chemotactic factor of anaphylaxis (ECF-A), *Nature (Lond.)* **289**:71–73.

Capron, M., Bazin, H., Joseph, M., and Capron, A., 1981b, Evidence for IgE-dependent cytotoxicity by rat eosinophils, *J. Immunol.* **126**:1764–1767.

Capron, A., Ameisen, J. C., Joseph, M., Auriault, C., Tonnel, A. B., and Caen, J., 1985, New functions for platelets and their pathological implications, *Int. Arch. Allergy Appl. Immunol.* **77**: 107–114.

Chignard, M., LeCouedic, J. P., Vargaftig, B. B., and Benveniste, J., 1980, Platelet-activating factor (Paf-acether) secretion from platelets: Effects of aggregating agents, *Br. J. Hematol.* **46**: 455–464.

Chung, K. F., Aizawa, H., Becker, A. B., Frick, O., Gold, W. M., and Nadel, J. A., 1986a, Inhibition of antigen-induced airway hyperresponsiveness by a thromboxane synthetase inhibitor (OKY-046) in allergic dogs, *Am. Rev. Respir. Dis.* **134**:258–261.

Chung, K. F., Aizawa, H., Leikauf, G. D., Ueki, I. F., Evans, T. W., and Nadel, J. A., 1986b, Airway hyperresponsiveness induced by platelet-activating factor: Role of thromboxane generation, *J. Pharmacol. Exp. Ther.* **236**:580–584.

Chung, K. F., Becker, A. B., Lazarus, S. C., Frick, O. L., Nadel, J. A., and Gold, W. M., 1985a, Antigen-induced airway hyperresponsiveness and pulmonary inflammation in allergic dogs, *J. Appl. Physiol.* **58**:1347–1353.

Chung, K. F., Evans, T. W., Graf, P. D., and Nadel, J. A., 1985b, Modulation of cholinergic neurotransmission in canine airways by thromboxane mimetic U46619, *Eur. J. Pharmacol.* **117**:373–375.

Cines, D. B., van der Keyl, H., and Levinson, A. I., 1986, In vitro binding of an IgE protein to human platelets, *J. Immunol.* **136**:3433–3440.

Coleman, J. W., 1982, Neuraminidase- and benzalkonium chloride-dependent inhibition of basic peptide-induced rat mast cell secretion, *Immunol. Lett.* **5**:197–201.

Coleman, J. W., Huang, Q., and Stanworth, D. R., 1986, The mast cell response to substance P: Effects of neuraminidase, limulin, and some novel synthetic peptide antagonists, *Peptides* **7**: 171–175.

Coleridge, J. C. G., and Coleridge, H. M., 1984, Afferent vagal C fibre innervation of the lungs and airways and its functional significance, *Rev. Physiol. Biochem. Pharmacol.* **99**:1–110.

Colley, D. G., 1973, Eosinophils and immune mechanisms. I. Eosinophil stimulation promotor (ESP): A lymphokine induced by specific antigen or phytohemagglutinin, *J. Immunol.* **110**: 1419–1423.

Curschmann, H., 1883, Ueber bronchiolitis excsudativa und ihr verhaltmis zum asthma nervosum, *Dtsch. Arch. Klin. Med.* **32**:1.

Cuss, F. M., Dixon, C. M. S., and Barnes, P. J., 1986, Effects of inhaled platelet activating factor on pulmonary function and bronchial responsiveness in man, *Lancet* **2**:189–192.

Denjean, A., Arnoux, B., Masse, R., Lockhart, A., and Benveniste, J., 1983, Acute effects of intratracheal administration of platelet-activating factor in baboons, *J. Appl. Physiol.* **55**:799–804.

Deuel, T. F., Senior, R. M., Chang, D., Griffin, G. L., Heinrickson, R. L., and Keiser, E. T., 1981, Platelet factor 4 is chemotactic for neutrophils and monocytes, *Proc. Natl. Acad. Sci. USA* **78**:4584–4587.

Dunnill, M. S., 1960, The pathology of asthma, with special reference to changes in the bronchial mucosa, *J. Clin. Pathol.* **13**:27–33.

Eling, T. E., Danilowicz, R. M., Henke, D. C., Sivarajah, K., Yankaskas, J. R., and Boucher, R. C., 1986, Arachidonic acid metabolism by canine tracheal epithelial cells, *J. Biol. Chem.* **261**:12841–12849.

Elwood, R. K., Kennedy, S., Belzberg, A., Hogg, J. C., and Pare, P. D., 1983, Respiratory mucosal permeability in asthma, *Am. Rev. Respir. Dis.* **128**:523–527.

Empey, D. W., Laitinen, L. A., Jacobs, L., Gold, W. M., and Nadel, J. A., 1976, Mechanisms of bronchial hyperreactivity in normal subjects following upper respiratory tract infection, *Am. Rev. Respir. Dis.* **113**:131–139.

Engels, F., Oosting, R. S., and Nijikamp, F. P., 1985, Pulmonary macrophages induce deterioration of guinea-pig tracheal β-adrenergic function through release of oxygen radicals, *Eur. J. Pharmacol.* **111**:143–144.

Fabbri, L. M., Aizawa, H., Alpert, S. E., Walters, E. H., O'Byrne, P. M., Gold, B. D., Nadel, J. A., and Holtzman, M. J., 1984, Airway hyperresponsiveness and changes in cell counts in bronchoalveolar lavage after ozone exposure in dogs, *Am. Rev. Respir. Dis.* **129**:288–291.

Fabbri, L. M., Aizawa, H., O'Byrne, P. M., Bethel, R. A., Walters, E. H., Holtzman, M. J., and Nadel, J. A., 1985, An anti-inflammatory drug (BW755C) inhibits airway hyperresponsiveness induced by ozone in dogs, *J. Allergy Clin. Immunol.* **76**:162–166.

Fewtrell, C. M. S., Foreman, J. C., Jordan, C. C., Oehme, P., Renner, H., and Stewart, J. M., 1982, The effects of substance P on histamine and 5-hydroxytryptamine release in the rat, *J. Physiol. (Lond.)* **330**:393–411.

Filley, W. V., Holley, K. E., Kephart, G. M., and Gleich, G. J., 1982, Identification by immunofluorescence of eosinophil granule major basic protein in lung tissues of patients with bronchial asthma, *Lancet* **2**:11–16.

Flavahan, N. A., Aarhus, L. L., Rimele, T. J., and Vanhoutte, P. M., 1985, Respiratory epithelium inhibits bronchial smooth muscle tone, *J. Appl. Physiol.* **58**:834–838.

Ford-Hutchinson, A. W., Bray, M. A., Doig, M. V., Shipley, M. E., and Smith, M. J. H., 1980, Leukotriene B, a potent chemokinetic and aggregating substance released from polymorphonuclear leukocytes, *Nature (Lond.)* **286**:264–265.

Foreman, J. C., and Jordan, C. C., 1983, Histamine release and vascular changes induced by neuropeptides, *Agents Action* **13**:105–116.

Frick, O. L., German, D. F., and Mills, J., 1979, Development of allergy in children. I. Association with virus infections, *J. Allergy Clin. Immunol.* **63**:228–241.

Frigas, E., Loegering, D. A., and Gleich, G. J., 1980, Cytotoxic effects on the guinea-pig eosinophil major basic protein on tracheal epithelium, *Lab. Invest.* **42**:35–43.

Frigas, E., Loegering, D. A., Solley, G. O., Farrow, G. M., and Gleich, G. J., 1981, Elevated levels of the eosinophil granule major basic protein in the sputum of patients with bronchial asthma, *Mayo Clin. Proc.* **56**:345–353.

Fujimura, M., Sasaki, F., Mifune, J., and Matsuda, T., 1986, The role of arachidonic acid metabolites on bronchial hyperresponsiveness to acetylcholine in asthmatic patients, *Arerugi* **35**:953–959.

Fuller, R. W., Dixon, C. M. S., Cuss, F. M., and Barnes, P. J., 1987, Bradykinin-induced bronchoconstriction in humans. Mode of Action, *Am. Rev. Respir. Dis.* **135**:176–180.

Fuller, R. W., Dixon, C. M. S., Dollery, C. T., and Barnes, P. J., 1986, Prostaglandin D_2 potentiates airway responsiveness to histamine and methacholine, *Am. Rev. Respir. Dis.* **133:** 252–254.

Gerrard, J. W., Cockcroft, D. W., Mink, J. T., Cotton, D. J., Poonawala, R., and Dosman, J. A., 1980, Increased nonspecific bronchial reactivity in cigarette smokers with normal lung function, *Am. Rev. Respir. Dis.* **122:**577–582.

Gleich, G. J., Loegering, D. A., and Maldonado, J. E., 1973, Identification of a major basic protein in guinea-pig eosinophil granules, *J. Exp. Med.* **137:**1459–1471.

Goetzl, E. J., and Austen, K. F., 1975, Purification and synthesis of eosinophilotactic tetrapeptides of human lung tissue: Identification as the eosinophil chemotactic factor of anaphylaxis, *Proc. Natl. Acad. Sci. USA* **72:**4123–4127.

Goetzl, E. J., Phillips, M. J., and Gold, W. M., 1983, Stimulus specificity of the generation of leukotrienes by dog mastocytoma cells, *J. Exp. Med.* **158:**731–737.

Golden, J. A., Nadel, J. A., and Boushey, H. A., 1978, Bronchial hyperirritability in healthy subjects after exposure to ozone, *Am. Rev. Respir. Dis.* **118:**287–294.

Goldstein, I. M., Malmsten, C. L., Kindahl, H., Kaplan, H. B., Radmark, O., Samuelsson, B., and Weissmann, G., 1978, Thromboxane generation by human peripheral blood polymorphonuclear leukocytes, *J. Exp. Med.* **148:**787–792.

Gordon, T., Sheppard, D., McDonald, D. M., Distefano, S., and Scypinski, L., 1985, Airway hyperresponsiveness and inflammation induced by toluene diisocyanate in guinea pigs, *Am. Rev. Respir. Dis.* **132:**1106–1112.

Gresele, P., Todisco, T., Merante, F., and Nenci, G. G., 1982, Platelet activation and allergic asthma. (Correspondence.) *N. Engl. J. Med.* **306:**549.

Hamberg, M., Svensson, J., and Samuelsson, B., 1975, Thromboxanes: A new group of biologically active compounds derived from prostaglandin endoperoxides, *Proc. Natl. Acad. Sci. USA* **72:** 2994–2998.

Hanahan, D. J., Demopoulos, C. A., Liehr, J., and Pinckard, R. N., 1980, Identification of platelet activating factor isolated from rabbit basophils as acetyl glyceryl ether phosphorylcholine, *J. Biol. Chem.* **255:**5514–5516.

Hardy, C. C., Robinson, C., Tattersfield, A. E., and Holgate, S. T., 1984, The bronchoconstrictor effect of inhaled prostaglandin D_2 in normal and asthmatic men, *N. Engl. J. Med.* **311** 209–213.

Hartung, H. P., and Toyka, K. V., 1983, Activation of macrophages by substance P: induction of oxidative burst and thromboxane release, *Eur. J. Pharmacol.* **89:**301–305.

Hemker, D. P., and Aiken, J. W., 1981, Actions of indomethacin and prostaglandin E_2 and D_2 on nerve transmission on the nictitating membrane of the cat, *Prostaglandins* **22:**599–611.

Henderson, F. W., Clyde, W. A., Jr., Collier, A. M., Denny, F. W., Senior, R. J., Sheaffer, C. I., Conley, W. G. III, and Christian, R. M., 1979, The etiologic and epidemiologic spectrum of bronchiolitis in pediatric practice, *J. Pediatr.* **95:**183–190.

Henderson, W. R., Chi, E. Y., and Klebanoff, S. J., 1980, Eosinophil peroxidase-induced mast cell secretion, *J. Exp. Med.* **152:**265–279.

Henderson, W. R., Harley, J. B., and Fauci, A. S., 1984, Arachidonic acid metabolism in normal and hypereosinophilic syndrome human eosinophils: Generation of leukotrienes B_4, C_4, D_4 and 15-lipoxygenase products, *Immunology* **51:**679–686.

Henderson, W. R., Jorg, A., and Klebanoff, S. J., 1982, Eosinophil peroxidase-mediated inactivation of leukotrienes B_4, C_4, and D_4, *J. Immunol.* **128:**2609–2613.

Higashida, H., Nakagawa, Y., and Miki, N., 1984, Facilitation of synaptic transmission by prostaglandin D_2 at synapses between NG108-15 hybrid and muscle cells, *Brain Res.* **295:**113–119.

Hogg, J. C., 1981, Bronchial mucosal permeability and its relationship to airways hyperreactivity, *J. Allergy Clin. Immunol.* **67:**421–425.

Holtzman, M. J., Fabbri, L. M., O'Byrne, P. M., Gold, B. D., Aizawa, H., Walters, E. H., Alpert, S. E., and Nadel, J. A., 1983a, Importance of airway inflammation for hyperresponsiveness induced by ozone in dogs, *Am. Rev. Respir. Dis.* **127:**686–690.

Holtzman, M. J., Aizawa, H., Nadel, J. A., and Goetzl, E. J., 1983b, Selective generation of leukotriene B_4 by tracheal epithelial cells from dogs, *Biochem. Biophys. Res. Commun.* **114:** 1071–1076.

Holtzman, M. J., Fabbri, L. M., Skoogh, B.-E., O'Byrne, P. M., Walters, E. H., Aizawa, H., and Nadel, J. A., 1983c, Time course of airway hyperresponsiveness induced by ozone in dogs, *J. Appl. Physiol.* **55:**1232–1236.

Horn, M. E. C., Reed, S. E., and Taylor, P., 1979, Role of viruses and bacteria in acute wheezy bronchitis in childhood: A study of sputum, *Arch. Dis. Child.* **54:**587–592.

Hubscher, T., 1975, Role of the eosinophil in the allergic reactions. I. EDI—An eosinophil derived inhibitor of histamine release, *J. Immunol.* **114:**1379–1388.

Hulbert, W., Walker, D. C., Jackson, A., and Hogg, J. C., 1981, Airway permeability to horseradish peroxidase in guinea pigs: The repair phase after injury by cigarette smoke, *Am. Rev. Respir. Dis.* **123:**320–326.

Hulbert, W. M., McLean, T., and Hogg, J. C., 1985, The effect of acute airway inflammation on bronchial reactivity in guinea pigs, *Am. Rev. Respir. Dis.* **132:**7–11.

Hunter, J. A., Finkbeiner, W. E., Nadel, J. A., Goetzl, E. J., and Holtzman, M. J., 1985, Predominant generation of 15-lipoxygense metabolites of arachidonic acid by epithelial cells from human trachea, *Proc. Natl. Acad. Sci. USA* **82:**4633–4637.

Iwamoto, I., Ra, C., Sato, T., Tomioka, H., and Yoshida, S., 1986, Thromboxane A_2 production in allergen-induced immediate and late asthmatic responses, *Arerugi* **35:**437–446.

Jacoby, D. B., Ueki, I. F., Widdicombe, J. H., Loegering, D. A., Gleich, G. J. and Nadel, J. A., 1987, Human eosinophil major basic protein stimulates prostaglandin E_2 production by dog tracheal epithelium, *Am. Rev. Respir. Dis.* **135:**A316.

Jacoby, D. B., Ueki, I. F., Widdicombe, J. H., Loegering, D. A., Gleich, G. J., and Nadel, J. A., 1988a, Effect of human eosinophil major basic protein on ion transport in dog tracheal epithelium, *Am. Rev. Respir. Dis.* **137:**13–16.

Jacoby, D. B., Tamaoki, J., Borson, D. B., and Nadel, J. A., 1988b, Influenza infection causes airway hyperresponsiveness by decreasing enkephalinase, *J. Appl. Physiol.* **64:**2653–2658.

Jensen, C., Henriksen, U., Dahl, B. T., Skov, P. S., and Norn, S., 1986, Influence of neuraminidase and *N*-acetylneuraminic acid on basophil histamine release in vitro, *Allergy* **41:**151–156.

Johnson, A. R., Ashton, J., Schulz, W. W., and Erdos, E. G., 1985, Neutral metalloendopeptidase in human lung tissue and cultured cells, *Am. Rev. Respir. Dis.* **132:**564–568.

Johnson, C. E., Belfield, P. W., Davis, S., Cooke, N. J., Spencer, A., and Davies, J. A., 1986, Platelet activation during exercise induced asthma: Effect of prophylaxis with cromoglycate and salbutamol, *Thorax* **41:**290–294.

Jong, E. C., and Klebanoff, S. J., 1980, Eosinophil-mediated mammalian tumor cell cytotoxicity: Role of the peroxidase system, *J. Immunol.* **124:**1949–1953.

Joseph, M., Capron, A., Ameisen, J.-C., Capron, M., Vorng, H., Pancre, V., Kusnierz, J.-P., and Auriault, C., 1986, The receptor of IgE on blood platelets, *Eur. J. Immunol.* **16:**306–312.

Kay, A. B., 1970, Studies on eosinophil leucocyte migration. II. Factors specifically chemotactic for eosinophils and neutrophils generated from guinea-pig serum by antigen–antibody complexes, *Clin. Exp. Immunol.* **7:**723–737.

Kennedy, S. M., Elwood, R. K., Wiggs, B. J. R., Pare, P. D., and Hogg, J. C., 1984, Increased airway mucosal permeability in smokers. Relationship to airway reactivity, *Am. Rev. Respir. Dis.* **129:**143–148.

Khalife, J., Capron, M., Grzych, J.-M., Bazin, H., and Capron, A., 1985, Fcγ receptors on rat eosinophils: Isotype-dependent cell activation, *J. Immunol.* **135:**2781–2784.

Kim, S. J., Lam, B., Godfrey, H. P., Wong, P. Y.-K., and Kikkawa, Y., 1987, Generation of lipoxins and short-chain aldehydes by rat alveolar macrophages, *Fed. Proc.* **46:**692A. (abst.).

Klavinskis, L. S., Patterson, S., Whiting, P. J., Willcox, H. N. A., Oxford, J. S., and Newsom-Davis, J., 1985, Infection of cultured human muscle cells by influenza virus, *J. Gen. Virol.* **66:** 2335–2345.

Knauer, K. A., Lichtenstein, L. M., Adkinson, N. F., and Fish, J. E., 1981, Platelet activation during antigen-induced airway reactions in asthmatic subjects, *N. Engl. J. Med.* **304:**1404–1407.

Laitinen, L. A., Heino, M., Laitinen, A., Kava, T. and Haahtela, T., 1985a, Damage of the airway epithelium and bronchial reactivity in patients with asthma, *Am. Rev. Respir. Dis.* **131:**599–606.

Laitinen, A., Partanen, M., Hervonen, A., Pelto-Huikko, M., and Laitinen, L. A., 1985b, VIP like immunoreactive nerves in human respiratory tract, *Histochemistry* **82:**313–319.

Lee, T.-C., Lenihan, D. J., Malone, B., Roddy, L. L., and Wasserman, S. I., 1984, Increased biosynthesis of platelet-activating factor in activated human eosinophils, *J. Biol. Chem.* **259:**5526–5530.

Leikauf, G. D., Ueki, I. F., Nadel, J. A., and Widdicombe, J. H., 1985a, Release of cyclooxygenase products from cultured epithelium derived from human and dog trachea, *Fed. Proc.* **44:**1920.

Leikauf, G. D., Ueki, I. F., Nadel, J. A., and Widdicombe, J. H., 1985b, Brdykinin stimulates Cl secretion and prostaglandin E_2 release by canine tracheal epithelium, *Am. J. Physiol.* **248:**F48–F55.

Leikauf, G. D., Ueki, I. F., Widdicombe, J. H., and Nadel, J. A., 1986, Alteration of chloride secretion across canine tracheal epithelium by lipoxygenase products of arachidonic acid, *Am. J. Physiol.* **250:**F47–F53.

Lellouch-Tubiana, A., Lefort, J., Pirotzky, E., Vargaftig, B. B., and Pfister, A., 1985, Ultrastructural evidence for extravascular platelet recruitment in the lung upon intravenous injection of platelet activating factor (Paf-acether) to guinea pigs, *Br. J. Exp. Pathol.* **66:**345–355.

Levine, J. D., Lam, D., Taiwo, Y. O., Donatoni, P., and Goetzl, E. J., 1986, Hyperalgesic properties of 15-lipoxygenase products of arachidonic acid, *Proc. Natl. Acad. Sci. USA* **83:**5331–5334.

Lewis, R. A., Soter, N. A., Diamond, P. T., Austen, K. F., Oates, J. A., and Roberts, L. J., 1982, II. Prostaglandin D_2 release after activation of rat and human mast cells with anti-IgE, *J. Immunol.* **129:**1627–1631.

Little, J. W., Hall, W. J., Douglas, R. G., Jr., Mudholkar, G. S., Speers, D. M., and Patel, K., 1978, Airway hyperreactivity and peripheral airway dysfunction in influenza A infection, *Am. Rev. Respir. Dis.* **118:**295–303.

Liu, M. C., Proud, D., Lichtenstein, L., MacGlashan, D. W., Jr., Schleimer, R. P., Adkinson, N. F., Jr., Kagey-Sobotka, A., Schulman, E. S., and Plaut, M., 1986, Human lung macrophage-derived histamine-releasing activity is due to IgE-dependent factors, *J. Immunol.* **136:**2588–2595.

Lotner, G. Z., Lynch, J. M., Betz, S. J., and Henson, P. M., 1980, Human neutrophil-derived platelet activating factor, *J. Immunol.* **124:**676–684.

Lowenhaupt, R. W., Silberstein, E. B., Sperling, M. I., and Mayfield, G., 1982, A quantitative method to measure human platelet chemotaxis using indium-111-oxine-labeled gel-filtered platelets, *Blood* **60:**1315–1352.

Lundberg, J. M., Brodin, E., and Saria, A., 1983, Effects and distribution of vagal capsaicin-sensitive substance P neurons with special reference to the trachea and lungs, *Acta Physiol. Scand.* **119:**243–252.

Lundberg, J. M., Brodin, E., Hua, X., and Saria, A., 1984b, Vascular permeability changes and smooth muscle contraction in relation to capsaicin-sensitive substance P afferents in the guinea-pig, *Acta Physiol. Scand.* **120:**217–227.

Lundberg, J. M., Franco-Cereceda, A., Hua, X., Hokfelt, T., and Fischer, J. A., 1985, Coexistence of substance P and calcitonin gene-related peptide-like immunoreactivities in sensory nerves in relation to cardiovascular and bronchoconstrictor effects of capsaicin, *Eur. J. Pharmacol.* **108:**315–319.

Lundberg, J. M., Hokefelt, T., Martling, C.-R., Saria, A., and Cuello, C., 1984a, Substance P-immunoreactive sensory nerves in the lower respiratory tract of various mammals including man, *Cell Tissue Res.* **235**:251–261.

Lundberg, J. M., and Saria, A., 1983b, Capsaicin-induced desensitization of airway mucosa to cigarette smoke, mechanical and chemical irritants, *Nature (Lond.)* **302**:251–253.

Lundberg, J. M., Martling, C.-R., and Saria, A., 1983c, Substance P and capsaicin-induced contraction of human bronchi, *Acta Physiol. Scand.* **119**:49–53.

Lundberg, J. M., Saria, A., Brodin, E., Rosell, S., and Folkers, K., 1983d, A substance P antagonist inhibits vagally induced increase in vascular permeability and bronchial smooth muscle contraction in the guinea pig, *Proc. Natl. Acad. Sci. USA* **80**:1120–1124.

Macconi, D., Morzenti, G., Livio, M., Morelli, C., Cassina, G., and Remuzzi, G., 1985, Acetyl glycerylphosphorylcholine aggregates human platelets through two distinct pathways, both dependent on arachidonic acid metabolism, *Lab. Invest.* **52**:159–168.

MacDermot, J., Kelsey, C. R., Waddell, K. A., Richmond, R., Knight, R. K., Cole, P. J., Dollery, C. T., Landon, D. N., and Blair, I. A., 1984, Synthesis of leukotriene B₄, and prostanoids by human alveolar macrophages: Analysis by gas chromatography/mass spectrometry, *Prostaglandins* **27**:163–179.

Maeno, K., and Kilbourne, E. D., 1970, Developmental sequence and intracellular sites of synthesis of three structural protein antigens of influenza A₂ virus, *J. Virol.* **5**:153–164.

Malmsten, C. L., Palmblad, J., Uden, A. M., Radmark, O., Engstedt, L., and Samuelsson, B., 1980, A highly potent stereospecific factor stimulating migration of polymorphonuclear leukocytes, *Acta Physiol. Scand.* **110**:449–451.

Mann, P. R., 1969, An electron microscope study of the relationship between mast cells and eosinophil leukocytes, *J. Pathol.* **98**:182–186.

Marasco, W. A., Showell, H. J., and Becker, E. L., 1981, Substance P binds to the formylpeptide chemotaxis receptor on the rabbit neutrophil, *Biochem. Biophys. Res. Commun.* **99**:1065–1072.

Martin, T. R., Altman, L. C., Albert, R. K., and Henderson, W. R., 1984, Leukotriene B₄ production by the human alveolar macrophage: A potential mechanism for amplifying inflammation in the lung, *Am. Rev. Respir. Dis.* **129**:106–111.

Martling, C.-R., Saria, A., Andersson, P., and Lundberg, J. M., 1984, Capsaicin pretreatment inhibits vagal cholinergic and non-cholinergic control of pulmonary mechanics in the guinea pig, *Arch. Pharmacol.* **325**:343–348.

McLaren, D. J., McKean, J. R., Olsson, I., Venges, P., and Kay, A. B., 1981, Morphological studies on the killing of schistosomula of *Schistosoma mansoni* by human eosinophil and neutrophil cationic proteins in vitro, *Parasite Immunol.* **3**:359–373.

McManus, L. M., Hanahan, D. J., Demopoulos, C. A., and Pinckard, R. N., 1980, Pathobiology of the intravenous infusion of acetyl glyceryl ether phosphorylcholine (AGEPC), a synthetic platelet-activating factor (PAF), in the rabbit, *J. Immunol.* **124**:2919–2924.

Metzger, W. J., Zavala, D., Richerson, H. B., Moseley, P., Iwamota, P., Monick, M., Sjoersma, K., and Hunninghake, G. W., 1987, Local allergen challenge and bronchoalveolar lavage of allergic asthmatic lungs, *Am. Rev. Respir. Dis.* **135**:433–440.

Migueres, J., Sallerin, F., and Didier, J., 1985, Viral and nonbacterial respiratory infections associated with exacerbation of asthma in the asthmatic adult, *Prog. Respir. Res.* **19**:200–204.

Murphy, K. R., Wilson, M. C., Irvin, C. G., Glezen, L. S., Marsh, W. R., Haslett, C., Henson, P. M., and Larsen, G. L., 1986, The requirement for polymorphonuclear leukocytes in the late asthmatic response and heightened airways reactivity in an animal model, *Am. Rev. Respir. Dis.* **134**:62–68.

Nakagawara, A., Nathan, C. F., and Cohn, Z. A., 1981, Hydrogen peroxide metabolism in human monocytes during differentiation in vitro, *J. Clin. Invest.* **68**:1243–1252.

Nakajima, M., and Toda, N., 1984, Neuroeffector actions of prostaglandin D₂ on isolated dog mesenteric arteries, *Prostaglandins* **27**:407–419.

Nathan, C. F., and Klebanoff, S. J., 1982, Augmentation of spontaneous macrophage-mediated cytolysis by eosinophil peroxidase, *J. Exp. Med.* **155**:1291–1308.

Naylor, B., 1962, The shedding of the mucosa of the bronchial tree in man, *Thorax* **17**:69–72.

O'Byrne, P. M., Walters, E. H., Aizawa, H., Fabbri, L. M., Holtzman, M. J., and Nadel, J. A., 1984a, Indomethacin inhibits the airway hyperresponsiveness but not the neutrophil influx induced by ozone in dogs, *Am. Rev. Respir. Dis.* **130**:220–224.

O'Bryne, P. M., Walters, E. H., Gold, B. D., Aizawa, H. A., Fabbri, L. M., Alpert, S. E., Nadel, J. A., and Holtzman, M. J., 1984b, Neutrophil depletion inhibits airway hyperresponsiveness induced by ozone exposure, *Am. Rev. Respir. Dis.* **130**:214–219.

O'Byrne, P. M., Dolovich, M., Dirks, R., Roberts, R. S., and Newhouse, M. T., 1984c, Lung epithelial permeability: Relation to nonspecific airway responsiveness, *J. Appl. Physiol.* **57**:77–84.

O'Byrne, P. M., Aizawa, H., Bethel, R. A., Chung, K. F., Nadel, J. A., and Holtzman, M. J., 1984d, Prostaglandin $F_{2\alpha}$ increases responsiveness of pulmonary airways in dogs, *Prostaglandins* **28**:537–543.

O'Byrne, P. M., Leikauf, G. D., Aizawa, H., Bethel, R. A., Ueki, I. F., Holtzman, M. J., and Nadel, J. A., 1985, Leukotriene B_4 induces airway hyperresponsiveness in dogs, *J. Appl. Physiol.* **59**:1941–1946.

O'Donnell, M. C., Ackerman, S. J., Gleich, G. J., and Thomas, L. L., 1983, Activation of basophil and mast cell histamine release by eosinophil granule major basic protein, *J. Exp. Med.* **157**: 1981–1991.

Oelz, O., Oelz, R., Knapp, H. R., Sweetman, B. J., and Oates, J. A., 1977, Biosynthesis of prostaglandin D_2. 1. Formation of prostaglandin D_2 by human platelets, *Prostaglandins* **13**: 225–234.

Orchard, M. A., Kagey-Sobotka, A., Proud, D., and Lichtenstein, L. M., 1986, Basophil histamine release induced by a substance from stimulated human platelets, *J. Immunol.* **136**:2240–2244.

Orehek, J., Douglas, J. S., and Bouhuys, A., 1975, Contractile responses to the guinea-pig trachea in vitro: Modification by prostaglandin synthesis-inhibiting drugs, *J. Pharmacol. Exp. Ther.* **194**:554–564.

Orehek, J., Massari, J. P., Gayrard, P., Grimaud, C., and Charpin, J., 1976, Effect of short-term, low-level nitrogen dioxide exposure on bronchial sensitivity of asthmatic patients, *J. Clin. Invest.* **57**;301–307.

Page, C. P., Paul, W., and Morley, J., 1984, Platelets and bronchospasm, *Int. Arch. Allergy Appl. Immunol.* **74**:347–350.

Palmblad, J., Gyllenhammar, H., Lindgren, J. A., and Malmsten, C. L., 1984, Effects of leukotrienes and F-met-leu-phe on oxidative metabolism of neutrophils and eosinophils, *J. Immunol.* **132**:3041–3045.

Palmer, J. B. D., Cuss, F. M. C., and Barnes, P. J., 1986, VIP and PHM and their role in nonadrenergic inhibitory responses in isolated human airways, *J. Appl. Physiol.* **61**:1322–1328.

Palmer, J. B. D., Cuss, F. M. C., Mulderry, P. K., Ghatei, M. A., Springall, D. R., Cadieux, A., Bloom, S. R., Polak, D. M., and Barnes, P. J., 1987, Calcitonin gene-related peptide is localised to human airway nerves and potently constricts human airway smooth msucle, *Br. J. Pharmacol.* **91**:95–101.

Palmer, R. M. J., Ferrige, A. G., and Moncada, S., 1987, Nitric oxide release accounts for the biological activity of endothelium-derived relaxing factor, *Nature (Lond.)* **327**:524–526.

Patterson, R., Bernstein, P. R., Harris, K. E., and Krell, R. D., 1984, Airway responses to sequential challenges with platelet activating factor and leukotriene D_4 in rhesus monkeys, *J. Lab. Clin. Med.* **104**:340–345.

Payan, D. G., Brewster, D. R., and Goetzl, E. J., 1983, Specific stimulation of human T-lymphocytes by substance P, *J. Immunol* **131**:1613–1615.

Payan, D. G., Brewster, D. R., and Goetzl, E. J., 1984, Stereospecific receptors for substance P on cultured human IM-9 lymphoblasts, *J. Immunol.* **133**:3260–3265.

Pecoud, A. R., Ruddy, S., and Conrad, D. H., 1981, Functional and partial chemical characterization of the carbohydrate moieties of the IgE receptor on rat basophilic leukemia cells and rat mast cells, *J. Immunol.* **126**:1624–1629.

Peters, S. P., MacGlashan, D. W., Jr., Schulman, E. S., Schleimer, R. P., and Lichtenstein, L. M., 1983, The production of arachidonic acide (AA) metabolites by purified human lung mast cells (HMC), *Fed. Proc.* **42**:1375.

Pullan, C. R., and Hey, E. N., 1982, Wheezing, asthma, and pulmonary dysfunction 10 years after infection with respiratory syncytial virus in infancy, *Br. Med. J.* **284**:1665–1668.

Rand, T. H., and Colley, D. G., 1983, Lymphokin-mediated regulation of murine eosinophils, in: *Immunobiology of the Eosinophil* (T. Yoshida and M. Torisu, eds.) pp. 13–26, Elsevier, New York.

Rosenfeld, M. G., Mermod, J. J., Amara, S. G., Swandson, L. H., Sawchenko, P. E., Rivier, J., Vale, N. N., and Evans, R. M., 1983, Production of a novel neuropeptide encoded by the calcitonin gene via tissue specific RNA processing, *Nature (Lond.)* **304**:129–135.

Ruff, M. R., Wahl, S. M., and Pert, C. B., 1985, Substance P receptor-mediated chemotaxis of human monocytes, *Peptides* **6**:107–111.

Schleimer, R. P., MacGlashan, D. W., Jr., Peters, S. P., Pinckard, R. N., Adkinson, N. F., Jr., and Lichtenstein, L. M., 1986, Characterization of inflammatory mediator release from purified human lung mast cells, *Am. Rev. Respir. Dis.* **133**:614–617.

Schulman, E. S., Liu, M. C., Proud, D., MacGlashan, D. W., Jr., Lichtenstein, L. M., and Plaut, M., 1985, Human lung macrophages induce histamine release from basophils and mast cells, *Am. Rev. Respir. Dis.* **131**:230–235.

Sekizawa, K., Tamaoki, J., Graf, P. D., Basbaum, C. B., Borson, D. B., and Nadel, J. A., 1987, Enkephalinase inhibitor potentiates mammalian tachykinin-induced contraction in ferret trachea, *J. Pharmacol. Exp. Ther.* **243**:1211–1217.

Seltzer, J., Bigby, B. G., Stulbarg, M., Holtzman, M. J., Nadel, J. A., Ueki, I. F., Leikauf, G. D., Goetzl, E. J., and Boushey, H. A., 1986, O_3-induced change in bronchial reactivity to methacholine and airway inflammation in humans, *J. Appl. Physiol.* **60**:1321–1326.

Serhan, C. N., Nicolaou, K. C., Webber, S. E., Veale, C. A., Dahlen, S. E., Puustinen, T. J., and Samuelsson, B., 1986, Lipoxin A stereochemistry and biosynthesis, *J. Biol. Chem.* **261**:16340–16345.

Shak, S., Perez, H. D., and Goldstein, I. M., 1983, A novel dioxygenation product of arachidonic acid possesses potent chemotactic activity for human polymorphonuclear leukocytes, *J. Biol. Chem.* **258**:14948–14953.

Shore, S., Collier, B., and Martin, J. G., 1987, Effect of endogenous prostaglandins on acetylcholine release from dog trachealis muscle, *J. Appl. Physiol.* **62**:1837–1844.

Sigal, E., Valone, F. H., Holtzman, M. J., and Goetzl, E. J., 1987, Preferential human eosinophil chemotactic activity of the platelet activating factor, 1-*o*-hexadecyl-2-acetyl-*sn*-glycerol-3-phosphocholine (AGEPC), *J. Clin. Immunol.* **7**:179–184.

Skidgel, R. A., Engelbrecht, A., Johnson, A. R., and Erdos, E. G., 1984, Hydrolysis of substance P and neurotensin by converting enzyme and neutral endoproteinase, *Peptides* **5**:769–776.

Snyderman, R., Wohlenberg, C., and Notkins, A. L., 1972, Inflammation and viral infection: Chemotactic activity resulting from the interaction of antiviral antibody and complement with cells infected with Herpes simplex virus, *J. Infect. Dis.* **126**:207–209.

Sydbom, A., 1982, Histamine release from isolated rat mast cells by neurotensin and other peptides, *Agents Actions* **12**:90–93.

Tamaoki, J., Sekizawa, K., Osborne, M. L., Ueki, I. F., Graf, P. D., and Nadel, J. A., 1987b, Platelet aggregation increases cholinergic neurotransmission in canine airways, *J. Appl. Physiol.* **62**:2246–2251.

Tamaoki, J., Sekizawa, K., Ueki, I. F., Graf, P. D., Nadel, J. A., and Bigby, T. D., 1987a, Short communication. Effect of macrophage stimulation on parasympathetic airway contraction in dogs, *Eur. J. Pharmacol.* **138:**421–425.

Tamaoki, J., Sekizawa, K., Graf, P. D., and Nadel, J. A., 1987c, Cholinergic neuromodulation by prostaglandin D_2 in canine airway smooth muscle, *J. Appl. Physiol.* **63:**1396–1400.

Tanaka, D. T., and Grunstein, M. M., 1986, Effect of substance P on neurally mediated contraction of rabbit airway smooth muscle, *J. Appl. Physiol.* **60:**458–463.

Thompson, J. E., Scypinski, L. A., Gordon, T., and Sheppard, D., 1986, Hydroxyurea inhibits airway hyperresponsiveness in guinea pigs by a granulocyte-independent mechanism, *Am. Rev. Respir. Dis.* **134:**1213–1218.

Tiku, K., Tiku, M. L., and Skosey, J. L., 1986, Interleukin 1 production by human polymorphonuclear neutrophils, *J. Immunol.* **136:**3677–3685.

Vanderhoek, J. Y., Bryant, R. W., and Bailey, J. M., 1980, Inhibition of leukotriene biosynthesis by the leukocyte product of 15-hyroxy-5,8,11,13-eicosatetraenoic acid, *J. Biol. Chem.* **255:**10064–10065.

Verhagen, J., Bruynzeel, L. B., Koedam, J. A., Wassink, G. A., de Boer, M., Terpstra, G. K., Kreukniet, J., Veldink, G. A., and Vliegenthart, J. F. G., 1984, Specific leukotriene formation by purified human eosinophils and neutrophils, *FEBS Lett.* **168:**23–28.

Wahlestedt, C., Bynke, G., and Hakanson, R., 1985, Pupillary constriction by bradykinin and capsaicin: Mode of action, *Eur. J. Pharmacol.* **106:**577–583.

Walsh, J. J., Dietlein, L. F., Low, F. N., Burch, G. E., and Mogabgab, W. J., 1960, Bronchotracheal response in human influenza, *Arch. Intern. Med.* **108:**376–388.

Walters, E. H., O'Byrne, P. M., Fabbri, L. M., Graf, P. D., Holtzman, M. J., and Nadel, J. A., 1984, Control of neurotransmission by prostaglandins in canine trachealis smooth muscle, *J. Appl. Physiol.* **57:**129–134.

Ward, P. A., Cohen, S., and Flanagan, T. D., 1972, Leukotactic factors elaborated by virus-infected tissues, *J. Exp. Med.* **135:**1095–1103.

Wardlaw, A. J., Moqbel, R., Cromwell, O., and Kay, A. B., 1986, Platelet-activating factor. A potent chemotactic and chemokinetic factor for human eosinophils, *J. Clin. Invest.* **78:**1701–1706.

Weiss, S., Robb, G. P., and Blumgart, H., 1929, The velocity of blood flow in health and disease as measured by the effect of histamine on the minute vessels, *Am. Heart J.* **4:**664–691.

Weiss, S. T., Tager, I. B., Munoz, A., and Speizer, F. E., 1985, The relationship of respiratory infections in early childhood to the occurrence of increased levels of bronchial responsiveness and atopy, *Am. Rev. Respir. Dis.* **131:**573–578.

Weksler, B. B., and Coupal, C. E., 1973, Platelet-dependent generation of chemotactic activity in serum, *J. Exp. Med.* **137:**1419–1429.

Weller, P. F., and Goetzl, E. J. (1980) The human eosinophil: Roles in host defense and tissue injury, *Am. J. Pathol.* **100:**793–820.

Weller, P. F., Lee, C. W., Foster, D. W., Corey, E. J., Austen, K. F., and Lewis, R. A., 1983, Generation and metabolism of 5-lipoxygenase pathway leukotrienes by human eosinophils: Predominant production of leukotriene C_4, *Proc. Natl. Acad. Sci. USA* **80:**7626–7630.

Welliver, R. C., 1983, Upper respiratory infections in asthma, *J. Allergy Clin. Immunol.* **72:**341–346.

Welliver, R. C., Kaul, A., and Ogra, P. L., 1979, Cell-mediated immune response to respiratory syncytial virus infection: Relationship to the development of reactive airway disease, *J. Pediatr.* **94:**370–375.

Welliver, R. C., Kaul, T. N., and Ogra, P. L., 1980, The appearance of cell-bound IgE in respiratory-tract epithelium after respiratory-syncytial-virus infection, *N. Engl. J. Med.* **303:**1198–1202.

Welliver, R. C., Wong, D. T., Sun, M., Middleton, E., Jr., Vaughan, R. S., and Ogra, P. L., 1981,

The development of respiratory syncytial virus-specific IgE and the release of histamine in nasopharyngeal secretions after infection, *N. Engl. J. Med.* **305**:841–846.

Zeiger, R. S., Yurdin, D. L., and Colten, H. R., 1976, Histamine metabolism. II. Cellular and subcellular localization of the catabolic enzymes, histaminase, and histamine methyl transferase, in human leukocytes, *J. Allergy Clin. Immunol.* **58**:172–179.

Zheutlin, L. M., Ackerman, S. J., Gleich, G. J., and Thomas, L. L., 1984, Stimulation of basophil and rat mast cell histamine release by eosinophil granula-derived cationic proteins, *J. Immunol.* **133**:2180–2185.

Chapter 13

Airway Hyperreactivity
Relationship to Disease States

James C. Hogg

Pulmonary Research Laboratory
St. Paul's Hospital
University of British Columbia
Vancouver, British Columbia, Canada V6Z 1Y6

I. INTRODUCTION

Airway hyperreactivity or hyperresponsiveness is a hallmark of the asthmatic state (Parker *et al.*, 1965; Cade and Pain, 1971; Cockcroft *et al.*, 1977a,b). However, it can also be demonstrated in subjects without asthma in whom the airways have been injured (Islam *et al.*, 1972; Orehek *et al.*, 1976; Golden *et al.*, 1978; Cockcroft *et al.*, 1979) and in patients in whom airway inflammatory processes have been induced by infection (Empey *et al.*, 1976; Little *et al.*, 1978). The presence of hyperreactivity is difficult to establish from the case history (Dales *et al.*, 1987) but can be documented by measuring excessive airway narrowing to a stimulus that has no effect on normal airways. The sequence of events that cause narrowing of the airways was recently reviewed by Moreno *et al.* (1986) and is summarized in Fig. 1. The first step in airway narrowing is the activation of airway smooth muscle, which then contracts and shortens to reduce the lumen and increase resistance. The purpose of this chapter is to provide an overview of these events and attempt to show that airway hyperresponsiveness can result from abnormal function at any step in this process.

II. GRADED DOSE–RESPONSE CURVE

The response of the airways to an inhaled stimulus can be measured in several ways. A solution of drug such as methacholine or histamine is usually generated in an aerosol, and the subject is asked to inhale a predetermined

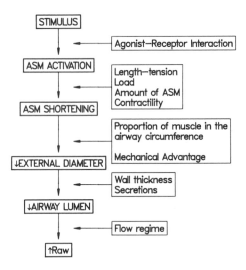

FIGURE 1. Airway hyperresponsiveness occurs as a result of excessive airway narrowing for a given stimulus. The outline shows the sequence of events leading to airway narrowing and the text discusses the possibility of excessive airway narrowing occurring as a result of abnormal function at each of these steps. (From Moreno *et al.*, 1987, with permission.)

amount. An airway response such as the maximum volume of air that can be exhaled in 1 sec (FEV$_1$), airway resistance or airway conductance is measured before and after increasing doses of the drug. A dose–response curve is then constructed and a predetermined point (i.e., the dose that produces a 20% drop in FEV$_1$ or 20% increase in airway resistance) is taken as a cutoff point. Subjects who show this change at low doses of the drug are said to be hyperresponsive, while subjects who fail to show any change at the highest doses administered are considered to have normal airway responsiveness. This type of graded dose–response curve differs substantially from the dose–response curve of isolated airway tissue.

Analysis of graded dose–response curves (Goldstein *et al.*, 1974) of an isolated tissue containing smooth muscle is based on the law of mass action and is derived from the classic Michaelis–Menton equation (Goldstein *et al.*, 1974). The equation gives the velocity V of an enzyme reaction as a function of substrate concentration S, the enzyme substrate dissociation constant K_m and maximum velocity V_{max}:

$$V = [V_{max} (S)]/[K_m + (S)] \tag{1}$$

The analogy for the drug–receptor interaction occurring on smooth muscle is that if a drug X combines with a receptor R, it forms a complex RX:

$$R + X \underset{K_1}{\overset{K_2}{\rightleftharpoons}} RX \qquad \text{where at equilibrium} \qquad K_2/K_1 = KX \qquad (2)$$

The response at any point can be compared with the maximum response at which all the receptors are occupied RT:

$$\Delta/\Delta_{max} = RX/RT = X/(KX + X)$$

Note that

$$\Delta/\Delta_{max} = \tfrac{1}{2} \quad \text{when } KX = X$$

and that

$$\Delta = (\Delta_{max} X)/(KX + X)$$

has the same form as Eq. (1). Under these circumstances, the sensitivity of the tissue to the agonist is defined as KX, which is the dose required to produce one half the maximal response, and its reactivity is defined as Δ_{max}/KX.

The problem with applying this approach to the measurement of airway responsiveness in an intact living subject is that it is seldom possible to study the complete dose–response curve without unacceptable risks. Furthermore, measurement of airway narrowing observed *in vivo* is influenced by more than muscle function alone (See Fig. 1). Indeed, failure to find correlations between airway responsiveness measured *in vivo* with the *in vitro* dose–response curves (Vincenc *et al.*, 1983; Armour *et al.*, 1984; Taylor *et al.*, 1985) of airway strips prepared from the same lungs shows that excess airway narrowing can occur when airways smooth muscle function is within normal limits.

III. SMOOTH MUSCLE ACTIVATION

Several attempts have been made to link airway hyperreactivity with abnormal smooth muscle activation (Kalsner, 1974; Antonissen *et al.*, 1979; Westfall, 1981). The mediators that activate receptors on smooth muscle can be deliberately administered to provoke a response or be released from cells present in the airways. The smooth muscle is also influenced by mediators such as ACh, which are released from efferent nerve endings, and by tachykinins, thought to be released by antidromic traffic in the sensory nerves (Lembeck and Gamse, 1982). Other mediators can be released by injury to the epithelial surface, where disruption of the cell membrane can lead to activation of phospholipases and generation of arachidonic acid metabolites that include the leukotrienes and prostaglandins (Holtzmann, 1982). All these mediators interact in a complex way to determine the amount of smooth muscle activation.

Barnes (1986a) recently reviewed the growing literature on the activation of smooth muscle by mediators and reviewed the evidence that airway reactivity might result from the smooth muscle having too few receptors for agents that relax it or too many receptors for agents that contract it. Unfortunately, he found little direct evidence to support either of these hypotheses. The alternate hypothesis that there are a normal number of receptors for smooth muscle agonists and that an increased local concentration of the agonist leads to a greater response through a greater receptor occupancy is slightly more viable. The increased concentration required for this hypothesis could be achieved by increasing either epithelial or vascular permeability or by decreasing bronchial blood flow so that the locally produced mediators are not removed. However, the evidence for increased local drug concentration in the tissue surrounding the smooth muscle is largely indirect.

Nadel (1965) was the first to postulate that airway hyperresponsiveness might be related to epithelial damage and this concept has stimulated a great deal of work directed toward determining the exact mechanism for the increased reactivity. He suggested that the mechanism for the excess smooth muscle spasm might be related to a lowering of the threshold of the irritant reflexes that would produce bronchoconstriction by increasing vagus nerve activity. Studies from our laboratory (Hogg, 1981) suggested that this increased sensitivity might be attributable to damaged tight junctions that would increase the exposure of the nerve endings. This suggested the possibility that a change in epithelial permeability following acute airways injury would permit more of an inhaled agonist to reach the irritant receptors and/or the smooth muscle. More recently, it has been suggested that the increased responsiveness is related to an axon reflex (Barnes, 1986b) whereby stimulation of sensory nerve endings results in a release of tachykinins such as substance P near the smooth muscle. It has also been postulated (Holtzmann, 1982; Holtzmann et al., 1983) that damage to the epithelial cells results in a release of chemotactic agents that caused neutrophils to migrate into the mucosa and release mediators that are responsible for the excessive airways narrowing. Although the neutrophil could be the cause of the airway hyperresponsiveness in some situations (Auburn et al., 1984), it definitely is not the cause in others (Thompson et al., 1986). In many clinical and experimental situations in which hyperresponsiveness exists, it is difficult to determine exactly what is responsible for the excessive airways narrowing.

IV. SMOOTH MUSCLE CONTRACTION

Several studies have reported an increase in the amount of muscle in the airways of patients with asthma or asthmatic bronchitis (Dunnill et al., 1969).

This is thought to be attributable to an absolute increase in the amount of muscle rather than an apparent increase due to bronchoconstriction (Huber and Koessler, 1922; Hurd and Hussain, 1973). Despite a substantial number of *in vitro* studies of muscle contraction in airways strips prepared from human lungs, relatively few studies of strips have been prepared from asthmatic airways. One study (Schellenberg and Foster, 1984) showed that the airways from an asthmatic boy had an increased maximum contraction, and a recent study reported similar findings in airways from patients with chronic obstructive pulmonary disease (COPD) (De Jongste *et al.*, 1987). Unfortunately, neither study ruled out the possibility that the increased maximum contraction was not the result of an increased amount of airways muscle rather than an excessive muscle contraction.

The length–tension relationship for smooth muscle defines the conditions under which muscle shortening occurs (Stevens-Newman, 1987). In the face of a large enough preload, the force generated by the contraction process will be dissipated by increasing tension in the muscle without shortening it. This means that if the muscle contracts against a preload, it must first generate enough force to move the load before it can shorten. Conversely, when no preload is placed on the muscle, it might shorten to 20–30% of its resting length. In the first situation, there will be no airway narrowing and no increase in resistance, but in the second, the airways will narrow maximally and an increase in resistance of up to 600-fold can be predicted in airways in which the muscle completely encircles the lumen (Macklem, 1985; Moreno *et al.*, 1985).

In the trachea, the smooth muscle is located in the posterior membranous sheath and occupies only ~30% of the airways circumference. In the peripheral airways, the smooth muscle is arranged in a helical spiral that completely encircles the airway lumen. This means that the same degree of muscle shortening will have a much greater effect on the lumen of a bronchiole than it will on the trachea. The effect of muscle shortening is also determined by the geometrical arrangement of the muscle around the lumen because the spiral arrangement means that some shortening will be used up in decreasing the length of the airway rather than in narrowing the lumen.

The amount that the muscle shortens will also be determined by the loads that are placed on it. Studies by Moreno *et al.* (1986) showed that softening the airway cartilage increases airway reactivity; James *et al.* (1987) showed that the cartilage provides a preload that prevents muscle shortening. The muscle in the intrapulmonary airways has an additional load placed on it by the lung parenchyma and this load increases with lung volume. When these preloads are reduced by either softening the airway cartilage (Moreno *et al.*, 1986) or breathing at low lung volume (Dirg *et al.*, 1987), there is a substantial increase in airway responsiveness. This suggests that it is possible to increase the responsiveness of the airway by altering the loads placed upon the muscle without necessarily changing the intrinsic properties of the muscle itself (Moreno *et al.*, 1985).

V. WALL THICKNESS AND SECRETIONS

The airway lumen may also be induced to narrow when an inflammatory exudate appears on the surface of the peripheral airways lumen (Fig. 2). This surface is normally covered by surfactant (Macklem *et al.*, 1970), which has a low surface tension; when it is replaced by the higher surface tension of the exudate, the lumen will tend to decrease in size. This narrowing of the airways lumen can be further aggravated by an increase in the luminal contents and thickening of the airway wall.

The most striking evidence for increased wall thickness and excess secretion is found in postmortem studies of patients in whom death was attributed to status asthmaticus (Dunnill *et al.*, 1969; Cardell, 1956; Houston *et al.*, 1953). As similar findings have been reported from autopsy studies of patients with well-documented asthma who died of other causes (Sobonya, 1984), it is reasonable to assume that they are present in patients suffering from chronic asthma. Postmortem studies show that in some cases of asthma, the airway lumen can be filled with tenacious viscid plugs. These plugs are most apparent in the segmental and subsegmental airways but can extend all the way to the respiratory bronchioles. They are formed from the inflammatory exudate and contain mucus, plasma proteins, migrating inflammatory cells, and sloughed epithelium; they can be extremely tenacious. These plugs sometimes form casts of the airways that are coughed up and appear in sputum as the so-called Curschman spirals (Naylor, 1972). The granules released from the eosinophils in the exudate form elongated crystalline structures called Charcot–Leyden crystals, which can also be identified in the sputum. The sputum may also contain compact clusters of airway epithelial cells known as Creola bodies after a patient (Creola Jones) from whom they were first demonstrated (Naylor, 1972). In addition to the luminal changes reflected in the patient's sputum, the airway epithelium and submucosa are also grossly abnormal. In some areas, the basement membrane is bare because of the epithelial sloughing; in other areas, the repair process results in a covering of either goblet or squamous cells rather than the normal pseudostratified respiratory epithelium. The basement membrane of the epithelium is thickened as a result of collagen deposition and increased deposition of immunoglobulins (Igs) rather than of true thickening of the true basal lamina (McCarter and Vazquez, 1966). The tracheobronchial mucus glands are increased in size, and the submucosa is increased in amount as a result of edema and infiltration with migrating inflammatory cells, particularly the eosinophil (Dunnill *et al.*, 1969).

The tissue swelling produced by the exudation of fluid and cells as well as the vascular congestion and hypertrophy of muscle and glands are all of a liquid nature and are therefore incompressible. This means that even if the smooth muscle shortens by a normal amount, the thickened mucosa will amplify the effect of the muscle contraction on the airway lumen (Orehek, 1983; Freedman,

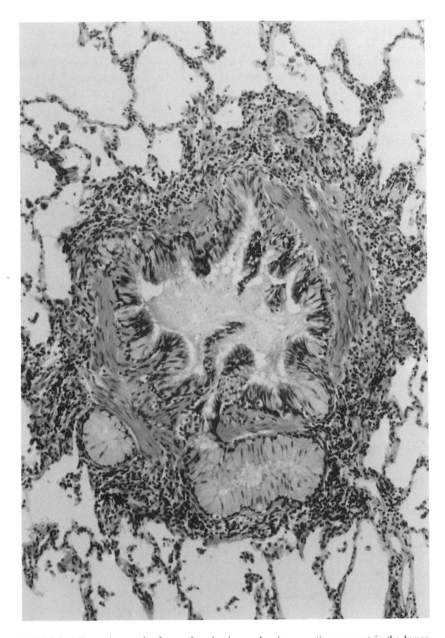

FIGURE 2. Photomicrograph of an asthmatic airway showing secretions present in the lumen, increased wall thickness, excessive smooth muscle, and the infiltrate of chronic inflammatory cells. (Courtesy of Dr. Alan James.)

1972). Moreno *et al.* (1986) calculated that an increase of only 10% in airway wall thickening, smooth muscle shortening, and degree to which the smooth muscle encircles the lumen has a profound effect on the dose–response curve. This means that although the reversal of the smooth muscle contraction may relieve the airway obstruction, it may not reverse the basic pathological process responsible for the asthma in the first place.

VI. SUMMARY

Airway hyperresponsiveness is an important feature of the asthmatic state, which can also be found when diseases other than asthma are present. It seems quite unlikely that the mechanism for the excess airway narrowing is the same in these divergent situations. For example, the acquired airway hyperresponsiveness that appears and disappears following an acute inhalational injury (Golden *et al.*, 1978) and infection (Little *et al.*, 1978) could be caused by epithelial injury, allowing an increased agonist to reach the smooth muscle, while the hyperresponsiveness seen in patients with chronic severe asthma may be caused by submucosal thickening produced by chronic inflammation. Similarly, mechanisms of the increased airway responsiveness that develops in chronic cigarette smokers (Mullen *et al.*, 1986) may not necessarily be the same as that seen following acute airway injury or in chronic asthma.

Airway hyperresponsiveness can result from factors that include variation in the concentration of mediators around the muscle, the state of the receptors on the muscle, the loads placed on the muscle by the tissues surrounding it, the contractibility of the muscle, and the coupling of muscle shortening to the submucosa, where increases in the tissue thickness can enhance the effect of muscle contraction on the airway lumen. Although one of these mechanisms may dominate the pathology in a particular clinical setting, it seems highly unlikely that a single mechanism will explain the cause of the airway narrowing in every case. The challenge of the future will be to devise experiments in which the sequence of events leading to the excess airway narrowing can be studied in a controlled fashion so that the step responsible for the excessive airway narrowing can be understood and eliminated in these separate forms of abnormal airway function.

VII. REFERENCES

Antonissen, L. A., Mitchell, R. W., Kroeger, E. A., Kepron, W., Sey, K. S., and Stevens, N. L., 1979, Mechanical alterations in airway smooth muscle in a canine asthmatic model, *J. Appl. Physiol.* **46**:681–687.

Armour, C. L., Lazar, N. M., Schellenberg, R. R., Taylor, S. M., Chan, N., Hogg, J. C., and Pare, P. D., 1984, A comparison of *in vivo* and *in vitro* human airways reactivity to histamine, *Am. Rev. Respir. Dis.* **129**:907–910.

Auburn, P. M., Walters, E., Gold, B., et al., 1984, Neutrophil depletion inhibits airway hyper-responsiveness induced by ozone, *Am. Rev. Respir. Dis.* **130:**214–219.

Barnes, P. J., 1986a, Asthma as an axone reflex, *Lancet* **1:**242–245.

Barnes, P. J., 1986b, Neural control of human airways in health and disease, *Am. Rev. Respir. Dis.* **134:**1289–1314.

Cade, J. F., and Pain, M. C., 1971, Role of bronchial reactivity in the etiology of asthma, *Lancet* **2:** 86–88.

Cardell, B. S., 1956, Pathological findings in deaths from asthma, *Intern. Arch. Allergy* **9:**189.

Cockroft, D. W., Killian, D. N., Mellon, J. J. A., and Hargreave, F. E., 1977a, Bronchial reactivity to inhaled histamine: A method and clinical survey, *Clin. Allergy* **7:**235–243.

Cockroft, D. W., Ruffin, R. E., Dolovich, J., and Hargreave, F. E., 1977b, Allergen-induced increase in non-allergic bronchial reactivity, *Clin. Allergy* **7:**503–513.

Cockcroft, D. W., Cotton, D. J., and Mink, J. T., 1979, Non-specific bronchial hyperactivity after exposure to western red cedar, *Am. Rev. Respir. Dis* **119:**505–510.

Dales, R. E., Ernst, P., Hanley, J. A., Baptista, R. N., and Becklake, M., 1987, The prediction of airways reactivity from responses to a standard respiratory symptom questionnaire, *Am. Rev. Respir. Dis.* **135:**817–821.

De Jongste, J. C., Mones, H., Block, R., Banta, I. L., Frederiksz, A. P., and Kerrebijn, K., 1987, Increased in vitro histamine responses in human small airways smooth muscle from patients with COPD, *Am. Rev. Respir. Dis.* **135:**549–553.

Dirg, D. J., Martin, J. C., and Macklem, P. T., 1987, Effects of lung volume on maximal meth-acholine induced bronchoconstriction in humans, *J. Appl. Physiol.* **62:**1324–1330.

Dunnill, M. S., Mezarella, G. R., and Anderson, J. A., 1969, A comparison of the quantitative anatomy of the bronchi in normal subjects and status asthmaticus in chronic bronchitis and in emphysema, *Thorax* **24:**176–179.

Empey, D. W., Laitinen, L. A., Jacobs, L., Gold, W. M., and Nadel, J. A., 1976, Mechanisms of bronchial hyperreactivity in normal subjects after upper respiratory tract infection, *Am. Rev. Respir. Dis.* **113:**131–139.

Freedman, B. J., 1972, The functional geometry of the bronchi, *Bull. Physiopathol. Respir.* **8:**545– 551.

Golden, J. A., Nadel, J. A., and Boushey, H. A., 1978, Bronchial hyperactivity in healthy subjects after exposure to ozone, *Am. Rev. Respir. Dis.* **118:**287–294.

Goldstein, A., Arrano, L., and Kalman, S. M., 1974, in: *Principles of Drug Action: The Basis of Pharmacology,* 2nd ed., pp. 82–111, Wiley, New York.

Hogg, J. C., 1981, Bronchial mucosal permeability and its relation to airways hyperreactivity, *J. Allergy Clin. Immun.* **67:**421–425.

Holtzmann, M. J., 1982, Pathophysiology of asthma: An overview of mechanisms of bronchial hyperreactivity, in: *Bronchial Hyperreactivity* (J. Morley, ed.), pp. 5–17, Academic, London.

Holtzmann, M. J., Fabre, L. M., Auburn, P. M., et al., 1983, Importance of airway inflammation for hyperresponsiveness induced by ozone, *Am. Rev. Respir. Dis.* **127:**686–690.

Houston, J. C., de Nevasquez, S., and Tounce, J. R., 1953, A clinical and pathological study of fatal cases of status asthmaticus, *Thorax* **8:**207.

Huber, H. L., and Koessler, K. K., 1922, The pathology of bronchial asthma, *Arch. Intern. Med.* **36:**89.

Hurd, B. E., and Hussain, S., 1973, Hyperplasia of bronchial muscle and asthma, *J. Pathol.* **110:** 319.

Islam, M. S., Vastag, E., Ulmer, W. T., 1972, Effect of noxious stimulants on bronchial reactivity, *Bull. Physiopathol. Respir.* **8:**509–518.

James, A. L., Pare, P. D., Moreno, R. H., and Hogg, J. C., 1987, Quantitative measurement of smooth muscle shortening in the isolated pig trachea. *J. Appl. Physiol.* **63:**1360–1365.

Kalsner, S., 1974, A new approach to the measurement and classification of forms of supersensitivity of autoanomic effect of responses, *Br. J. Pharmacol.* **51:**424–434.

Lembeck, F., and Gamse, R., 1982, Substance P and perhipheral sensory processes, in: *Substance P in the Nervous System* (R. Porter, and M. O'Conner, eds.), pp. 35–49, Pitman, London.

Little, J. N., Hall, W. J., Douglas, R. G., Jr., Mudholkar, G. S., Speers, D. M., and Patel, K., 1978, Airway hyperreactivity in peripheral airway dysfunction in influenza A infection, *Am. Rev. Respir. Dis.* **118:**295–303.

Macklem, P. T., 1985, Bronchial hyper-responsiveness, *Chest* **87:**1585–1595.

Macklem, P. T., Parker, D. F., and Hogg, J. C., 1970, The stability of peripheral airways, *Respir. Physiol.* **8:**191–203.

McCarter, J. H., and Vazquez, J. J., 1966, Bronchial basement membrane in asthma, *Arch. Pathol. Lab. Med.* **82:**328.

Moreno, R. H., Dalhby, R., Hogg, J. C., and Pare, P. D., 1985, Increased airway responsiveness caused by airway cartilage softening in rabbits, *Am. Rev. Respir. Dis.* **131:**A288.

Moreno, R. H., Hogg, J. C., and Pare, P. D., 1986, Mechanisms of airway narrowing, *Am. Rev. Respir. Dis.* **113:**1171–1180.

Mullen, B. J., Wiggs, B. R., Wright, J. L., Hogg, J. C., and Pare, P. D., 1986, Nonspecific airways reactivity in cigarette smokers, *Am. Rev. Respir. Dis.* **133:**120–125.

Nadel, J. A., 1965, Structure and functional relationship in airways, *Med. Thor.* **22:**231.

Naylor, B., 1972, The shedding of the mucosa of the bronchial tree in asthma, *Thorax* **17:**69.

Orehek, J., 1983, The concept of airway "sensitivity" and "reactivity," *Eur. J. Respir. Dis.* **64** (Suppl. 131):27–48.

Orehek, J., Massari, J. P., Gayrard, T., Grimaud, C., and Sharpin, J., 1976, Effect of short term, low level nitrogen dioxide exposure on bronchial sensitivity as asthmatic subjects, *J. Clin. Invest.* **5:**7301–7307.

Parker, C. D., Bilbo, R. E., and Reed, C. E., 1965, Methacholine aerosol as a test for bronchial asthma, *Arch. Intern. Med.* **115:**452–458.

Schellenberg, R. R., and Foster, A., 1984, In vitro responses to human asthmatic airways and pulmonary vascular smooth muscle, *Int. Arch. Allergy Clin. Immunol.* **75:**237–241.

Sobonya, R. E., 1984, A concise clinical study of quantitative structural alterations in long-standing allergic asthma, *Am. Rev. Respir. Dis.* **132:**89.

Stevens-Newman, L., 1987, Airway smooth muscle, *Am. Rev. Respir. Dis.* **135:**960–975.

Taylor, S. M., Pare, P. D., More, C. L., Hogg, J. C., and Schellenberg, R. R., 1985, Airway reactivity in chronic obstructive pulmonary disease, *Am. Rev. Respir. Dis.* **132:**30–35.

Thompson, J. E., Scypinski, L. A., Gorden, T., and Sheppard, D., 1986, Hydroxyurea inhibits airways hyperresponsiveness in guinea pigs by a granulocyte independent mechanism, *Am. Rev. Respir. Dis.* **123:**1213–1216.

Vincenc, K. S., Black, J. L., Yan, K., Armour, C. L., Donnelly, P. D., and Woolcock, A. J., 1983, Comparison of *in vivo* and *in vitro* responses to histamine in human airways, *Am. Rev. Respir. Dis.* **128:**875–879.

Westfall, D. P., 1981, Supersensitivity of smooth muscle, in: *Smooth Muscle: An Assessment of Current Knowledge* (E. Bulbringe, A. F. Braiding, and E. W. Jones, eds.), pp. 285–309, University of Texas Press, Austin.

Chapter 14

Current Concepts of the Pathophysiology of Allergic Asthma

Ruth Jacobs and Michael Kaliner

Allergic Diseases Section
National Institute of Allergy and Infectious Diseases
National Institutes of Health
Bethesda, Maryland 20205

I. INTRODUCTION

The word *asthma* was derived from the Greek word for "panting," or "breathlessness" and thus might be considered a description of a symptom rather than a diagnosis. Although medical advances of the nineteenth and twentieth centuries have established asthma as a diagnostic entity, it is recognized that this disease is heterogeneous and difficult to classify. Asthma can be defined clinically as a recurrent disease causing intermittent wheezing, breathlessness, and cough with sputum production. Pathologically, asthma is a disease of airflow obstruction caused by combinations of bronchial smooth muscle contraction, mucosal edema and inflammation, and viscous mucin secretion. The disease involves large and small airways but not alveoli, and the processes leading to airflow obstruction are reversible, either spontaneously or in response to medication.

Allergy is the most frequent cause of asthma, accounting for up to 90% of asthmastics. The basic events in allergic reactions are the development of immunoglobin E (IgE) antibodies directed at ordinarily harmless antigens (e.g., pollens, dust, or danders); fixation of these IgE antibodies to the surface of mast cells found in the connective tissue throughout the body but in highest concentration in the mucous membranes and skin; stimulation of mast cell; degranulation upon exposure to the sensitizing antigen (as occurs during pollen exposure); release of a diverse group of chemical mediators into the connective tissue, where they initiate events leading to the allergic reaction; and the subsequent development of asthma, if the event occurs in the airways. There are a number of other causes of asthma, summarized in Table I. The differential diagnosis of wheezing is presented in Table II.

II. ALLERGIC ASTHMA

A. Diagnosis

One should suspect allergic asthma when (1) there is a family history of allergic disease; (2) the clinical presentation shows seasonal exacerbations; (3) there is concomitant allergic rhinitis, eczema, or other allergic disease; (4) mild to moderate eosinophilia is present (300–1000 mm^3) or eosinophilia in sputum is observed; or (5) the patient is under 40 years of age. Skin testing confirms the suspicion of IgE directed against the incriminated allergen but does not confirm the cause-and-effect relationship. Allergen bronchial provocation will be positive in subjects with positive skin tests and a history of seasonal asthma but is rarely indicated as a diagnostic test. IgE levels are useful but not diagnostic. About 60% of allergic asthmatic subjects have IgE levels above 300 ng/ml, while the remaining 40% are within the normal range.

B. Epidemiology

The basic pathophysiologic events leading to asthma may be modified by the patient's age, sex, emotional stability, recent exposure to viral infections,

TABLE I
Causes of Asthma[a]

Allergy
    Allergen-induced asthma
    Allergic bronchopulomonary aspergillosis
Infections
    Bronchiolitis
    Upper respiratory infections
Industrial–occupational exposure
    Irritants
    Allergens
Chemical or drug ingestion
    Nonsteroidal anti-inflammatory drugs (NSAIDs)
    Sulfiting agents
    β-adrenergic antagonists
Exercise
Vasculitis
Idiopathic (intrinsic)

[a]Modified from Kaliner and McFadden (1988).

TABLE II
Differential Diagnosis of Asthma[a,b]

Nonasthmatic conditions associated with wheezing
    Pulmonary embolism
    Cardiac failure
    Foreign bodies[b]
    Tumors in central airways
    Aspiration (gastroesophageal reflux)[b]
    Carcinoid syndrome
    Chondromalacia[b]
    Löffler syndrome
    Bronchiectasis
    Tropical eosinophilia
    Hyperventilation syndrome
    Laryngeal edema
    Laryngeal or tracheal obstruction[b]
    Factitious wheezing
    α-Antitrypsin deficiency
    Immotile cilia syndrome, Kartagener syndrome[b]
    Bronchopulmonary dysplasia[b]
    Bronchiolitis, croup[b]

Overlapping disease
    Chronic bronchitis and emphysema
    Cystic fibrosis[b]

[a]Modified from Kaliner and McFadden (1988).
[b]Especially important in differential diagnosis in children.

allergen exposure, work or home environment, and genetic makeup. Prevalence estimates for asthma in children range from nearly 0 in Gambia to more than 25% among Tokelau Islanders living in New Zealand (Godfrey, 1975; Waite *et al.*, 1980). The disease is rare among adults in Sweden (0.2%) and quite common in adults in Arizona (7.9%) (Dodge and Burrows, 1980; Lebowitz *et al.*, 1975; J. M. Smith, 1983). Epidemiology studies in the United States suggest a prevalence of asthma of 3–6.6% (Broder *et al.*, 1974; Dodge and Burrows, 1980; Lebowitz *et al.*, 1975; Wilder, 1973). According to 1986 census data, up to 18 million Americans may have asthma.

Asthma may be more common among blacks in the United States than among whites. Both the U.S. Public Health Survey and a study in Baltimore showed a higher prevalence of asthma in block schoolchildren than among their white counterparts (Wilder, 1973), even after adjusting for differences in socioeconomic status (Mak *et al.*, 1982; Wilder, 1973). Hiss (1966), however, was unable to find any differences in asthma prevalence between white and black children in Houston. The higher rates among blacks, if true, cannot be explained on racial grounds alone. Studies of tribal groups in nonwhite developing countries demonstrate a low prevalence of asthma, but document major increases in asthma in the same tribal groups when they move from rural to urban areas (Cookson and Makoni, 1980; Godfrey, 1975; Van Niekerk *et al.*, 1979). In the United States, asthma is uncommon among American Indians, Eskimos, and, perhaps, Mexican-Americans.

Allergic asthma is most common in childhood and early adulthood. In a study performed in Arizona (Dodge and Burrows, 1980), the incidence of asthma was highest in young children, decreased in late adolescence, and increased again in the early adult years. In the young, the onset of asthma was strongly associated with allergy skin-test reactivity. Persons in whom asthma developed after age 40, however, frequently had prior symptoms of chronic bronchial irritation, blurring the distinction between asthma and chronic bronchitis. Hence, the high prevalence reported in older adults in some studies may reflect imprecise application of diagnostic criteria.

C. Natural History, Morbidity, and Mortality

The natural history of asthma is not well established. Cross-sectional and longitudinal studies support the view that the recurrent episodes of obstruction associated with this illness do not lead to permanent lung dysfunction. Thus, asthma differs from other obstructive airway disease (e.g., chronic bronchitis or emphysema), in that its course is not relentlessly progressive.

In general, the onset of asthma in childhood has a somewhat better prognosis than when the disease begins in adulthood. In Rackemann and Edward's study of childhood asthma (1952), 50% of the patients lost their disease over the 20-year follow-up period, and another 25% remained well by avoiding known exacerbating factors. A remission of at least 3 years can be expected in 25–30% of children with asthma (Blair, 1977). Nonetheless, most persons who appear to have outgrown their asthma still have abnormal airway reactivity and may experience recurrences as adults (Martin *et al.*, 1980).

The death rate from asthma is low, averaging approximately 0.3 deaths per 10,000 persons. Typically, 2000–3000 deaths—mostly among the very young and the elderly—are reported yearly (National Institute of Allergy and Infectious Diseases Task Force Report, 1979). Data from 1983 and 1984 suggest that the death rate may be increasing slightly for reasons that are not yet clear. Blacks, particularly inner-city blacks, have an increased death rate compared with whites, with a risk factor approaching threefold.

Unlike mortality, the morbidity associated with asthma is staggering. It is one of the most common chronic diseases of childhood and is the leading cause among chronic illnesses of school absences. Asthma results in an estimated 27 million patient visits, 6 million lost work-days, and 90.5 million days of restricted activity per year. According to pharmaceutical industry figures, $362 million was spent for asthma medications in 1983.

D. Inheritance

"Asthmatics beget asthmatics." When differentiating asthma from other obstructive airways diseases, it is always relevant to ask whether family members experience the same symptoms. If the asthmatic is atopic, familial concordance is very likely to be present, while nonallergic asthmatics have less tendency for genetic transmission. Cooke and Veer (1916) recognized that 48.4% of allergic subjects had a positive family history, in contrast to 14% for 76 control subjects. Schwartz (1952) compared the family histories of 191 asthmatic patients with those of 200 matched controls and found a significantly increased incidence of allergy in the close family members of his patients. The conclusion was that asthma is a genetic disease.

In addition to these studies, Van Arsdale and Motulsky (1959) demonstrated that if both parents are atopic, 58% of the offspring were likely to be atopic, while if only one parent was atopic, only 20% of the offspring were allergic. Only 6% of offspring of nonatopic parents were allergic. Not just the inheritance of atopy is genetic, but also the manifestation of atopy. Thus, asthmatics have asthmatic children, while rhinitics have children with rhinitis (Gerrard, 1976).

III. PATHOPHYSIOLOGY OF ALLERGIC ASTHMA

On inhalation antigen challenge, the allergic asthmatic may display an immediate response reflected by the development of airway obstruction, an immediate followed by late obstructive response or rarely an isolated late response (Bhagat et al., 1985) (Fig. 1). An immediate asthmatic response is usually characterized by pulmonary obstruction beginning within minutes after exposure to an antigen, peaking within 10–30 min, and spontaneously resolving with 1–2 hr. A late asthmatic response has its onset 2–4 hr after exposure, peaks within 5–12 hr, and may cause symptoms persisting for one to several days.

Pathological studies of the immediate- and late-phase allergic reactions in the airways are unavailable in humans but have been studied in experimental animals, including the sheep and rabbit. In the rabbit airway, tissue edema and dilation of blood vessel are prominent features of the immediate allergic response, followed by the development of an acute inflammatory infiltrate over 4–6 hr. The inflammation and accompanying edema are the prominent feature of the late-phase reaction in the rabbit (Wilson and Larson, 1986).

Pathological findings at autopsy in humans dying of status asthmaticus shows changes reflective of both the acute injury as well as years of chronic disease. At autopsy, lungs from patients dying from asthma are hyperinflated and tend not to collapse postmortem, reflecting the presence of fixed airway obstruction due to plugs of mucus filling the segmental and subsegmental airways and bronchioles. The mucus plugs contain large numbers of epithelial cells, eosino-

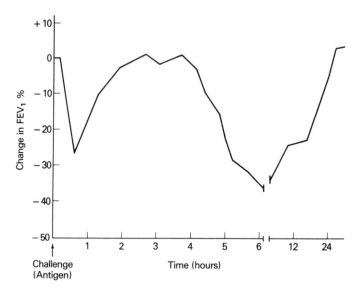

FIGURE 1. Pulmonary responses to antigen challenge. Inhalational bronchial provocation with specific antigen in atopic subjects provokes allergic reactions that include an immediate airway obstruction apparent within 15–30 min that usually resolves by 60–120 min and a late-phase reaction that develops over 4–8 hr after challenge. The airflow obstruction in the late-phase reaction is maximal by 8–12 hr and usually resolves by 24–48 hr. The initial airflow obstruction is caused by a combination of airway edema, mucus secretion, and muscle spasm, while the late-phase obstruction is caused by edema and cellular inflammation.

phils, and neutrophils as well as Charcot–Leydin crystals, Creola bodies, and Curschman's spirals. The submucous glands are hypertrophied, and the goblet cells are hyperplastic. The presence of an increased amount of albumin in the airway mucus found in asthmatics suggests an ongoing inflammatory process with transudation of serum proteins. In addition, the basement membrane thickness is increased from 7 μm in normals to an average of 17.45 μm in asthmatics (Callerame *et al.*, 1971). Inflammation and edema of the lamina propria of the segmental and subsegmental airways along with edema is prominant while the airway smooth muscle may be hypertrophied. The presence of concertina-like infoldings of the airways suggests that the smooth muscle might have been contracted *in vivo*.

A. Antigen Entry

The actual path of entry of antigen into the lung is yet to be determined. Ragweed pollen is about 20 μm in diameter, tree pollens vary from 20 to 60 μm, and grass pollens are mostly 30–40 μm. In order for particles in the inhaled airstream to reach the airways, they must be less than 5 μm in size. Intact pollen grains are unable to penetrate to the airway, but recent studies indicate that fragments of pollen can and do reach the airways (Agrawal *et al.*, 1983).

Pseudostratified columnar epithelium covers the major bronchi, thereafter becoming simple cuboidal. The cells are held together by terminal bars known as tight (zonula occludens) and intermediate (zonula adherens) junctions extending around the lateral surface of cells adjacent to the airway lumen. The terminal bars provide a normally impermeable barrier to foreign substances. However, these junctions may be disrupted by a variety of factors, including tobacco smoke (Richardson *et al.*, 1976; Boucher *et al.*, 1980); activation of mast cells in the bronchial lumen; viral infection leading to epithelial disruption; or by airway inflammation as occurs after either ozone exposure or as part of the late-phase allergic reaction. Thus, mechanisms exist to facilitate the transport of potential antigens from the bronchial lumen, across the epithelium, and into the lamina propria, where most mast cells are found. In addition, potentially antigenic substances in excess of 40,000 M_r may be directly transported by epithelial cells without requiring any disruption of the epithelial tight junctions (Richardson *et al.*, 1976).

B. Pulmonary Mast Cells

Of the cells found in the airway lumen of humans, 0.5–2% are mast cells. Antigen stimulation of luminal mast cells causes increased epithelial permeability to antigen in guinea pigs (Richardson *et al.*, 1973). Most pulmonary mast cells are found in the loose connective tissue beneath the airway basement membrane (Lamb and Lumsden, 1982), near submucosal blood vessels (Kawanami *et al.*, 1979; Trotter and Orr, 1973; Casale and Kaliner, 1984), in muscle bundles (Casale and Kaliner, 1984), and in the intra-alveolar septa (Casale and Kaliner, 1984; Fox *et al.*, 1981). Approximately 40% are located in pulmonary parenchyma and 60% in the airways. In alveolar tissue there may be an average of 350 mast cells per mm^2, and the mast cells may comprise up to 2% of the alveolar wall (Fox *et al.*, 1981).

C. Mast Cell Stimulation and Mediator Release

After antigen-induced activation, a variety of mast cell mediators are released. Currently recognized mast cell-derived mediators are listed in Table III. These chemical mediators are either preformed and rapidly released from mast cells or are a stable component of the granule matrix that persists in the connective tissue for some time after degranulation. The pathological consequences of mediator release in the airway include both immediate- and late-phase asthmatic reactions. These mediators are too diverse a group of mast cell mediators to be discussed individually as to their possible roles in asthma. Therefore, it is easier to discuss these mediators in the context of pathophysiologic events responsible for causing the airflow obstruction.

TABLE III
Mast Cell-Derived Mediators[a]

Preformed, rapidly released under physiologic conditions
  Histamine
  Eosinophil chemotactic factors of anaphylaxis (ECF-A)
  Neutrophil chemotactic factors (NCF)
  Kininogenase
  Arylsulfatase A
  Serotonin[b]
  Exoglycosidases[b] (β-hexosaminidase, β-D-Galactosidase, β-glu-
    curonidase)

Secondary or newly generated mediators
  Superoxide and other reactive oxygen species
  Leukotrienes C_4, D_4, E_4 (previously known as SRS-A)
  Prostaglandins
  Monohydroxyeicosatetraenoic acids (HETEs)
  Hydroperoxyeicosatetraenoic acids (HPETEs)
  Hydroxyheptadecatrienoic acid (HHT)
  Thromboxanes
  Prostaglandin-generating factor of anaphylaxis (PGF-A)
  Adenosine
  Bradykinin platelet-activating factor (PAF)[b]

Granule-associated mediators
  Heparin or other proteoglycan
  Tryptase
  Chymotryptic proteinase
  Arylsulfatase B
  Inflammatory factors of anaphylaxis (IF-A)[b]
  Peroxidase[b]
  Superoxide dismutase[b]

[a]Modified from Kaliner and McFadden (1988).
[b]Demonstrated in nonhuman mast cells.

IV. PATHOLOGICAL PROCESSES IN ALLERGIC ASTHMA

A. Bronchospasm

It is difficult to conceive of any mechanism of airway obstruction other than spasm of the muscle layers encircling the airways that would be capable of the rapid reversal seen clinically in allergic asthmatics undergoing an immediate phase response to antigen challenge. Of the mast cell-derived mediators, histamine, leukotrienes C, D, and E (LTC, LTD, and LDE), prostaglandins PGG_2, $PGF_{2\alpha}$, and PGD_2 and thromboxane A_2 (TXA_2), platelet activating factor (PAF), and bradykinin are capable of causing bronchial smooth muscle constriction.

Two cellular receptors for histamine have been identified, designated H_1

and H_2. Histamine causes airway obstruction through stimulation of H_1 receptors on muscle fibers. Analysis of histamine-induced constriction *in vivo* demonstrates that histamine acts directly on muscles (Rosenthal *et al.*, 1977) and, in addition, may have a vagally mediated reflex parasympathetic action (Yu *et al.*, 1972).

The biologically active moiety previously referred to as slow-reacting substances of anaphylaxis (SRS-A) is now known to be composed of leukotrienes C, D, and E. Leukotrienes are a series of closely related conjugated trienes derived from arachidonic acid through the lipoxygenase pathway. *In vivo* experiments comparing the effects of SRS-A on airway resistance and lung compliance, as well as comparisons of the effects of LTC and LTD on tracheal and parenchymal strips, indicate that these agents are more active in contracting small airways and peripheral lung tissue than in large airways (Spannhake *et al.*, 1981). Studies on human bronchial muscle (Dahlén *et al.*, 1980) indicate that LTC and LTD are 1000 times more potent than histamine and 500 times more potent than $PGF_{2\alpha}$ on a molar basis. Cyclo-oxygenase inhibitors may enhance allergic tracheal contraction by diverting arachidonic acid through the lipoxygenase pathway toward the production of SRS-A or related bronchoconstrictive leukotrienes (Paterson *et al.*, 1981; Burka and Paterson, 1980; Hitchcock and Kokolis, 1981) or of LTD suggested that the leukotrienes were 3800 times more potent than histamine in this action. These studies indicate that in normal humans, LTC and LTD are the most potent bronchoconstrictor substances yet described. Their 15- to 20-min duration of action is consistent with a possible role in producing IgE-mediated bronchoconstriction (Weiss *et al.*, 1982).

Prostaglandins are generated during the course of anaphylaxis in human (Platshon and Kaliner, 1978) and guinea pig lung (Dawson *et al.*, 1976). Prostaglandins PGG_2, PGF_2, and $PGGD_2$, and TXA_2 produce smooth muscle contraction. PGE and PGI_2 produce smooth muscle relaxation. The balance between these opposing influences may affect bronchial tone and play a role in allergic bronchospasm.

The appearance of histamine in lung supernatants after immunological challenge is closely paralleled by the generation of $PGF_{2\alpha}$ and PGE (Kaliner, 1980). Histamine through stimulation of H1 receptors is responsible for about 50% of the prostaglandins generated during anaphylaxis (Platshon and Kaliner, 1978). A novel mediator of lung anaphylaxis, termed prostaglandin-generating factor of anaphylaxis (PGF-A), is an oligopeptide of 1450 daltons that induces the production of $PGF_{2\alpha}$, PGE, TXB_2 and 5-, 12-, and 15-hydroxyeicostatetraenoic acid (HETE) from human lung parenchyma and airways. Experimental evidence suggests that this factor is not preformed (like histamine) but newly synthesized or rapidly generated by the anaphylactic event (like leukotrienes) (Steel and Kaliner, 1981).

Platelet-activating factor causes contraction of parenchymal lung strips, as well as large airway contractions and is a possible important contributor to allergen-induced bronchospasm. Its chemical structure is 1-alkyl-acetyl-*sn*-glycero-3-phosphocholine. PAF is newly synthesized by rabbit basophils and rat and mouse marrow-derived mucosal mast cells in culture (but not by human basophils) during activation (Betz *et al.*, 1980; Mencia-Huerta and Benveniste, 1979). Its role in human anaphylaxis has not been entirely delineated.

Finally, a kininogenase released from partially purified human lung mast cells in a dose-dependent fashion by anti-IgE has been reported (Proud *et al.*, 1985). The kininogenase that is released may participate in IgE-mediated asthma by the production of kinins. Released kinins may contract bronchial smooth muscle, increase vascular permeability, and mediate PG formation.

B. Mucosal Edema

Edema of airway mucosa is caused by increased capillary permeability with leakage of serum proteins into interstitial areas. Histamine, leukotrienes, PGE, PAF, and bradykinin are all capable of causing increased capillary permeability.

C. Mucus Secretion

Mucus secretion is a feature of both immediate- and late-phase reactions in asthma. The primary sources of the mucus component of these secretions are the submucosal glands and goblet cells. Parasympathetic fibers innervate submucous glands, stimulating smooth muscle contraction and glandular secretion. Asthma is characteristically associated with hyperplasia of goblet cells that, in contrast to mucous glands, appear devoid of innervation (Kaliner, 1980).

Pathological examinations of fatal cases of status asthmaticus almost always reveal diffuse secretions of mucus, which appear to contribute significantly to the airway obstruction (Messer *et al.*, 1960). The precise mechanisms responsible for increased mucus production are unclear, although the association between allergic pulmonary reactions and mucorrhea suggests that immediate hypersensitivity reactions may lead to the release or generation of secretagogues. Analysis of the relationship between allergic pulmonary reactions and mucorrhea suggests that immediate hypersensitivity reactions may lead to the release or generation of secretagogues. Analysis of the relationship between allergic pulmonary reactions and the release of mucous glycoproteins indicates that the order of potency of the mediators of anaphylaxis on mucus secretion is LTD4 $\geq$ LTC4 > HETEs > $PGF_{2\alpha}=PGD_2=PGI_2=PGE_1=PGA_2$ > histamine stimulation (Shelhamer *et al.*, 1982). In addition, stimulation of cyclic guanosine monophosphate (cGMP) formation in human airways by cholinergic and α-adrenergic neurohormones leads to increased mucous glycoprotein release (Shelhamer *et al.*, 1980a).

Finally, macrophages may also play a role in mucus secretion by producing a low-molecular-weight product (2400 daltons) during phagocytosis (Marom *et al.*, 1984). This molecule is not lipid or a prostaglandin and has been termed macrophage mucus secretagogue.

D. Late-Phase Reactions

Although the identity and precise role of the mast cell mediators involved in the late-phase reaction (LPR) are controversial and under active investigation, there is no controversy as to the role of the mast cell in initiating the process. Cellular infiltrates are the hallmark of the late-phase reactions. Mediators released or generated by the mast cell lead to an influx of neutrophils and eosinophils at the site of mast cell degranulation (Lemanske and Kaliner, 1982). The initial polymorphonuclear infiltrate is then replaced by a more chronic inflammatory reaction characterized by mononuclear cells (Lemanske et al., 1983). Mediators that are candidates for causing LPR include inflammatory factors of anaphylaxis, several chemotactic factors capable of attracting polymorphonuclear leukocytes, leukotriene B4, PAF, and probably other unrecognized mediators as well (Tables IV and V).

TABLE IV
Pathological Changes in Asthma and the Mediators Possibly Responsible[a]

Pathological changes	Mediators responsible
Bronchial smooth muscle contraction	Histamine (H_1 response)
	LTC_4, LTD_4, LTE_4
	Prostaglandins and thromboxane A_2
	Bradykinin
	Platelet activating factor (PAF)
Mucosal edema	Histamine (H_1 response)
	LTC_4, LTD_4, LTE_4
	PGE
	Bradykinin
	PAF
Cellular infiltration (airway hyperreactivity)	Inflammatory factors of anaphylaxis (IF-A)
	Eosinophil chemotactic factors
	Neutrophil chemotactic factors
	HETEs
	LTB_4
	PAF
Mucus secretion	Histamine (H_2 response)
	Prostaglandins
	HETEs
	LTC_4, LTD_4, LTE_4
Desquamation	O_2, H_2O_2, OH^-
	Proteolytic enzymes
Basement membrane thickening	O_2^- proteolytic enzymes

[a]Modified from Kaliner and McFadden (1988).

TABLE V
Putative Mediators Causing Late-Phase Reaction[a]

Inflammatory factors of anaphylaxis	Chemotactic *in vitro*
Low molecular weight	Attract PMN *in vivo* at less than nano-
High molecular weight (appears to	gram quantities
be the same molecule as neu-	Confirmed in rodents and monkeys
trophil chemotactic activity)	but not in humans
Neutrophil chemotactic activity (NCA)	Chemotactic *in vitro*
	Found in human plasma during LPR
Platelet activating factor	Cause inflammation *in vivo* in rodents
	and humans
LTB$_4$	Chemotactic for human PMN *in vitro*
	Cause inflammation *in vivo* in humans

[a]LPR, late-phase reaction; PMN, polymorphonuclear cell.

Clinically, LPR is always characterized by swelling, often of an angioedematous rather than urticarial nature. In the airways, this swelling is poorly responsive to bronchodilators, suggesting that smooth muscle spasm is not an important component. Assessment of venous plasma in asthmatics experiencing LPR after antigen or exercise provocations (Nagy *et al.*, 1982; Atkins *et al.*, 1984) demonstrates that LPR is accompanied by a significant release of histamine. Nasal lavage fluids during nasal LPR contain histamine and other mediators of allergy but not PGD$_2$ (Naclerio *et al.*, 1985). It has been suggested that basophils may be the source of histamine release, as basophils are unable to generate PGD$_2$ but are capable of releasing the other mediators detected in nasal lavage (Naclerio *et al.*, 1985). In support of this possibility, basophils have been detected in nasal secretions after allergic responses but are not generally seen in the lamina propria or even in the epithelial layers of the nose (Okuda and Otsuka, 1977; Hastie *et al.*, 1979; Friedman and Kaliner, 1985).

Alternatively, inflammatory cells attracted to the site of mast cell degranulation may cause renewed mast cell mediator release. White described human neutrophil-derived histamine releasing activity (termed HRA-N) generated spontaneously from neutrophils (White and Kaliner, 1985). HRA-N is capable of causing degranulation of rat basophilic tumor maintained in culture (the RBL tumor line), in human basophils, in mouse bone marrow-derived mast cells, and in guinea pig skin mast cells *in vivo* (White *et al.*, 1986). The activity of HRA-N is independent of the presence of IGE, is non-cytotoxic, and is calcium dependent. In addition to HRA-N, eosinophil-derived peroxidase (Henderson *et al.*, 1980) and major basic protein (Zheutlin *et al.*, 1984) can also stimulate mast cell degranulation, as can lymphocyte (Thueson *et al.*, 1979) and mononuclear cell-derived (Schulman *et al.*, 1985) releasing factors. Thus, there now exists evidence suggesting that inflammatory cells attracted to the site of mast cell degranulation may cause renewed mast cell mediator release.

One might conceive of a cycle of mast cell degranulation, inflammatory cell infiltration, renewed mast cell degranulation, and so on. Controlling influences that dampen this cycle must exist, as LPR is usually self-limited to 1 or 2 days.

E. Desquamation of Surface Epithelium, Thickening of the Basement Membrane, and Goblet Cell Hyperplasia

Denudation of airway epithelial surfaces with appearance of epithelial clumps (Creola bodies) in expectorated secretions accompanies severe asthma. The denuded epithelial surfaces may be replaced by goblet cells, resulting in goblet cell hyperplasia. Mechanisms possibly responsible for these changes may include production of superoxide anion by mast cells undergoing degranulation (Henderson and Kaliner, 1978), which may lead to the production of H_2O_2, OH-, singlet oxygen, and other highly destructive oxygen radicals. These molecules are able to damage cell membranes and may thereby contribute to denudation of epithelium. Granule matrix-derived proteolytic enzymes may also help release epithelial cells from the basal layer by weakening intercellular bonds.

In addition, the eosinophilic infiltration associated with the late-phase allergic reaction may contribute to epithelial desquamation. Bronchoalveolar lavage of patients with late asthmatic reactions showed a significant eosinophilia in patients experiencing the late-phase reaction compared with control groups. Eosinophil cationic protein–albumin ratios were elevated in the bronchoalveolar lavage fluid from patients with late asthmatic reactions (De Monchy *et al.*, 1985). Filley and co-workers (1982) demonstrated large amounts of eosinophil major basic protein (MBP) deposited in the airways of patients dying in status asthmaticus. MBP has cytotoxic properties and in guinea pigs is cytotoxic to the epithelium (Frigas *et al.*, 1980). MBP could therefore be one factor that contributes to the characteristic epithelial shedding of asthma (Laitinen, 1985).

V. AIRWAY INFLAMMATION AND HYPERREACTIVITY

Bronchial hyperreactivity is a fundamental component of asthma and is found universally in asthmatics. Asthmatics develop airway obstruction after natural exposure to cold air, exercise, irritating chemicals, laughing or coughing, or laboratory provocations with histamine, methacholine, $PGF_{2\alpha}$, adenosine, ultrasonic mist, serotonin, bradykinin, or leukotrienes. The severity of asthma is directly proportional to the degree of increased bronchial responsiveness, and some degree of abnormality may remain long after asthma symptoms have disappeared. Airway hyperreactivity appears to be a prerequisite for the development of asthma, but can be present in normal subjects who never become asthmatic. Five percent of normal subjects are highly reactive to methacholine, and 8% have medium positive responses (Townley and Bewtra, 1983). A sizable portion of asthmatic family members have abnormal airway responsiveness in the absence of clinical asthma and, if followed, are found to be at increased risk of the eventual development of asthma. In fact, 18% of normal nonatopic individuals with a family history of asthma (Townley and Bewtra, 1983) had medium positive responses to methacholine. In a 4- to 7-year follow-up of five families, including 56 subjects (Townley, 1982), four subjects developed *de novo* asthma. Each of these subjects had developed an abnormal methacholine bronchial provocation response before the onset of asthma. We can conclude that the presence of airway reactivity is a necessary setting for asthma to develop, but the presence of airway reactivity itself is not sufficient to produce clinical asthma.

Both epithelial damage and airway inflammation, such as that found following the late-phase reaction, have been implicated in the bronchial hyperreactivity that accompanies allergic asthma (Cockcroft *et al.*, 1984). *De novo* production of airway hyperreactivity has been studied using ozone exposure. Ozone can cause airway hyperreactivity in about 50% of exposed human subjects or dogs. In animals, ozone-induced airway hyperreactivity is associated with an acute neutrophil-rich inflammatory response (Holtzman *et al.*, 1983; Snapper and Brigham, 1984). Introduction of neutropenia prevents ozone-induced airway hyperreactivity, suggesting that the neutrophil is a critical determinant. Treatment with nonsteroidal anti-inflammatory drugs (NSAIDs) to prevent PG production also prevents the induction of ozone-hyperreactivity (O'Byrne *et al.*, 1984). This finding would suggest that acute airway inflammation in response to ozone leads to neutrophil infiltration and associated airway hyperreactivity and that PG formation is essential for the hyperreactivity to occur. The degree of airway hyperreactivity induced by ozone exposure appears to reflect the development of epithelial inflammation.

A further role for epithelial damage in the pathogenesis of bronchial hyperreactivity is suggested by the dramatic increases in airway reactivity that may occur following certain viral infections. The common cold (Empey *et al.*, 1976) as well as croup (Zach *et al.*, 1980), rhinovirus (Halpern *et al.*, 1983), and influenza (Minor *et al.*, 1976) infections can induce airway hyperreactivity that persists for days and weeks.

The most likely mechanism by which epithelial damage or inflammation could induce hyperreactivity involves increased responsiveness of sensory nerves in the epithelium. Opening of tight junctions in airway epithelium has been found to occur after cigarette smoke, methacholine, histamine, and antigen exposure. Vagal sensory nerve endings lie beneath the tight junctions (Widdicomb, 1977), and exposure of these endings to irritants can cause exaggerated vagal reflex reactions.

The immediate airway response to inhaled allergen appears to be a function of two components: the degree of nonspecific airway reactivity, and the responsiveness of mast cells to allergen. Asthmatic patients with very responsive (irritable) airways require only a small degree of mast cell activation (as measured by increases in venous histamine and neutrophil chemotactic factor) to produce a major airway response, whereas subjects with less abnormal airway reactivity require up to threefold greater mediator release to produce a similar airway response (Holgate et al., 1985).

VI. OTHER MODULATORS OF AIRWAY RESPONSIVENESS

Vagal influences, abnormal autonomic nervous system responsiveness, nonimmunological mast cell secretagogues, and modulation of the various arachidonic acid metabolites (PGs, HETEs, and leukotrienes) may participate in determining the degree of airway reactivity. Pulmonary innervation consists of parasympathetic, adrenergic, and noncholinergic nonadrenergic efferent fibers. In the human, the predominant influence is parasympathetic. Resting bronchial tone is under parasympathetic control. Reflex cholinergically mediated bronchospasm can be elicited by emotional stress, gastric or esophageal irritation, or nasosinopulmonary reflexes. Instillation of dilute HCl onto the lower esophagus regularly causes bronchospasm in asthmatics with symptomatic gastroesophageal reflux (Mansfield and Stein, 1978; Spaulding et al., 1982). Kratschmer (1928) showed that chemical irritation of the nasal mucosa can led to bronchoconstriction in the cat or rabbit. Exposure of the nasal mucosa of normal subjects to silica particles for 150 sec causes significant increases in airway resistance (Kaufman and Wright, 1969). Pretreatment of these subjects with atropine prevents these responses, suggesting that the operative mechanism involves vagally mediated cholinergic reflexes. More recently, cold challenge to the nose resulted in increased airway resistance (Nolte and Berger, 1983), while nasal histamine challenges caused a greater than 20% drop in FEV_1 in 50% of subjects (Yan and Salome, 1983). In one study, treatment of concomitant sinus disease in asthmatics led to subjective improvement in 28 of 33 subjects and to reduction of steroid requirement in 15 of 18 subjects (Slavin, 1984).

Patients with allergic asthma have been found to have abnormal autonomic nervous system responses. Allergic asthma has been associated with β-adrenergic hyporeactivity (Shelhamer et al., 1980b) and with cholinergic (L. J. Smith et al., 1980; Makino et al., 1970) and α-adrenergic (Henderson et al., 1979) hypersensitivity. Both β-adrenergic antagonists (propranolol) (McNeil, 1964) and α-adrenergic agonists (Anthracite et al., 1977; Snashall et al., 1978) produce bronchoconstriction in asthmatic patients but not in normal patients. The etiology of this autonomic imbalance is unclear. Some in vitro studies have demonstrated reduced numbers of β-adrenergic receptors on lymphocyte membranes from asthmatics; these lymphocytes have been found to generate less cyclic adenosine monophosphate (cAMP) in response to β-adrenergic agonists (Brooks *et al.*, 1979; Kariman, 1980; Parker and Smith, 1973). Other studies have shown that reduced leukocyte β-adrenergic responsiveness may occur only after continuous treatment with β-adrenergic agonists (Conolly and Greenacre, 1976). Antibodies directed at β-adrenergic receptors have been found in the serum of 5% asthmatics, and the presence of these antibodies strongly corresponds to the degree of β-adrenergic responsiveness expressed in these patients. A few normal subjects as well as nonasthmatic atopics also have low titers of these autoantibodies (Fraser *et al.*, 1981).

The precise contribution of these autonomic nervous system abnormalities to asthma is unclear. However, the functional imbalance of the autonomic nervous system is likely to influence three areas of the asthmatic's response: (1) bronchial smooth muscle dilation and response to β-adrenergic stimulation would be reduced, while cholinergically and α-adrenergically mediated constriction would be augmented; (2) mast cell mediator release, generally suppressed by β-adrenergic stimulation, would be resistant to β-adrenergic agonists, while both cholinergic and α-adrenergic enhancement of mediator release would be exaggerated; and (3) increased mucus secretion in response to α-adrenergic and cholinergic stimulation would be augmented, while sodium and water fluxes into tracheobronchial secretions in response to β-adrenergic stimulation would be reduced.

A number of compounds are known mast cell secretagogues. What is unknown is their precise role in modulating airway hyperreactivity. Compounds inducing noncytotoxic degranulation of mast cells *in vivo* include the anaphylatoxins C3A, C4A, and C5A; adenosine triphosphate; hypoxia; releasing factors produced by neutrophils, eosinophils, lymphocytes, and macrophages; muscle-depolarizing agents such as succinyl choline and D-tubocurare; enzymes found in bee and snake venoms; and neuropeptides, including the opiates.

In recent years, studies have suggested a role for neuropeptides as mast cell secretagogues. Neuropeptides may influence pulmonary function either by directly affecting glandular secretion, smooth muscle cell contraction, blood vessel dilation and permeability, and leukocyte function or by indirectly modulating secretion of mediators from mast cells or basophils. Substance P (SP), vasoactive intestinal polypeptide (VIP), neurotensin (NT), and somatostatin (SOM) are among the neuropeptides that have been studied. In humans, substance P-like immunoreactivity is present in airway sensory nerve fibers (Pernow, 1983; Lundberg *et al.*, 1984). Vasoactive VIP-like immunoreactive nerves have been observed in the smooth muscle layers of the airways from secondary bronchi to small bronchioles and in bronchial glands. VIP-reactive nerve fibers are also found surrounding nerve cell bodies in the microganglia of the lower respiratory tract. In addition, VIP-positive nerve cell bodies are present in the ganglia of trachea and of major bronchi (Dey *et al.*, 1981; Ghatei *et al.*, 1982; Laitinen, 1985). Somatostatin, a hormone that also functions as a neurotransmitter in the gut, has also been identified in mammalian airways.

Substance P is a potent smooth muscle contractile and vasodilatory factor. Nanomolar concentrations of SP evoke maximal contraction of pulmonary as well as gastrointestinal (GI) smooth muscle by acting directly through stereospecific receptors that recognize the carboxy-terminal portion of the peptide (Hanley and Iversen, 1980). Intravenous or aerosol administration of SP at pM concentration to humans elicits flushing, diaphoresis, rhinorrhea, hypotension, and pulmonary reactions ranging from coughing to bronchospasm (Lundberg *et al.*, 1983). By contrast, VIP induces bronchodilation, both directly and by antagonizing effects of histamine on human airways (Morice *et al.*, 1983).

The study of neuropeptides and their influence on the lung is in its infancy. The published data on the influence of neuropeptides on mast cells mostly involve rat mast cells, although some human and primate work has been reported. Because the results of studies for many neuropeptides are not yet confirmed, any conclusions must be drawn very cautiously, especially as evidence suggests that mast cell subtypes appear to respond differently to the same stimulus. Table VI provides an overview of *in vitro* studies performed to determine the effects of neutropeptides on mast cell release (Sertl and Kaliner, 1988).

VII. SUMMARY

Allergic asthma is a complex disease in which mast cell activation and mediator release play a pivotal role. The pathophysiological events set into motion by mast cell activation are modulated by heredity, underlying airway reactivity, airway inflammation, the presence or absence of epithelial destruction, the integrity of epithelial destruction, antigenic stimulation, the reactivity of

TABLE VI
Neuropeptides and Mast Cell Histamine Release[a,b]

Mast cell source	NT	SP	SST	VIP
Rat peritoneal	+/−	+	+	+/[c]
Rat intestinal	−	+	−	−
Rat pleural	+	+	?	?
Rat lung	?	?	?	?[c]
Rat basophilic leukemia	−	−	I/[c]	?
Monkey lung	−	+	−	−
Monkey intestinal	−	+	−	−
Mouse bone marrow-derived	?	?	I	?
Human skin	+	+	−	−
Human basophils	−	−	I/[c]	−

[a]From Sertl and Kaliner (1988).
[b]I, inhibition of histamine release.
[c]Contains peptide.

the subject to vagal reflexes, as well as the presence of abnormal autonomic nervous system reactivity, and possibly by the existence of nonimmunological mast cell secretagogues. Although much work remains to be done, pathophysiological findings of both the immediate phase and late phase of allergic asthma can be related to known actions of mast cell mediators released after IgE and antigen activation of mast cells.

VIII. REFERENCES

Agrawal, M. K., Swanson, M. C., Reed, C. E., and Yunginger, J. W., 1983, Immunochemical quantitation of airborne short ragweed, alternaria, antigen E and alt 1 allergies: A two year prospective study, *J. Allergy Clin. Immunol.* **72**:40–45.

Anthracite, R. F., Vachon, L., and Knapp, P. H., 1971, Alpha-adrenergic receptors in the human lung, *Psychosomat. Med.* **33**:481–488.

Atkins, P. C., Bedard, P. M., Zweiman, B., Dyer, J., and Kaliner, M., 1984, Increased antigen-induced local and systemic mediator release in rhinitis subjects with pulmonary symptoms in the pollen season, *J. Allergy Clin. Immunol.* **73**:341–347.

Betz, S. J., Lotner, G. Z., and Henson, P. M., 1980, Generation and release of platelet-activating factor PAF from enriched preparations of rabbit basophils; failure of human basophils to release PAF, *J. Immunol.* **125**:2749–2755.

Bhagat, R. G., Strunk, R. C., and Larsen, G. L., 1985, The late asthmatic response, *Ann. Allergy* **54**:297–302.

Blair, H., 1977, Natural history of childhood asthma, *Arch. Dis. Child.* **52**:613–619.

Boucher, R. C., Johnson, J., Inone S., Hulbert, W., and Hogg, J. C., 1980, The effect of cigarette smoke on the permeability of guinea pig airways, *Lab. Invest.* **43**:94–100.

Broder, I., Barlow, P. P., and Horton, R. J. M., 1962a, The epidemiology of asthma and hay fever in a total community, Tecumseh, Michigan, *J. Allergy* **33**:513–523.

Broder, I., Barlow, P. P., and Horton, R. J. M., 1962b, The epidemiology of asthma and hay fever in a total community, Tecumseh, Michigan, *J. Allergy* **33**:524–531.

Broder, I., Higgins, M. W., Mathews, K. P., and Keller, J. B., 1974, Epidemiology of asthma and allergic rhinitis in a total community, Tecumseh, Michigan, *J. Allergy Clin. Immunol.* **53**:127–138.

Brooks, S. M., McCowan, K., Bernstein, I. L., Altenau, P., and Peagler, J., 1979, Relationship between numbers of beta-adrenergic receptors in lymphocytes and disease severity in asthma, *J. Allergy Clin. Immunol.* **63**:401–406.

Burka, J. F., and Paterson, N. A. M., 1980, Evidence for lipoxgenase pathway involvement in allergic tracheal contraction, *Prostaglandins* **19**:499–515.

Callerame, M. L., Condemi, J. J., Bohrod, M. G., Vaughn, J. H., 1971, Immunologic reactions of bronchial tissues in asthma, *N. Engl. J. Med.* **284**:459–64.

Casale, T., and Kaliner, M., 1984, Allergic reactions in the respiratory tract, in: *Immunology of the Lung* (J. Beinenstock, ed.), pp. 326–341, McGraw-Hill, New York.

Cockcroft, D. W., Hoeppner, V. H., and Werner, G. D., 1984, Recurrent nocturnal asthma after bronchoprovocation with western red cedar sawdust: association with acute increase in non-allergic bronchial responsiveness, *Clin. Allergy* **14**:61–68.

Conolly, M. E., and Greenacre, J. K., 1976, The lymphocyte beta-adrenoceptor in normal subjects and patients with bronchial asthma. The effect of different forms of treatment on receptor function, *J. Clin. Invest.* **58**:1307–1316.

Cooke, R. A., and Veer, V. A., 1916, Human sensitization, *J. Immunol.* **1**:201–218.

Cookson, J. B., and Makoni, G., 1980, Prevalence of asthma in Rhodesian Africans, *Thorax* **35:** 833–837.

Dahlen, S. E., Hedqvist, P., Hammarström, S., and Samuelsson, B., 1980, Leukotrienes are potent constrictors of human bronchi, *Nature (Lond.)* **288:**484–486.

Dawson, W., Boot, J. R., Cockerell, A. F., Mallen, D. N. B., and Osbourne, D. J., 1976, The release of novel prostaglandins and thromboxanes after immunologic challenge of guinea pig lung, *Nature (Lond.)* **262:**699–703.

De Monchy, J. G. R., Kauffman, H. F., Venge, P., Koëter, Jansen, H. M., Sluiter, H. J., and deVries, K., 1985, Bronchoalveolar eosinophilia during allergen-induced late asthmatic reactions, *Am. Rev. Respir. Dis.* **131:**373–376.

Dey, R. D., Shannon, W. A., and Said, S. I., 1981, Localization of VIP-immuno-reactive nerves in airways and pulmonary vessels of dogs, cats and human subjects, *Cell Tissue Res.* **220:**231– 238.

Dodge, R. R., and Burrows, B., 1980, The prevalence and incidence of asthma and asthma-like symptoms in a general population sample, *Am. Rev. Respir. Dis.* **122:**567–575.

Empey, D. W., Laitinen, L. A., Jacobs, L., Gold, W. M., and Nadel, J. A., 1976, Mechanisms of bronchial hyperreactivity in normal subjects after upper respiratory tract infections, *Am. Rev. Respir. Dis.* **113:**131–139.

Filley, W. V., Holley, K. E., Kephart, G. M., and Bleich, G. J., 1982, Identification of eosinophil granule major basic protein in lung tissues of patients with bronchial asthma, *Lancet* **2:**11–15.

Fox, B., Bull, T. B., and Guz, A., 1981, Mast cells in the human alveolar wall: An electron microscopic study, *J. Clin. Pathol.* **34:**1333–1342.

Fraser, C., Venter, J., and Kaliner, M., 1981, Autonomic abnormalities and autoantibodies to beta adrenergic receptors, *N. Engl. J. Med.* **305:**1165–1170.

Friedman, M. M., and Kaliner, M., 1985, In situ degranulation of human nasal mucosal mast cells: Ultrastructural features and cell–cell interactions, *J. Allergy Clin. Immunol.* **76:**70–82.

Frigas, E., Loegering, D. A., and Gleich, G. J., 1980, Cytotoxic effects of the guinea pig eosinophil major basic protein on tracheal epithelium, *Lab. Invest.* **42:**35–43.

Gerrard, J. W., 1976, The familial incidence of allergic disease, *Ann. Allergy* **36:**10–15.

Ghatei, M. A., Sheppard, M. N., O'Shaughnessy, D. J., Adrian, T. E., McGregor, G. P., Polak, J. M., and Bloom, S. R., 1982, Regulatory peptides in the mammalian respiratory tract, *Endocrinology* **111:**1248–1354.

Godfrey, R. C., 1975, Asthma and IgE levels in rural and urban communities of the Gambia, *Clin. Allergy* **5:**201–207.

Galpern, S. A., Eggleston, P. A., Hendley, J. O., Suratt, P. M., Groshel, D. H. M., and Waltney, J. M., Jr., 1983, Pathogenesis of lower respiratory tract symptoms in experimental rhinovirus infection, *Am. Rev. Respir. Dis.* **128:**806–810.

Hanley, M. R., and Iversen, L. L., 1980, Substance P receptors, in: *Neurotransmitter Receptors.* Part I. *Receptors and Recognition* (S. J. Ehna and H. I. Yamamura, eds.), pp. 71–95, Chapman and Hall, London.

Hastie, R., Chir, B., Heroy, J. H., and Levy, D. A., 1979, Basophils, leukocytes, and mast cells in human nasal secretions and scraping studied by light microscopy, *Lab. Invest.* **40:**554–561.

Henderson, W. R., and Kaliner, M., 1978, Immunologic and nonimmunologic generation of superoxide from mast cells and basophils, *J. Clin. Invest.* **61:**187–196.

Henderson, W. R., Shelhamer, J. H., Reingold, D. B., Smith, L. J., Evans, R., and Kaliner, M., 1979, Alpha-adrenergic hyperresponsiveness in asthma: Analysis of vascular and pupillary responses, *N. Engl. J. Med.* **300:**642–647.

Henderson, W. R., Chi, E. Y., and Klebanoff, S. J., 1980, Eosinophil peroxidase-induced mast cell secretion, *J. Exp. Med.* **152:**265–279.

Hiss, A. E., 1966, Asthma among school children, *J. School Health* **316:**353–356.

Hitchcock, M., and Kokolis, N. A., 1981, Arachidonic acid metabolism and modulation of in vitro anaphylaxis by 5, 8, 11, 14-eicosatetraynoic acid and 9a,12a-octadecadiynoic acid, *Br. J. Pharmacol.* **72:**689–695.

Holgate, S. T., Howarth, P. H., Lee, T. H., Durham, S. R., Nagakura, T. A., and Kay, A. B., 1985, Airway hyperreactivity and mediatory release in clinical models of asthma, *J. Allergy Clin. Immunol.* **75:**140 (abst.).

Holtzman, M. J., Fabbri, L. M., O'Byrne, P. M., Gold, B. D., Aizawa, H., Walters, E. H., Alpert, S. E., and Nadel, J. A., 1983, Importance of airway inflammation for hyperresponsiveness induced by ozone, *Am. Rev. Respir. Dis.* **127:**686–690.

Kaliner, M. A., 1980, Mast cell-derived mediators and bronchial asthma, in: *Airway Reactivity* (E. F. Hargreave, ed.), pp. 175–188, Astra Pharmaceuticals Canada, Mississauga.

Kaliner, M., Eggleston, P., and Mathews, K., 1988, Rhinitis and asthma, *JAMA.*

Kariman, K., 1980, Beta-adrenergic receptor binding in lymphocytes from patients with bronchial asthma, *Lung* **158:**41–51.

Kaufman, J., and Wright, G. W., 1969, The effect of nasal and nasopharyngeal irritation on airway resistance in man, *Am. Rev. Respir. Dis.* **100:**626–630.

Kawanami O., Ferrans, V. J., Fulmer, J. D., and Crystal, R. G., 1979, Ultrastructure of pulmonary mast cells in patients with fibrotic lung disorders, *Lab. Invest.* **40:**717–735.

Kratschmer, F., 1928, Effect on respiratory, blood pressure and carotid pulse of various inhaled and insufflated vapors when stimulating one cranial nerve and various combinations of cranial nerves, *Am. J. Physiol.* **87:**319–325.

Laitinen A., 1985, Autonomic innervation of the human respiratory tract as revealed by histochemical and ultrastructural methods, *Eur. J. Respir. Dis.* **661**(Suppl. 140):1–42.

Lamb, D., and Lumsden, A., 1982, Intra-epithelial mast cells in human airway epithelium: Evidence for smoking induced changes in their frequency, *Thorax* **37:**334–342.

Lebowitz, M. D., Kudson, R. J., and Burrows, B., 1975, Tucson epidemiologic study of obstructive lung diseases, *Am. J. Epidemiol.* **102:**137–152.

Lemanske, R. F., and Kaliner, M., 1982, Mast cell-dependent late phase reactions, *Clin. Immunol. Rev.* **1:**547–580.

Lemanske, R. F., Guthman, D. A., Oertel, H., Barr, L., and Kaliner, M., 1983, The biologic activity of mast cell granules. VI. The effect of vinblastine-induced neutropenia on rat cutaneous late phase reactions, *J. Immunol.* **130:**2837–2842.

Lundberg, J. M., Martlin, C. R., and Saria A., 1983, Substance P and capsaicin-induced contraction of human bronchi, *Acta Physiol. Scand.* **119:**49–53.

Lundberg, J. M., Hökfelt, T., Martlin, C. R., Saria, A., and Cuello, C., 1984, Substance P immunoreactive sensory nerves in the lower respiratory tract of various mammals including man, *Cell Tissue Res.* **235:**251–261.

Mak, H., Johnson, P., and Abbey, H., 1982, Prevalence of asthma and health service utilization of asthmatic children in an inner city. *J. Allergy Clin. Immunol.* **70:**367–372.

Makino, S., Ouelette, J. J., Reed, C. E., and Fishel, C., 1970, Correlation between increased bronchial response to acetylcholine and diminished metabolic and eosinopenic responses to epinephrine in asthma, *J. Allergy Clin. Immunol.* **46:**178–189.

Mansfield, L. E., and Stein, M. R., 1978, Gastroesophageal reflux and asthma: A possible reflex mechanism, *Ann. Allergy* **41:**224–226.

Marom, Z., Shelhamer, J. H., and Kaliner, M., 1984, Human pulmonary macrophage-derived mucus scretagogue, *J. Exp. Med.* **189:**844–860.

Martin, A. J., Landau, L. I., and Phelan, P. D., 1980, Lung function in young adults who had asthma in childhood, *Am. Rev. Respir. Dis.* **122:**609–616.

McNeill, R. S., 1964, Effects of a beta-adrenergic blocking agent, propranolol, on asthmatics, *Lancet* **2:**1101–1102.

Mencia-Huerta, J. M., and Benveniste, J., 1979, Platelet activating factor and macrophages. I. Evidence for the release from rat and mouse peritoneal macrophages and not from mastocytes, *Eur. J. Immunol.* **9:**409–415.

Messer, J. W., Peters, G. A., and Bennett, W. A., 1960, Causes of death and pathologic findings in 304 cases of bronchial asthma, *Dis. Chest* **38**:616–624.

Minor, T. E., Dick, E. C., Baker, J. W., Ouellette, J. J., Cohen, M., and Reed, C. E., 1976, Rhinovirus and influenza type A infections as precipitants of asthma, *Am. Rev. Respir. Dis.* **113**:149–153.

Morice, A., Unwin, R. J., and Sever, P. S., 1983, Vasoactive intestinal peptide causes bronchodilation and protects against histamine-induced bronchoconstriction in asthmatic subjects, *Lancet* **2**: 1225–1230.

Naclerio, R. M., Proud, D., Togias, A., Adkinson, N. F., Kagey-Sobotka, A., Meyers, D. A., Plaut, M., Norman, P. S., and Lichtenstein, L., 1985, Inflammatory mediators in late antigen-induced rhinitis, *N. Engl. J. Med.* **313**:65–70.

Nagy, L., Lee, T. H., and Kay, A. G., 1982, Neutrophil chemotactic activity in antigen-induced late asthmatic restrictions, *N. Engl. J. Med.* **306**:497–501.

National Institute of Allergy and Infectious Diseases Task Force Report on Asthma and Allergic Diseases, 1979, NIH Publication 79–387.

Nolte, D., and Berger, D., 1983, On vagal bronchoconstriction in asthmatic patients by nasal irritation, *Eur. J. Respir. Dis.* **64**(Suppl. 128):110–115.

O'Byrne, P. M., Walters, E. H., Aizawa, H., Fabbri, L. M., Holtzman, J. M., and Nadel, J. A., 1984, Indomethacin inhibits the airway hyperresponsiveness but not the neutrophil influx induced by ozone in dogs, *Am. Rev. Respir. Dis.* **130**:220–224.

Okuda, M., and Otsuka, H., 1977, Basophilic cells in allergic nasal secretions, *Arch. Oto-Rhino-Laryngol.* **214**:283–289.

Parker, C., and Smith, J. W., 1973, Alterations in cyclic adenosine monophosphate metabolism in human bronchial asthma. I. Leukocyte responsiveness to beta-adrenergic agents, *J. Clin. Invest.* **52**:48–59.

Paterson, N. A. M., Burka, J. F., and Craig, I. D., 1981, Release of slow reacting substance of anaphylaxis from dispersed pig lung cells: Effect of cyclo-oxygenase and lipoxygenase inhibitors, *J. Allergy Clin. Immunol.* **67**:425–434.

Pernow, B., 1983, Substance P, *Pharmacol. Rev.* **35**:86–128.

Platshon, L., and Kaliner, M., 1978, The effect of the immunologic release of histamine upon human lung cyclic nucleotide levels and prostaglandin generation, *J. Clin. Invest.* **62**:1113–1121.

Proud, D., Schulman, E. S., MacGlashan, D. W., Pierce, J. V., and Newball, H. H., 1985, IgE-mediated release of a kininogenase from purified human lung mast cells, *Am. Rev. Respir. Dis.* **132**:405–408.

Rackmann, F. M., and Edwards, M. C., 1952, Asthma in children: A follow-up study of 688 patients after an interval of twenty years, *N. Engl. J. Med.* **246**:815–823.

Richardson, J. B., Hogg, J. C., Bouchard, T., and Hall, D. H., 1973, Localization of antigen in experimental bronchoconstriction in guinea pigs, *J. Allergy Clin. Immunol.* **52**:172–181.

Richardson, J. B., Bouchard, I., and Ferguson, C. C., 1976, Uptake and transport of exogenous proteins by respiratory epithelium, *Lab. Invest.* **35**:307–314.

Rosenthal, R. R., Norman, P. S., Summer, W. R., and Permutt, S., 1977, Role of the parasympathetic system in antigen-induced bronchospasm, *J. Appl. Physiol.* **42**:600–606.

Schulman, E. J., Liu, R. C., Proud, D., Macglashan, D. W., Lichtenstein, L., and Plaut, M., 1985, Human lung macrophages induced histamine release from basophils and mast cells, *Am. Rev. Respir. Dis.* **131**:230–235.

Schwartz, M., 1952, Heredity in bronchial asthma, *Acta Allergol.* **5**(Suppl. 2):1–288.

Stertl, K., and Kaliner, M. A., 1988, The influence of neurohormones and neuropeptides on mast cells, in: *The Airways: Neural Control in Health and Disease* (M. Kaliner and P. Barnes, eds.), pp. 447–466, Marcel Dekker, New York.

Shelhamer, J. H., Marom, Z., and Kaliner, M., 1980a, Immunologic and neuropharmacologic stimulation of mucous glycoprotein release from human airways in vitro, *J. Clin. Invest.* **66**: 1400–1408.

Shelhamer, J. H., Metcalfe, D. D., Smith, L. J., and Kaliner, M., 1980b, Abnormal beta adrenergic responsiveness in allergic subjects: Analysis of isoproterenol-induced cardiovascular and plasma cyclic adenosine monophosphate responses, *J. Allergy Clin. Immunol.* **66**:52–60.

Shelhamer, J. H., Marom, Z., Sun, F., Bach, M. K., and Kaliner, M., 1982, The effects of arachinoids and leukotrienes on the release of mucus from human airways, *Chest* **81**:36S–37S.

Slavin, R. G., 1984, Sinus disease and asthma, *ENT J.* **63**:49–50.

Smith, J. M., 1983, Epidemiology and natural history of asthma, allergic rhinitis, and atopic dermatitis (eczema), in: *Allergy: Principles and Practice,* 2nd ed., (E. Middleton, C. E. Reed, and E. F. Ellis, eds., pp. 771–810, C. V. Mosby, St. Louis, Missouri.

Smith, L. J., Shelhamer, J. H., and Kaliner, M., 1980, The cholinergic nervous system and immediate hypersensitivity. II. An analysis of pupillary responses, *J. Allergy Clin. Immunol.* **66**:374–378.

Snapper, J. R., and Brigham, K. L., 1984, Inflammation and airway reactivity, *Exp. Lung Res.* **6**:83–89.

Snashall, P. D., Boother, F. A., and Sterling, G. M., 1978, The effect of alpha-adrenergic receptor stimulation on the airways of normal and asthmatic man, *Clin. Sci. Mol. Med.* **54**:283–289.

Spannhake, E. W., Hyman, A. L., and Kadowitz, P. J., 1981, Bronchoactive metabolites of arachidonic acid and their role in airway function, *Prostaglandins* **22**:1013–1026.

Spaulding, H. S., Mansfield, L. E., Steen, M. R., Sellner, J. C., and Gremillion, D. E., 1982, Further investigation on the association between gastroesophagial reflux and bronchoconstriction, *J. Allergy Clin. Immunol.* **69**:516–521.

Steel, L., and Kaliner, M., 1981, Prostaglandin-generating factor of anaphylaxis—Identification and isolation, *J. Biol. Chem.* **256**:12692–12698.

Thueson, D. O., Speck, L. S., Lett-Brown, M. A., and Grant, J. A., 1979, Histamine releasing activity (HRA). I. Production by mitogen- or antigen-stimulated human mononuclear cells, *J. Immunol.* **123**:626–632.

Townley, R. G., 1982, Receptors and Nonspecific Bronchial Reactivity, in: *Proceedings of the Eleventh International Congress of Allergology and Clinical Immunology* (J. W. Kerr and M. A. Ganderton, eds.), pp. 197–203, Macmillan, London.

Townley, R., and Bewtra, A., 1983, Airway Reactivity to Methacholine in Asthma Family Members and Twins, in: *Provocative Challenge Procedures: Bronchial, Oral, Nasal, and Exercise,* Vol. I (S. L. Spector, ed.), pp. 177–185, CRC Press, Boca Raton, Florida.

Trotter, C. M., and Orr, T. S. C., 1973, A fine structure study of some cellular components in allergic reactions, *Clin. Allergy* **3**:411–425.

Van Arsdale, P. P., and Motulsky, A. G., 1959, Frequency and heritability of asthma and allergic rhinitis in college students, *Acta. Genet. Statist. Med.* **9**:101–114.

Van Niekerk, C. H., Weinberg, E. G., Shore, S. C., De V Heese, H., and Van Schalkwyk, D. J., 1979, Prevalence of asthma: A comparative study of urban and rural Xhosa children, *Clin. Allergy* **9**:319–324.

Waite, D. A., Eyles, E. F., and Tonkin, S. L., 1980, Asthma prevalence in Tokelauan children in two environments, *Clin. Allergy* **10**:71–75.

Weiss, J. W., Drazen, J. M., Coles, N., McFadden, E. R., Weller, P., Corey, E. J., Lewis, R., and Austen, K. F., 1982, Bronchoconstriction effects of leukotriene C in humans, *Science* **216**:196–198.

White, M. V., and Kaliner, M. A., 1985, Neutrophil induced mast cell degranulation, *J. Allergy Clin. Immunol.* **75**:175 (abst.).

White, M. V., Kaliner, M. A., and Baer, H., 1986, Stimulated neutrophils release a histamine releasing factor, *J. Allergy Clin. Immunol.* **77**:132 (abst.).

Widdicomb, J. G., 1977, Some experimental models of acute asthma, *J. R. Coll. Physicians Lond.* **11**:141–155.

Wilder, C. S., 1973, Prevalance of selected chronic respiratory conditions, *Vital Health Stat.* **10**:1–49.

Wilson, M. C., and Larsen, G. L., 1986, Gaining control over the late asthmatic response. *J. Respir. Dis.* **7**(8):51–60.

Yan, K., and Salome, C., 1983, The response of the airways to nasal stimulation in asthmatics with rhinitis, *Eur. J. Respir. Dis.* **64**(Suppl.):105–108.

Yu, D. Y. C., Galant, S. P., and Gold, W. M., 1972, Inhibition of antigen-induced bronchoconstriction by atropine in asthmatic patients, *J. Appl. Physiol.* **32**:832–843.

Zach, M. S., Schnall, R. P., and Landau, L. I., 1980, Upper and lower airway hyperreactivity in recurrent croup, *Am. Rev. Respir. Dis.* **121**:979–983.

Zheutlin, L. M., Ackerman, S. J., Gleich, G. J., and Thomas, L. L., 1984, Stimulation of basophil and mast cell histamine release by eosinophil granule-derived cationic proteins, *J. Immunol.* **133**:2100–2185.

Chapter 15

Exercise-Induced Bronchoconstriction

Joseph F. Souhrada

John B. Pierce Foundation Laboratory
and
School of Medicine
Yale University
New Haven, Connecticut 06519

I. DEFINITION OF EXERCISE-INDUCED BRONCHOCONSTRICTION

Shortness of breath during exercise is frequently observed in patients with different types of cardiac or pulmonary disease. The shortness of breath and bronchoconstriction that develop after short-term exercise is a characteristic clinical response that signifies airway hyperreactivity in patients with bronchial asthma. It is known as either exercise-induced asthma or exercise-induced bronchoconstriction (EIB), or both. Since this entity is characterized by bronchoconstriction, which can be objectively measured, and as the term asthma has a broader meaning, it is more appropriate to call this phenomenon exercise-induced bronchoconstriction.

In clinical medicine, the definition of airway hyperreactivity is usually based on clinical history, the result of inhalation challenge (specific antigen, histamine, or methacholine), or the result of exercise testing. Although airway hyperreactivity also known as airway hyperresponsiveness, is usually found in patients with bronchial asthma, it can also be detected in subjects with different forms of airway injury such as during and following upper airway viral infection (Empey *et al.*, 1976; Parker *et al.*, 1965), following influenza vaccine (Ouellette and Reed, 1965), after ozone exposure (Silverman, 1979), and possibly after exposure to other types of pollutants (Jaeger *et al.*, 1979; Orehek *et al.*, 1976). Increased airway reactivity has also been reported in patients with pulmonary tuberculosis (Laitinen and Kokkola, 1974) and chronic bronchitis (Simonsson, 1965). In adult patients with bronchial asthma, when the presence of bronchitic component can be reasonably assumed, the presence of EIB was also demonstrated (Simonson *et al.*, 1972; Chryssanthopoulos *et al.*, 1979). Finally, EIB has been reported in young children who showed features related to asthma, such as wheezy bronchitis (Konig et al., 1972), and in patients with cystic fibrosis (Godfrey and Mearns, 1971).

A. Response of Normal Subjects to Exercise

In normal humans, short-term exercise either causes bronchodilation (Konig et al., 1972) or no measurable change in pulmonary function (Godfrey and Mearns, 1971). During exercise, Mansfield et al. (1979) found that exercise in healthy children causes an immediate decrease in the respiratory resistance followed by a return to baseline levels at the end of exercise. Similarly, Lefcoe (1969) observed that exercise improved the pulmonary function of normal subjects. In another study (Kattan et al., 1978) however, short-term exercise in healthy children induced a minimum change in pulmonary function. Changes in pulmonary function during and after exercise in normal subjects are probably related to the activation of the sympathetic nervous system and the consecutive effect of blood-borne catecholamines on airways smooth muscle.

B. Clinical Presentation of Exercise-Induced Bronchoconstriction

The incidence of EIB among asthmatic and atopic nonasthmatic children is 63% and 41%, respectively (Kawabori et al., 1976). It has been reported that 75% of children with asthma have EIB (Jones, 1966). Because of a lack of workload standardization, duration and type of exercise, and the methodology to determine the magnitude of airway obstruction, comparison of data between different investigators is difficult.

In susceptible persons, the degree of EIB is generally proportional to the severity and type of exercise and to its duration. The temperature and humidity of inspired air during physical activity are also important factors. Although there is a wide range of responses to exercise in asthmatics, a typical bronchoconstrictive response is usually detected after exercise, achieving maximum at 7 to 15 min postexercise (Anderson et al., 1975). Respiratory difficulties may include cough, shortness of breath, chest tightness, and wheezing. Only a few patients seem to develop bronchoconstriction during exercise, and in some, continuation of exercise may actually bring relief of dyspnea as well as bronchoconstriction. In most asthmatics, when postexercise bronchoconstriction is not too severe, a spontaneous recovery of symptoms is usually observed 30–35 min after exercise. The most characteristic change in pulmonary function is a decrease in maximal expiratory flow rates as measured by spirometry or flow–volume relationship. A decrease of 15–20% in FEV_1 as compared with the baseline FEV_1 suggests the presence of bronchoconstriction. Usually an increase in static lung volumes, such as residual volume (RV) or functional residual capacity (FRC) can also be detected leading to the increase in the resting respiratory position and a significant lung hyperinflation. If measurement of airway resistance (R_{aw}) is available, a significant increase in this value and a decrease in the specific conductance (SG_{aw}) is also observed.

Until recently, EIB was associated only with an immediate (i.e., 7–15 min postexercise) bronchomotor response. It has been shown, however, that after complete recovery from EIB, which spontaneously occurred 2.5 hours postexercise, a second fall in the pulmonary function can be present 3.5–6 hr postexercise (Bierman, 1984; Iikura *et al.*, 1985). In some patients, this late response is even greater than an immediate fall in pulmonary function. Lee and co-workers reported that asthmatics who had both exercise-induced early and late responses had two peaks in serum level of neutrophile chemotactic factor that correlated with these two episodes of bronchoconstriction (Lee *et al.*, 1983; Nagy *et al.*, 1982). It is also of interest that the rate of spontaneous recovery from immediate (early) response was slower in patients with dual responses, suggesting that this variable may predict development of the late-phase response in EIB. Recent data, however, suggest that a late response observed after exercise might be a nonspecific epiphenomenon (Zawadski *et al.*, 1988).

As one would expect, there are also abnormalities in other functional characteristics. The alveolar–arterial oxygen gradient is increased both during and after exercise in asthmatics with EIB, a finding that suggests deterioration of the ventilation–perfusion relationship (Katz *et al.*, 1971). P_{CO_2}, however, seems to remain within baseline range (Katz *et al.*, 1971). During a postexercise period, a deterioration in a gas exchange is usually observed (Vassallo *et al.*, 1972). The pH value of arterial blood has been reported to decrease after exercise (Silverman *et al.*, 1972). This finding may be related to the higher lactate levels present in asthmatics at the completion of exercise.

II. MECHANISM OF EXERCISE-INDUCED BRONCHOCONSTRICTION

A. Heat and Water Exchange in Upper Airways

In humans, one of the most important functions of the nose is to modify inspired air. In addition to removing small particles and some gases (e.g., sulfur dioxide), the nose provides an important place for the adjustment of inspired air to the body temperature and its humidification. Because the saturation of air with water depends on the temperature, more water vapor is required for maintaining full saturation as the inspired air is warmed to 37°C. This nasal air-conditioning function is highly efficient under resting conditions when the level of ventilation varies from 8 liters/min to 15 liters/min. Even during exercise (up to certain ventilatory levels), a man is normally capable of breathing through the nose, because during exercise a special reflex mechanism constricts vessels in the nose, which in turn decreases nasal airflow resistance. As the level of exercise and ventilation increases even more, increasing nasal resistance raises the work of breathing, and then breathing is done through the mouth only. Consequently, the air conditioning of inspired air through the nose is eliminated. Under these conditions, air inhaled into the upper airways will be cooler and drier. Thus, during physical work, in a particularly cold environment, the temperature and humidity of the upper airways can be compromised. At these higher levels of ventilation, a significant cooling and drying of mucosa can occur even in more peripheral airways. McFadden *et al.* (1982) showed that during quick breathing most of the heating of the inspired air took place in the upper airways as expected. However, during hyperventilation and inhalation of subfreezing temperature air, a temperature decrease could be detected, even in small peripheral airways. This finding suggests that the conditioning of inspired air is a continuous process most likely involving a majority of the tracheobronchial tree.

B. Role of Minute Ventilation

Hyperventilation *per se* has been implicated as a possible causative factor in EIB, but data on this subject conflict (Godfrey *et al.*, 1973; Buckley *et al.*, 1974; Herxheimer, 1946). The controversy might have arisen in part because investigators have not always considered the humidity and temperature of inspired air during exercise and/or hyperventilation (McFadden *et al.*, 1977). This may be important, as it has been shown that warm and humid air supplied during exercise to asthmatic patients can prevent the development of EIB (Chen *et al.*, 1976; Bar-Or *et al.*, 1977). In addition, it has been demonstrated that the severity of EIB is decreased in asthmatic children who are instructed to use only nasal breathing during exercise. By contrast, oral breathing during exercise was found to potentiate EIB (Shturman-Ellstein *et al.*, 1978).

In carefully controlled studies, it had been shown that significant bronchoconstriction, comparable in severity to that observed after moderate treadmill exercise, can be induced in asthmatics by voluntary isocapnic hyperventilation of 3-min and 10-min duration. In both hyperventilation and exercise, nasal breathing inhibited the bronchoconstriction, whereas mouth breathing potentiated the bronchoconstrictive response. It is of interest that 10 min of voluntary isocapnic hyperventilation did not represent a greater bronchoconstrictive stimulus than did 10 min of exercise or 3 min of isocapnic hyperventilation (Zeballos *et al.*, 1978).

The results of this and other studies suggest that hyperventilation can be considered an underlying mechanism of EIB, when investigated under controlled conditions. Thus, this study confirms the observation of other investigators who suggested that hyperventilation is an important factor in EIB (Herxheimer, 1946; Shturman-Ellstein *et al.*, 1978; Zeballos *et al.*, 1978; Chang-Yeung *et al.*, 1971; Rebuck and Reed, 1968; Simonsson *et al.*, 1967; Stanescu and Teculescu, 1970).

C. Consequence of Heat Loss and Water Loss

1. Effect of Water Loss

Anderson (1983) suggested that the loss of water from airways during exercise may change the osmotic environment of irritant receptors or mast cells present in the vicinity of ephithelial cells, triggering bronchoconstriction. Consecutive studies have shown that at least in some asthmatics, the loss of water from airways *per se* is the most important stimulus in EIB (Anderson, 1984). EIB may occur when expired air temperature during exercise is close to body temperature and higher then that normally observed at rest. It was concluded that evaporative water loss from the airway mucosa could induce a transient change in the osmolarity of the epithelial fluid. In turn, the increase in the osmolarity may induce a release of mediators from inflammatory cells or epithelial cells or may initiate a stimulus in nerve endings (Hahn, 1984; Anderson *et al.*, 1985; Anderson, 1985).

2. Initiation of the Vagal Reflex

Numerous nerve endings have been described superficially between cylindrical epithelial cells of the airways. In the upper airways they are called cough receptors, and in the lower large airways they are known as irritant receptors. Their afferent and efferent pathways are connected to the vagus and airway smooth muscle, respectively. Stimulation of these receptors will cause bronchoconstriction. Cooling or drying of the mucosal surface could stimulate these receptors and cause bronchoconstriction. Bronchoconstriction can stimulate stretch receptors, which could initiate a vagal positive feedback loop, which in turn may cause a further potentiation of bronchoconstriction. Indeed, it has been shown that inhalation of cold air in asthmatic subjects will cause immediate bronchoconstriction. This bronchoconstriction can be prevented by the administration of atropine (Sheppard *et al.*, 1982).

If during exercise these receptors are stimulated by relatively cold and dry air, causing heat and water loss from airways, one would expect that in most patients with bronchial asthma, EIB could be prevented with atropine pretreatment. However, only in a few patients has atropine pretreatment prevented EIB. It is likely that in these studies insufficient doses of antimuscarinic agents were administered. Conversely, a local anesthesia of the upper airways in asthmatic patients before exercise has been shown to attenuate the development of bronchoconstriction significantly. Since the same anesthesia also eliminates the cough and gag reflexes, it was speculated that stimulation of these receptors may participate in the development of EIB at least in some patients with bronchial asthma (McNally *et al.*, 1979; Enright *et al.*, 1980).

3. Direct Effect of Cooling on Airway Smooth Muscle

Even though good evidence exists showing that cooling-induced bron-choconstriction could stimulate irritant receptors and that this response can be attenuated by atropine treatment, one can speculate that cooling could also affect other structures located in the wall of airways, including airway smooth muscle cells. Several recent studies have shown that changes in the temperature can have a direct effect on airway smooth muscle cells. Thus, a decrease in temperature caused a depolarization of airway smooth muscle cells and potentiated both electrical and contractile responses to histamine (Souhrada and Souhrada, 1981) and acetylcholine (ACh) (Murlas *et al.*, 1982).

Subsequent studies have shown that quick cooling of airway smooth muscle preparations in the range of temperatures of 22–29.5°C caused a biphasic re-sponse, i.e., contraction followed by relaxation. A consecutive rewarming in-duced a large contraction of airway smooth muscle cells. These responses were temperature dose dependent (Souhrada *et al.*, 1983). Similar findings were re-ported by Park *et al.* (1982), who observed contraction of isolated guinea pig trachea after cooling to 26°C.

These data suggest that quick temperature changes may represent an impor-tant stimulus in airways. This cooling- and rewarming-induced contraction of airway smooth muscle cells may furthermore stimulate stretch receptors, and consequent vagal response may lead to an even greater contraction of airway smooth muscle. Alternatively, one may speculate if the cooling-induced inhibi-tion of the electrogenic Na^+ pump (Souhrada and Souhrada, 1981), which leads to the increased synthesis of inhibitory prostaglandins (Coburn and Soltoff, 1977; Soltoff *et al.*, 1979), is unaltered in asthmatic airways. The alteration of this mechanism could explain an excessive sensitivity of asthmatic airways to inhalation of cool air and lack of this phenomenon in normal airways.

4. Role of α-Adrenergic Mechanism

It has been suggested that in the asthmatic lung, a functional imbalance between β- and α-receptors in the tracheobronchial tree could be responsible for the presence of airway hyperreactivity. The administration of α-blocking agents has been shown to improve the symptoms experienced by asthmatics. Further-more, it was reported that α-adrenergic blocking agents could attenuate bron-choconstriction response following exercise (Gross *et al.*, 1974; Beil and De-Kock, 1978; Bleecker *et al.*, 1983).

In a carefully designed study, Walden and associates (1984) reported that α-adrenergic system may partially mediate airway response to exercise and cold air hyperventilation. These workers also showed that this mechanism is not involved in airway response of asthmatics to histamine and ragweed antigen. Therefore, it was speculated that exercise- and cold-induced bronchospasm may be partially mediated by α-adrenergic neural mechanisms perhaps through sensitization of α-receptors during airway cooling. These data are supported by *in vitro* studies of human bronchi isolated from patients with respiratory diseases demonstrating that the α-receptor mechanism in those tissues is increased (Kneussel and Richardson, 1978).

5. Mediator Release

When repeated exercise challenge was done with the same patients, it was observed that the magnitude of the response to exercise decreased with the repetition of the test. This so-called refractoriness was suggested to be due to the depletion of the stored mediators, and it was speculated that it would take some time for these mediators to be replenished. Therefore, numerous attempts have been made to demonstrate the possible presence of different mediators in plasma following the exercise and to correlate the mediator (its presence or concentration) with decreased pulmonary function.

a. Histamine. Many investigators studied changes in the level of the circulating plasma histamine both during and after exercise in asthmatics. The results of these studies are contradictory. Some of these data could be explained by technical difficulties of measuring small changes in plasma concentration of histamine. Recent data employing sensitive radioenzymatic techniques confirmed that plasma histamine in asthmatics is increased after exercise (Anderson *et al.,* 1981; Barnes and Brown, 1981). It is likely that histamine is released from tissue mast cells, present on the surface of mucosa, or from circulating basophils.

Conversely, it was also emphasized that the specificity of mediators for mast cells should be defined before conclusions can be drawn on changes in these mediators in the plasma during exercise (Nagakura *et al.,* 1983). It was pointed out that in the event of histamine increase, plasma levels may be caused by the release of histamine from basophils. Therefore, the histamine increase described in plasma after EIB may be due to accompanying basophilia or alternatively to the sample handling. In fact, Howarth and Holgate (1983) suggested that it is unlikely that mast cell degranulation is directly involved in the pathogenesis of EIB, emphasizing a different time course of events between exercise and antigen challenge.

b. Prostaglandins. Anderson and co-workers (1976) failed to find any involvement of $PGF_{2\alpha}$ (or its metabolites) in EIB. If mast cell degranulation is involved in the mechanism of EIB, one would expect $PGF_{2\alpha}$ metabolites to be present in the plasma of asthmatics after exercise.

c. Neutrophil Chemotactic Factor. In 1977 Atkins and co-workers (1977) identified a high-molecular-weight heat-stable neutrophil chemotactic factor

(NCF) in the plasma of asthmatic patients after antigen inhalation challenge. No changes in NCF, however, were found after methacholine inhalation challenge. Later it was shown that in asthmatics, treadmill exercise can cause a three- to fivefold increase in plasma NCF (Lee *et al.,* 1982; Kay and Lee, 1983). Peak levels of NCF correlated well with the maximal decrease in the pulmonary function. Furthermore, NCF increased during exercise-induced late asthmatic response, an observation similar to that reported after antigen inhalation challenge. NCF could participate in early recruitment of neutrophils. The release of intracellular constituents from neutrophils that infiltrate bronchi might participate in the phenomenon of airway hyperreactivity. Indeed, Papageorgio and co-workers (1983) showed that after exercise in asthmatics, inflammatory cells are activated possibly as a result of the release of NCF and that exercise induced a kinetic increase in the expression of neutrophil C3b receptor (Lee *et al.,* 1984).

   d. Release of Other Mediators and Hormones.   It has been reported that exercise in atopic asthmatics can induce a release of glucagon (Ahonen *et al.,* 1983). These investigators speculated that the rise of glucagon in plasma is a defense mechanism against EIB. Recent reports, however, seem to suggest that a variety of additional mediators may be involved in the mechanism of EIB. Thus, Neuman *et al.* (1983) reported a consistent elevation of the plasma kallikrein levels following an exercise challenge of asthmatics. A very interesting finding reported by Johnson and associates (1986) suggests that in asthmatics, exercise can increase plasma concentration of platelet factor 4 and β-thromboglobulin, a rise that preceded a fall in the pulmonary function. These findings indicate that exercise in asthmatics can induce a rise in platelet release products similar to that observed in bronchoconstriction induced by a specific antigen inhalation challenge. Indeed, a selective 5-HT-blocking agent had a partial inhibitory effect on EIB (So *et al.,* 1985).

III. SUMMARY

   The results of clinical studies suggest that EIB is caused by multiple factors that occur in response to the loss of water and heat from the surface of airways. These mechanisms are summarized in Fig. 1. It is likely that the initiating stimulus is both water and heat loss, which leads to the cascade of events, some interrelated, leading to the bronchoconstriction (Nadel, 1984). Furthermore, it is possible that in different patients with bronchial asthma, the importance of these mechanisms may vary (Lee and Anderson, 1985). The confirmation of this hypothesis may require additional work.

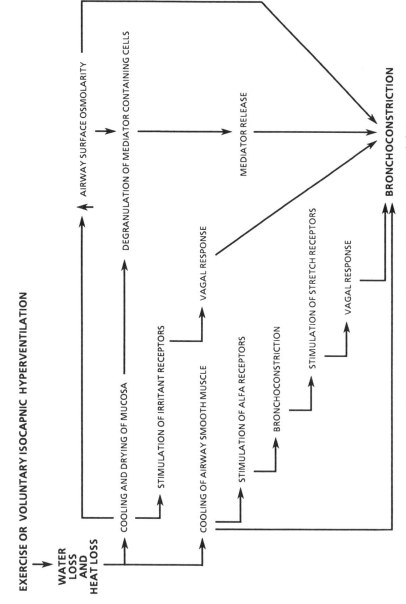

FIGURE 1. Mechanisms participating in exercise-induced bronchoconstriction.

IV. REFERENCES

Ahonen, A., Sovijarvi, A. R. A., and Muittari, A., 1983, Plasma glucagon response during exercise-induced asthma, *Eur. J. Respir. Dis.* **64**:166–171.

Anderson, S. D., 1983, Recent advances in the understanding of exercise-induced asthma, *Eur. J. Respir. Dis.* **64**:225–236.

Anderson, S. D., 1984, Is there a unifying hypothesis for exercise-induced asthma?, *J. Allergy Clin. Immunol.* **73**:660–665.

Anderson, S. D., Silverman, M., Konig, P., and Godfrey, S., 1975, Exercise induced asthma, *Br. J. Dis. Chest* **69**:1–17.

Anderson, S. D., Poser, R., Smith, I. D., and Temple, D., 1976, Exercise-related changes in plasma levels of 15-keto-13, 14-dihydro-prostaglandin F_2 and noradrenaline in asthmatic and normal subjects,*Scand. J. Respir. Dis.* **57**:41–48.

Anderson, S. D., Bye, P. T. P., Schoeffel, R. E., Seale, J. P., Taylor, K. M., and Ferris, L., 1981, Arterial plasma histamine levels at rest and during and after exercise in patients with asthma: Effect of terbutaline aerosol, *Thorax* **36**:259–267.

Anderson, S. D., 1985, Issues in exercise-induced asthma, *J. Allergy Clin. Immunol.* **76**:764–772.

Anderson, S. D., Schoeffel, R. E., Black, J. L., and Daviskas, E., 1985, Airway coding and the stimulus to exercise-induced asthma—A re-evaluation, *Eur. J. Respir. Dis.* **67**:20–30.

Atkins, P. C., Nordan, M., Weiner, H., and Zweim, B., 1977, Release of neutrophil chemotactic activity during immediate hypersensitivity reactions in humans, *Arch. Intern. Med.* **86**:415–420.

Barnes, P. J., and Brown, M. J., 1981, Venous plasma histamine in exercise and hyperventilation-induced asthma in man, *Clin. Sci.* **61**:159–162.

Bar-Or, O., Neuman, L., and Dothan, R., 1977, Effects of drug and humid climates on exercise-induced asthma in children and adolescents, *J. Allergy Clin. Immunol.* **60**:163–168.

Beil, M., and DeKock, M. A., 1978, Role of alpha-adrenergic receptors in exercise-induced bronchoconstriction, *Respiration* **35**:78–86.

Bierman, C. W., 1984, A comparison of late reactions to antigen and exercise, *J. Allergy Clin. Immunol.* **73**:654–659.

Bleecker, E. R., Chahal, K. S., Mason, P., and Permutt, S., 1983, The effect of alpha adrenergic blockage on non-specific airway reactivity and exercise-induced asthma, *Eur. J. Respir. Dis.* **64**:258–264.

Buckley, J. M., Souhrada, J. F., and Kopetzky, M. T., 1974, Detection of Airway obstruction in exercise-induced asthma, *Chest* **66**:244–251.

Chang-Yeung, M. M. W., Vyas, M. N., and Gryzbowski, S., 1971, Exercise-induced asthma, *Am. Rev. Respir. Dis.* **104**:915–923.

Chen, W. Y., Horton, D. J., and Souhrada, J. F., 1976, Respiratory tract and water loss and exercise-induced asthma, *Physiologist* **19**:152 (abstr.).

Chryssanthopoulos, C., Maksud, M. G., Funahashi, A., Hoffmann, R. G., and Barboriak, J. J., 1979, An assessment of cardiorespiratory adjustments of asthmatic adults to exercise, *J. Allergy Clin. Immunol.* **63**:321–327.

Coburn, R. F., and Soltoff, S., 1977, Na^+-K^+-ATPase inhibition stimulates PgE release in guinea pig taenia coli, *Am. J. Physiol.* **232**:C191–C195.

Empey, D. W., Laitinen, L. A., Jacobs, L., Gold, W. M., and Nadel, J. A., 1976, Mechanisms of bronchial hyperactivity in normal subjects after upper respiratory tract infection, *Am. Rev. Respir. Dis.* **113**:131–139.

Enright, P. L., McNally, J. F., and Souhrada, J. F., 1980, Effect of lidocaine on the ventilatory and airway responses to exercise in asthmatics, *Am. Rev. Respir. Dis.* **122**:823–828.

Godfrey, S., and Mearns, M., 1971, Pulmonary function and response to exercise in cystic fibrosis, *Arch. Dis. Child.* **46**:144–152.

Godfrey, S., Silverman, M., and Anderson, S. A., 1973, Problems of interpreting exercise-induced asthma, *J. Allergy Clin. Immunol.* **52:**199–209.

Gross, G. N., Souhrada, J. F., and Farr, R. S., 1974, Long-term treatment of an asthmatic patient using phentolamine, *Chest* **66:**397–401.

Hahn, A., Anderson, S. D., Morton, A. R., Black, J. L., and Fitch, K. D., 1984, A reinterpretation of the effect of temperature and water content of the inspired air in exercise-induced asthma, *Am. Rev. Respir. Dis.* **130:**575–579.

Herxheimer, H., 1946, Hyperventilation asthma, *Lancet* **1:**87–89.

Howarth, P. H., and Holgate, S. F., 1983, Exercise, asthma, and mast cells, *Lancet* **1:**822 (abstr.).

Iikura, Y., Inui, H., Nagakura, T., and Tak H. Lee, 1985, Factors predisposing to exercise-induced late asthmatic responses, *J. Allergy Clin. Immunol.* **75:**285–289.

Jaeger, M. J., Tribble, D., and Wittig, H. J., 1979, Effect of 0.5 ppm sulfur dioxide on the respiratory function of normal and asthmatic subjects, *Lung* **156:**119–224.

Johnson, C. E., Belfield, P. W., Davis, S., Cooke, N. J., Spencer, A., and Davies, J. A., 1986, Platelet activation during exercise-induced asthma: Effect of prophylaxis with chlomoglycate and solbutabol. *Thorax* **41:**290–298.

Jones, R. S., 1966, Assessment of respiratory function in the asthmatic child, *Br. Med. J.* **2:**972–975.

Kattan, M., Keens, T. G., Mellis, C. M., and Levison, H., 1978, The response to exercise in normal and asthmatic children, *J. Pediatr.* **92:**718–721.

Katz, R. M., Whipp, B. J., Heimlich, E. M., and Wasserman, K., 1971, Exercise-induced bronchospasm, ventilation and blood gases in asthmatic children, *J. Allergy* **47:**148–158.

Kay, A. B., and Lee, T. H., 1983, Mediators of hypersensitivity in exercise induced asthma, *Eur. J. Respir. Dis.* **64:**237–241.

Kawabori, S., Pierson, W. E., and Conquest, C. W., 1976, Incidence of exercise-induced asthma in children, *J. Allergy Clin. Immunol.* **58:**447–455.

Kneussel, M. P., and Richardson, J. B., 1978, Alpha-adrenergic receptors in human and canine tracheal and bronchial smooth muscle, *J. Appl. Physiol.* **45:**307–311.

Konig, P., Godfrey, S., and Abrahamov, A., 1972, Exercise-induced bronchial lability in children with a history of wheezy bronchitis, *Arch. Dis. Child.* **47:**578–580.

Laitinen, L. A., and Kokkola, K., 1974, Bronchial reactivity to histamine in pulmonary tuberculosis, *Scand. J. Respir. Dis.* **89**(Suppl.):201–205.

Lee, T. H., and Anderson, S. D., 1985, Heterogeneity of mechanism in exercise-induced asthma, *Thorax* **40:**481–489.

Lee, T. H., Nagy, L., Nagakura, T., Walport, M. J., and Kay, A. B., 1982, Identification and partial characterization of a exercise-induced neutrophil chemotactic factor in bronchial asthma, *J. Clin. Invest.* **69:**889–899.

Lee, T. H., Nagakura, T., Papageorgiou, N., Iikura, Y., and Kay, A. B., 1983, Exercise-induced late asthmatic reactions with neutrophil chemotactic activity, *N. Engl. J. Med.* **308:**1502–1505.

Lee, T. H., Nagakura, T., Papageorgio, N., Cromwell, O., Iikura, Y., and Kay, A. B., 1984, Mediators in exercise-induced asthma, *J. Allergy Clin. Immunol.* **73:**634–639.

Lefcoe, N. M., 1969, The time course of maximum ventilatory performance during and after moderately heavy exercise, *Clin. Sci.* **36:**47–52.

Mansfield, L., McDonnell, J., Morgan, W., and Souhrada, J. F., 1979, Airway response in asthmatic children during and after exercise, *Respiration* **38:**135–143.

McFadden, E. R., Jr., Sterns, D. S., Ingram, R. H., Jr., and Leith, D. E., 1977, Relative contributions of hypocapnia and hyperpnea as mechanisms in a post-exercise asthma, *J. Appl. Physiol.* **42:**22–27.

McFadden, R. E., Jr., Denison, D. M., Walker, J. F., Assoufi, B., Peacock, A., and Sopwith, T., 1982, Direct recording of the temperature in the tracheobronchial tree in normal man, *J. Clin. Invest.* **69:**700–705.

McNally, J. F., Jr., Enright, P., Hirsh, J. E., and Souhrada, J. F., 1979, The attenuation of exercise-induced bronchoconstriction by oropharyngeal anesthesia, *Am. Rev. Resp. Dis.* **119**:247–252.

Murlas, C., Tencati, J., Mahutte, K., Nadel, J. A., and Roberts, J. M., 1982, Direct effect of cooling on human bronchial smooth muscle, *Am. Rev. Resp. Dis.* **125**:222a.

Nadel, J. A., 1984, Inflammation and asthma, *J. Allergy Clin. Immunol.* **73**:651–653.

Nagakura, T., Lee, T. H., Assoufi, B. K., Denison, D. M., Newman-Taylor, A. J., and Kay, A. B., 1983, Neutrophil chemotactic factor in exercise and hyperventilation-induced asthma, *Am. Rev. Resp. Dis.* **128**:294–296.

Nagy, L., Lee, T. H., and Kay, A. B., 1982, Neutrophil chemotactic activity in antigen-induced late asthmatic reactions, *N. Engl. J. Med.* **306**:496–501.

Neuman, I., Inbar, O., and Creter, D., 1983, The kinin system in exercise-induced asthma, *Am. Allergy* **53**:351–354.

Ouellette, J. J., and Reed, E., 1965, Increased response of asthmatic subjects to methacholine after influenza vaccine, *J. Allergy* **36**:558–563.

Orehek, J., Massari, J. P., Gayrard, P., Grimaud, C., and Charpin, J., 1976, Effect of short-term, low-level nitrogen dioxide exposure on bronchial sensitivity of asthmatic patients, *J. Clin. Invest.* **57**:301–307.

Papageorgio, N., Durham, S. R., Walsh, G. M., Carroll, M., Lee, T. H., and Kay, A. B., 1983, Complement receptor enhancement as evidence of neutrophil activation after exercise-induced asthma, *Lancet* **2**:1220–1222.

Park, M. K., Hayashi, S., Wise, F. M., and Robotham, J. L., 1982, Mechanism of cooling-induced contraction of isolated tracheal smooth muscle, *Fed. Proc.* **41**:1356a.

Parker, C. D., Bilbo, R. E., and Reed, C. E., 1965, Methalcholine aerosol as test for bronchial asthma, *Arch. Int. Med.* **115**:452–460.

Rebuck, A. S., and Reed, J., 1968, Exercise-induced asthma, *Lancet* **2**:429–431.

Sheppard, D., Epstein, J., Holtzman, M. J., Nadel, J. A., and Boushey, H. A., 1982, Dose-dependent inhibition of cold air-induced broncho-constriction by atropine, *J. Appl. Physiol.* **53**:169–174.

Shturman-Ellstein, R., Zeballos, R. J., Buckley, J. M., and Souhrada, J. F., 1978, The beneficial effect of nasal breathing on exercise-induced bronchoconstriction, *Am. Rev. Resp. Dis.* **118**:65–73.

Silverman, M., Anderson, S. D., and Walker, S. R., 1972, Metabolic changes preceeding exercise-induced bronchoconstriction, *Br. Med. J.* **1**:207–209.

Silverman, F., 1979, Asthma and respiratory irritants (ozone), *Environ. Health Perspect.* **29**:131–136.

Simonsson, B. G., 1965, Clinical and physiological studies on chronic bronchitis. III. Bronchial reactivity to inhaled acetylcholine, *Acta Allergol.* **20**:325–348.

Simonsson, B. G., Jacobs, F. J., and Nadel, J. A., 1967, Role of autonomic nervous system and the cough reflex in the increased responsiveness of airway in patients with obstructive airway disease, *J. Clin. Investig.* **46**:1812–1820.

Simonsson, B. G., Skoogh, B. E., and Ekstrom-Jodal, B., 1972, Exercise-induced airways constriction, *Thorax* **27**:169–180.

Soltoff, S., Carcillo, T., and Coburn, R. F., 1979, Zero extracellular K^+ and prostaglandine release in the guinea pig taenia coli, *Biochim. Biophys. Acta.* **557**:217–229.

So, S. Y., Lam, W. K., and Kwan, S., 1985, Selective 5-HT_2 receptor blockage in exercise-induced asthma, *Clin. Allergy* **15**:371–386.

Souhrada, M., and Souhrada, J. F., 1981, The direct effect of temperature on airway smooth muscle, *Resp. Physiol.* **44**: 311–323.

Souhrada, J. F., Presley, D., and Souhrada, M., 1983, Mechanisms of the temperature effect on airway smooth muscle, *Resp. Physiol.* **53**:225–237.

Stanescu, D. C., and Teculescu, D. B., 1970, Exercise and cough-reduced asthma, *Respiration* **27:** 377–402.

Vassallo, C. L., Gee, J. B. L., and Domm, B. M., 1972, Exercise-induced asthma: observations regarding hypocapnia and acidosis, *Am. Rev. Resp. Dis.* **105:**42–49.

Walden, S. M., Bleecker, E. R., Chahal, K., Britt, J., Mason, P., and Permutt, S., 1984, Effect of alpha-adrenergic blockage on exercise-induced asthma and conditioned cold air, *Am. Rev. Resp. Dis.* **130:**357–362.

Zawadski, D. K., Lenner, K. A., and McFadden, E. R., Jr., 1988, Re-examination of the late asthmatic response to exercise, *Am. Rev. Resp. Dis.* **137:**837–841.

Zeballos, R. J., Shturman-Ellstein, R., McNally Jr., J. F., Hirsh, J. E., and Souhrada, J. F., 1978, The role of hyperventilation in exercise-induced bronchoconstriction, *Am. Rev. Resp. Dis.* **118:** 877–884.

Index